EVOLVING TOMORROW

ADVANCE PRAISE FOR THE BOOK

"Cutter presents us with a wonderfully stimulating book. Rather than simply yet another basic introduction to evolutionary processes with some highlights of promises from modern genetic engineering, the author elegantly weaves in personal narratives, natural history examples, and inspiring thought experiments. Who wouldn't want to read a step-by-step account of how to build a dragon, for example? Importantly, he also has the reader consider consequences of indiscriminate applications of genetic engineering, such as the effects of our own biases and assumptions around normativity."

Mohamed Noor, Professor of Biology and Interim Vice Provost for Academic Affairs, Duke University.

"*Evolving Tomorrow* guides readers on an awe-inspiring journey through the spectacular potential of modern DNA technologies. We now possess God-like powers beyond anything humans have had before. We could end malaria by tweaking the DNA of mosquitoes to make them go extinct. We could bring back the animals we've wiped off the face of the Earth, like mammoths, woolly rhinos, and cave lions. We may even be able to create fire-breathing dragons. But should we? Cutter explores these themes, and others, as he sees them on the front lines of DNA science, presenting them clearly in the context of evolutionary biology. This book is a must-read for anyone curious about the potential of genetic technologies to control the very fate of nature, itself. And if you happen to be someone who wants to build a dragon from scratch, then it's a good book for you, too!"

Daniel Riskin, Adjunct Professor, University of Toronto Mississauga, TV Host, and Producer.

"Asher Cutter arms the reader with a robust understanding of the evolutionary process and then asks them to apply this knowledge to consider how we, humans, are a new kind of evolutionary force. *Evolving Tomorrow* is an excellent primer for anyone hoping to understand how today's biotechnologies - and the way we choose to use them - have the power to change all of life on Earth."

Beth Shapiro, Professor, University of California, Santa Cruz. Author of *How to Clone a Mammoth* and *Life As We Made It*.

"Cutter has written a highly enjoyable and thought-provoking book about the ways in which timeless biological processes and ever more powerful human tinkering are, together, shaping the evolutionary future. A must read for anyone who cares about the future of humanity and the rest of life."

Rob Dunn, Professor, North Carolina State University.

Evolving Tomorrow

Genetic Engineering and the Evolutionary Future of the Anthropocene

Asher D. Cutter

Department of Ecology & Evolutionary Biology, University of Toronto

OXFORD

UNIVERSITY PRESS

OXFORD
UNIVERSITY PRESS

Great Clarendon Street, Oxford, OX2 6DP,
United Kingdom

Oxford University Press is a department of the University of Oxford.
It furthers the University's objective of excellence in research, scholarship,
and education by publishing worldwide. Oxford is a registered trade mark of
Oxford University Press in the UK and in certain other countries

Published in the United States of America by Oxford University Press
198 Madison Avenue, New York, NY 10016, United States of America

British Library Cataloguing in Publication Data
Data available

Library of Congress Control Number: 2022952187

ISBN 978–0–19–887452–2

DOI: 10.1093/oso/9780198874522.001.0001

Printed and bound by
CPI Group (UK) Ltd, Croydon, CR0 4YY

Links to third party websites are provided by Oxford in good faith and
for information only. Oxford disclaims any responsibility for the materials
contained in any third party website referenced in this work.

For my most precious experiments in evolution, Beatrix and Oona, and all the kin of their generation: Fletcher, Lucy, Damon, Jasper, Eddie Arlene, Hildegard, and Arlo.

Contents

Preface

I can promise you that life today, despite millennia of profound evolutionary change, will continue to evolve tomorrow. Some of that evolution to species' features has nothing to do with humanity. Some of it arises as an incidental byproduct of our existence. And some evolution results directly from intentional intervention by people. This book will scrutinize how all of that evolutionary change can happen, how modern-day genetic engineering can influence it, and what consequences of these forces we can expect to see in the future of the Anthropocene era.

Part 1 of this book, *The hearts of nature*, explores how living organisms change over time. Over these first 10 chapters, we'll unearth nature's evolutionary forces and how these forces interplay with genetics. Part 1 explains the intricacies of genetic biotechnologies like CRISPR-Cas9 genome editing, and how CRISPR-Cas9, when used to make gene drives, now permits humans to modify the genomes of every single individual in any species of our choosing. I make the case that this kind of genetic engineering, "genetic welding," unleashed into the wild, will present a new force of evolution. I show just how fast evolution can proceed to change a species, and we'll learn what it is that makes a species a species in the first place. I infuse the conceptual points with incredible tales of organismal biology: of melanic squirrels and blanched mice, of tomatoes and the opsin genes that let us see, of crocodiles and skunk musk, of chimeric sex and of the virtues of chickens. I draw on my own experience of living on three continents and conducting research in the lab and in the field of tropical habitats over the past few decades.

Part 2, *Evolutionary futures*, shifts attention toward the applications and implications of using evolutionary forces, and genetic engineering, to manipulate organisms and to create new species. We'll walk through the ecosystem consequences of species gain and loss in our current Anthropocene era. We'll deliberate the ins-and-outs of de-extinction with aurochs and woolly mammoths, of rewilding with wisent (European bison) and chestnuts, of "pest" control over invasive species like hippos, all as modern large-scale incarnations of ecological restoration. We'll ponder the figurative dragons that we would need to confront in applying genetic welding to create and introduce new species into ecosystems. From the view of genetics, humans are just another animal. In considering the ethical soup of genetic engineering of our own species, I highlight lessons from body modification and in vitro fertilization, identifying the potential for genetic welding of human genomes to unleash what I term *guerrilla eugenics*. I share my own internal conflicts in these multifaceted ethical discussions. We'll lay bare the competing and often internally inconsistent views of what is "wild" and what people desire from "nature," the oft-neglected considerations that are crucial in deciding how humans should deploy or prevent evolutionary engineering and genetic welding in ecosystems. What we decide to do as a global society will shape the kind of evolutionary change that will accumulate in the world of tomorrow and throughout the next millennia.

In preparing *Evolving Tomorrow*, I've had the good fortune of good colleagues and good supports. For expert consultation on my blue-moon questions, I am grateful to the kind and quick minds of Spencer C. H. Barrett, John Calarco, Belinda Chang, Marc T. J. Johnson, Luke Mahler, Shelby Riskin, F. Helen Rodd, Njal Rollinson, and Stephen Wright. I also benefited from Graham Coop's lucid insights into genetics delivered on a very snowy evening for the 2019 Darwin Day lecture at the University of Toronto. Thanks also go to Ashley Reynolds for assistance with the teaching collection of mammal bones in the Department of Ecology & Evolutionary Biology. I am fortunate for the favor and encouragement of careful and thoughtful readers of draft versions of chapters. In particular, I'm grateful to Belinda Chang, Zoe Clarke, Steven Coyne, Beatrix Cutter, Oona Cutter, Marc T. J. Johnson, Nicole Mideo, Shelby Riskin, Yee-Fan Sun, Jason Weir, and Stephen Wright. Thank you also to the Evolutionary Genetics Discussion Group and the members of my research team at the University of Toronto, for discussion of early versions of chapters, including Eniolaye Balogun, Maia Dall'Acqua, Daniel Fusca, Karl Grieshop, Vedika Jha, Katja Kasimatis, Else Mikkelsen, Christine Rehaluk, Rebecca Schalkowski, María Tocora, Athmaja Viswanath, Sabrina Zaidi, and Linyi Zhang. I'm very appreciative of my production editor, Katie Lakina, for keeping it all in order. I'm indebted to my editor, Ian Sherman, at Oxford University Press, for his encouragement and frank advice in working from the proposal through to book finalization, and to the collection of thoughtful anonymous reviewers of the initial book proposal who shared exceptional constructive feedback. One accidental blessing of the COVID-19 pandemic is that it gifted me the luxury of writing this book from home while surrounded by family. I am continually grateful for the love and support of Yee-Fan, Oona, and Beatrix, in all those moments when I needed it most.

My growth as a scientist was shaped by a long train of biological mentors who took great care in guiding my development, starting from when I was a teenager. Thank you, Leo P. Kenney, Norton H. Nickerson, Sara M. Lewis, Stephen H. Levine, Michael P. Ghiglieri, Christopher A. M. Reid, Leticia Avilés, Samuel Ward, Deborah Charlesworth, and Brian Charlesworth. While I am so very grateful for the personal contributions of these individual people to my particular scientific development, my debt runs deeper: to the entire community of the world's scientists, past and present. The open sharing of their findings and data, as well as the blossoming and fading of ideas across the decades as recorded in the scientific literature, made this book possible and all the more enjoyable to write. The inquisitiveness and rigor of scientific researchers in disciplines of all stripes has given every one of us this current understanding of the world.

In writing this book in Toronto, Ontario, I gratefully acknowledge this land that is now home to diverse peoples as the traditional territory of the Mississaugas of the Credit, the Anishnabeg, the Chippewa, the Haudenosaunee, the Wendat peoples, and their ancestors.

Abbreviations

BRCA1:   Breast cancer gene 1, found in human genomes with mutated gene variants that confer a high likelihood of developing breast or ovarian cancer in one's lifetime.

Cas9:   CRISPR-associated protein 9, encoded by genes found in bacteria and archaea that is involved in defense against phage viruses and that has been co-opted in biotechnology for genome editing.

CCR5:   C-C chemokine receptor type 5 gene, found in human genomes with a mutated gene variant associated with resistance to infection by the human immunodeficiency virus (HIV).

ClvR: Cleave and rescue gene drive, a kind of genetically engineered selfish genetic element system that acts to increase its abundance in a population as it transmits itself from generation to generation.

CRISPR: Clustered regularly interspersed short palindromic repeats, regions of bacterial and archaeal genomes that encode a library of phage viral DNA to target as part of their immune system, portions of which get expressed to form guide RNAs to work in concert with Cas9 proteins to cut invading viral DNA, a system that has been co-opted for biotechnological use in genetic engineering.

dCas9:   dead CRISPR-associated protein 9, a genetically modified version of Cas9 protein that disables the ability to cut DNA.

DMI: Dobzhansky-Muller incompatibility, a detrimental interaction between genes that gets revealed when interspecies hybrids develop.

DNA: Deoxyribonucleic acid, the chemical responsible for storing heritable material in animals, plants, fungi, bacteria, archaea, and DNA-based viruses.

GDO: Gene drive organism, a genetically modified organism that encodes a synthetic gene drive in its genome.

GE: Genetically engineered, synonym for genetically modified by virtue of alterations of the genome through directed molecular biology techniques, typically involving transgenic DNA.

GFP: Green fluorescent protein, a protein found naturally in the genomes of some jellyfish species that has been co-opted for use in experimental molecular biology due to its ability to emit light when exposed to UV illumination.

GMO: Genetically modified organism, an organism whose genome includes changes from genetic engineering, usually of transgenic origin.

GPCR: G-protein-coupled receptor, a kind of protein that is part of a large family of proteins in animal genomes that localize to cell membranes and often have functions related to transduction of information into the cell, including sensation of light or odor.

HDR: Homology directed repair, a highly accurate cellular mechanism that fixes damaged DNA by using a similar sequence of DNA as the template for repair, usually using a homologous chromosome under natural circumstances but that can also use a template of DNA provided by a researcher when conducting genome editing.

HGT: Horizontal gene transfer, the movement of genetic material from one species to another species, usually in the context of natural microbial mechanisms of DNA exchange.

HHGE: Heritable human genome editing, genetic engineering applied to human cells that will lead to transmission of the genetic changes across generations.

IUCN: International Union for Conservation of Nature, a nongovernmental organization with observer status within the United Nations and expertise in environmental issues.

IVF: In vitro fertilization, a medical and veterinary technique for humans and other mammals to facilitate fusion of gametes in a Petri dish outside the body for subsequent transfer to the womb of a host female to enable gestation of the resulting embryo and fetus.

MC1R: Melanocortin-1 receptor protein, a type of protein encoded in genomes of mammals and other vertebrates with a key role in the pathway responsible for production of melanin pigments.

mRNA: Messenger ribonucleic acid, the intermediary molecule that results from transcription of a gene in the genome that will then get translated into a protein.

MRSA: Methicillin-resistant *Staphylococcus aureus*, a pathogenic bacterium responsible for staph infections that is not treatable by many antibiotics.

NHEJ: Nonhomologous end joining, a mechanism of DNA repair that is prone to introduce insertion or deletion changes to the DNA sequence in the course of fixing breaks in DNA.

PGC: Primordial germ cell, a kind of stem cell that will enable development of sperm cells or egg cells in the adult organism.

PGD: Preimplantation genetic diagnosis, a genetic test performed on a particular gene at an early stage of embryogenesis before inserting the embryo into a womb via IVF.

PGS: Preimplantation genomic screen, genetic analysis of the entire genome of an early stage embryo before IVF transfer into a womb.

RNA: Ribonucleic acid, a kind of biomolecule that can contain heritable information (as for RNA viruses), serve as an intermediate in the transmission of heritable information in the course of producing proteins (as for mRNA), or function as a bioactive component of cells to influence expression of genes or other cellular activities (such as guiding Cas9 proteins to genomic locations, in the case of guide RNAs).

RNAi: Ribonucleic acid interference, an experimental technique that enables a researcher to inhibit the expression of a gene.

SCNT: Somatic cell nuclear transfer, a key experimental procedure used to clone mammals that involves insertion of the cell nucleus and DNA from a body cell such as a skin cell into an enucleated egg cell.

TARE: Toxin-antidote recessive embryo gene drive, a particular type of genetically engineered selfish genetic element that is capable of increasing its abundance in the genomes of a population from one generation to the next.

UV: Ultraviolet light, a form of electromagnetic radiation with a wavelength between 10 nm and 400 nm, that is not visible to humans but can be visible to other organisms with appropriate short-wavelength-sensitive opsin proteins.

About the Author

Asher D. Cutter is a professor of Ecology and Evolutionary Biology at the University of Toronto, Canada. A former Fulbright Scholar, Cutter trained at Tufts University, James Cook University, the University of Arizona, and the University of Edinburgh. He authored the textbook *A Primer of Molecular Population Genetics*, as well as nearly 100 scientific articles on the topics of genome evolution, population genetics, speciation, and the biology of *Caenorhabditis* nematode roundworms. Raised in the Boston, Massachusetts, area, Cutter lives in Toronto with his family.

Part 1

The hearts of nature

1

Nature in the raw, and cooked

Wandering the long beach of the Gold Coast of Australia on one sunny July afternoon, a shift of light caught my eye in the thin skim of surf on the sand. Leaning down, I couldn't see much, just an oval shimmer against the tan sand. When I cupped my hands to hold a palm full of water and sand grains, however, I discovered something that I'd never encountered before: an utterly transparent fish-shaped creature. As long as my palm was wide, its flat-sided body sported a pigmented dark eye on each side of its head, and it wriggled sedately between my fingers. It was absolutely stunning. I later learned that this creature is called a leptocephalus larva, the juvenile form of a marine eel. It was entirely transparent and alive, and the wispy bones of its skeleton were only just visible as faint pale creases, as if encased in thick wet cellophane.

I was amazed. I had only seen such incredible transparency in boneless creatures like jellyfish and nematode roundworms. Or, in the chemically preserved fish in jars that I prepped at the Smithsonian National Museum of Natural History one summer as a student. I'd cleared and stained their dead piscine bodies to make visible their skeleton of crimson bones and cerulean cartilage, courtesy of alizarin and alcian stains, the other tissues rendered transparent from a recipe of chemical acids and bases and enzymes. Not only had the existence of *living* transparent vertebrates—a see-through live animal with a backbone!—somehow evaded all of my prior experience, but the unexpectedness of the encounter in nature in the palm of my hand was a marvel (Figure 1.1).

As a biologist, perhaps I should not be surprised that I am constantly surprised. I am, though. I'm surprised and amazed by the multitude of shapes, sizes, colors, and behaviors of organisms that I come across year in and year out, of animals and plants and critters that are neither.

In the day to day of my biological research, though, it is strings of DNA that I think about most. I think about the space-time of genetics, of the swirl of the double helix of a chromosome as the sequence of chemical letters on its rungs go on to mutate and evolve. I think about how the DNA sequences that trail across my computer screen encode the ongoing tug of distinct and ancient forces of evolution, a record of what has evolved and of clues to where evolution will go next. I think about what experiments we might conduct to alter those DNA sequences, to create something new to test the limits of our understanding of biological principles. Just as I can't help but marvel at a new personal discovery on a beach, I marvel at the prospect of melding genetic engineering with the forces of evolution.

Evolving Tomorrow. Asher D. Cutter, Oxford University Press. © Asher D. Cutter (2023). DOI: 10.1093/oso/9780198874522.003.0001

Figure 1.1 *Transparent leptocephalus larva of a marine eel (Queensland, Australia), one of the marvels of the living world that already exists.*
Photo by the author.

In the laboratory, we can create extraordinary creatures. In the laboratory, we can use genetic engineering to change a creature's DNA so that it expresses characteristics it never had before or that no creature ever had before. In my lab, we have worms like none found in nature: worms with genes that make them glow green or red or yellow when you shine a UV lamp onto them. This fact teases a bigger idea, a niggling question about life beyond the lab. Would the great outdoors also benefit from newly evolved creatures, populations of new and extraordinary species, what we might call hyperexotic species, produced with evolutionary engineering to be so remarkable that the world had never before seen their likeness?

I didn't always think about the DNA of genomes and how to mess about with them. I once was a little country boy who sat among the overgrown pokeweed in the sunny garden of my parents' backyard, smooshing their brilliant berries and following the slow trails of insects along their stems. I once was a kid who roamed the New England forest that ringed my childhood home at the end of a dirt road, snatching bullfrogs in swampy ponds and spying the pink sac-like flowers of lady slipper orchids—*Cypripedium acaule* boorishly renamed as "scrotum plants"—that sprouted from the rusty duff of pine needles in spring. Even as I got older, through university, it was the outer lives in the living world that I spent time with. I rubbed elbows with the plants and animals of vernal pools and bogs and mangrove swamps, I peeped in on the sex lives of beetles, I sought out the signs of life in Eocene fossil leaves and the ants of Australian tropical rainforests.

After I transplanted my life from New England to Arizona for a time, I began to learn more about the inner lives in the living world. At the end of that first year I lived in Tucson, 1999, for the first time in the world, the first animal got its entire genome sequenced. It's a little squiggle of a creature known as a nematode roundworm, often referred to as *C. elegans*. When you watch through the eyepieces of a microscope as these miniature animals swerve sinuously with quiet determination, you understand how they got their name: *elegans* derives from the Latin for *elegant*. Also, they are transparent. I was hooked. In the ensuing years that have now turned into decades, they've taught me how DNA makes an organism develop and how DNA changes, how to engineer genetic changes, and how those changes make an organism develop differently and behave differently.

I didn't give up on outer lives though. Understanding evolution means paying attention to both the unseen and the seen, both DNA and the magnificent diversity of forms and species all around us. For a Yankee New Englander, the urban incarnation of the Sonoran Desert of Arizona showed me an alien landscape. I began a prickly pear garden in my Tucson backyard. I exploited the astounding biological feat that many plants, and even some animals, are able to do: to grow an entirely new individual from a broken-off piece of another. Transplanting the trimmings of diverse varieties of *Opuntia* cactus that I scavenged in the alleys around my Tucson neighborhood—pale green and purple cladode pads,[1] cladodes covered in long whitish or yellowish spikes, enormous emerald-green cladodes nearly devoid of spines, cladodes nearly brown with the dense clusters of tawny prickles—there soon grew a spiny display of living plant-paddle biodiversity unlike anything I'd experienced in my hometown outside of Boston. They were my in-town homage to the saguaro, those 10-m-tall (33-foot-tall) botanical icons that stood in legions throughout the foothills of the mountains that ringed the edge of the city.

On some of my cacti, tiny white fluffballs pocked the surface. They looked like clusters of Q-tip tops that, at first, I thought might be a fungal infection. After a poke produced a smear of shocking purple-red brilliance, I realized that they must be cochineal bugs, a creature I'd heard about only in stories. Cochineal *Dactylopius coccus* are those magnificent so-called scale insects, parasites of *Opuntia* prickly pear, that produce copious quantities of carminic acid as a defense against predators. Humans harvest that carminic acid to make the intense red color of carmine dye, a.k.a. natural red #4 or ingredient E120. You might also know carmine from the ingredient list of all manner of FDA-approved cosmetics and foodstuffs, from lipstick to hotdogs, store-bought strawberry jam to the red veneer of imitation crab.

Do you ever happen upon nature in the raw to find something unexpected in the wild? Something beautiful or inexplicable, it might even be something tiny, something that compels a smile or a gasp or a laugh? The woven detail of a bird's nest dangling on a branch, the flash of the first firefly's light on a summer evening, fossil shells spied in a rock on a cliffside, a transparent animal caught in the surf on the beach. There are true wonders in this world right now, every minute. They have outer wonders and inner wonders, encoded in DNA. Perhaps we should—or perhaps we should not—go out of our way to make more wondrous animals. Could we? How would we? And, again, *should* we?

Evolution, genetic engineering, ecosystems, bioethics, and the state of the world—we'll need to sift through all of these things to answer those questions.

1.1   The nature of change

The world is awash in change. Decade over decade, the climate across the globe has grown higher average temperatures and greater extremes of temperature. Compared to preindustrial periods, the air has 50% more carbon dioxide, and the oceans are 30% more acidic, achieving levels unprecedented over the past 65 million years.[2] We did that. Entire species have gone extinct at such an astounding rate that some say that we are amidst the sixth-greatest mass extinction event that our planet has ever experienced. Estimates put the world on a pace of nearly 1000 species extinctions accumulating globally each year, a rate 1000 times higher than it was prehistorically.[3] Population declines and extinctions hit some types of organisms, and some places, harder than others. Invertebrates, large animals, amphibians, and the Indo-Pacific are especially hard-hit.[4] We did that, too. And at the same time, nonnative, invasive species of animals and plants arrive on the shores, islands, and waterways of distant lands and distant lakes, rivers, and seas. This spreading of life alongside human movement takes away lives, too, through extinctions mediated by predation, competition, and pathogenicity.[5] Yes, we also did that. We have come to depend on some of those nonnative species, however, for food or other goods. They fill the agricultural space that now occupies over 41% of the world's land.[6] Human-manufactured material, from buildings to soda bottles to the microplastic flecks that infiltrate the tissues of seemingly pristine arctic wildlife,[7] now exceed the combined mass of all living things on planet Earth, all 1.1 trillion metric tons of it.[8] We stupefyingly did that. These inglorious markers of change that leave geologically indelible tracks on the planet are the hallmarks of the Anthropocene era.[9]

The Anthropocene is the here-and-now epoch defined by humanity's profound influence on changing the shape of the natural world. I've listed some of the direct material changes, most of which you have probably heard about from news headlines. You can relate to their causes or consequences in some part of your day-to-day life. But they also impact the natural world indirectly: shifts in ways both subtle and profound to the ecological interactions that organisms have with one another and with the environments they live in. These kinds of impacts cause shifts to the natural selection pressures that sculpt the evolution of those species that manage to persist, that manage to evade extinction.

Humans caused all this change in about one million days. The knock-on effects on ecosystems and evolution of individual species, however, will follow the laws of life that have ruled the planet for roughly 3700 million years.[10] Humans nudge, or punch, the scales of the wild, tipping balances. But all the usual processes still apply in how nature responds: nutrient uptake and release; predation; competition between organisms for resources, for sun and space and food and mates; under-the-hood evolutionary processes like mutation and natural selection and genetic drift. All these things still operate in the same fundamental way in response to all the manhandling. With one potential exception. Mankind, or in this case womankind, given the 2020 awarding of the relevant Nobel Prize in Chemistry to Emmanuelle Charpentier and Jennifer Doudna,[11] has devised biotechnology with the capacity for humans to bend those laws of life.

The idea of bending the laws of life may make you nervous. The idea of carving the DNA of wild organisms with tools of genetic engineering may even inspire in you a feeling of horror at the hubris. Before indignity jumps up too high, however, take stock of the fact that we've been shaping the DNA of animals and plants and microbes for millennia. Human creation of domesticated animals and plants are the obvious cases—the horse and hog and dog, the Brussels sprout and Nashi pear and freestone peach. *Homo sapiens* is also an ecosystem engineer par excellence, meaning that our successes in making the world more to our liking also modifies the world for the innocent bystanders. Our Anthropocene exploits push the forces of evolution, driving change in response to new environmental conditions, habitat perturbation, elimination of fellow species, or creating overlap among species that had never before encountered one another. The evolutionary responses in the outdoors are, perhaps, slow enough for us individually to acclimate to, to accept as a new normal. This acclimation is how shifting baselines are born, from one generation to the next, casual acceptance of what is profoundly different from the generation before.

What to do? One option for civilization is to continue to acclimate. This kind of proposal is not necessarily as apathetic as it might seem. For example, by proactively establishing large regions for rewilding, we could let ecological and evolutionary processes proceed in a passive and noninterventionist way to promote biological diversity and holistic ecosystem functionality. The news out of the United Kingdom, these days, seems chock full of reports of former pasture land left to seed itself into forest without the nibbling of large herds of sheep and cattle. It then becomes a question of patience. We humans, however, are not known for our patience.

Another wing of thought holds that the symptoms of Earth's ills are so dire that the cure requires active treatment, not simple bed rest. That is, humans ought to take a heavy hand in establishing functioning ecosystems. To help keep the most catastrophic effects of rapid climate change at bay, for example, we could partake of solar geoengineering—altering atmospheric chemistry to block out the sun—as a temporary chemical parasol for the planet. Perhaps we even ought to model nature on the structure and composition of ecological landscapes prior to widespread human impacts, so-called Pleistocene trophic rewilding. This plan could introduce existing species into environments in which their ecological analogs are extinct, intentionally changing the biota with a cascade of ecological consequences. This plan could also introduce genetically engineered species, perhaps species resurrected—as for the so-called de-extinction of mammoths—or even brand-new species evolved de novo with the aid of biotechnological tools.

It is this last possibility that I would like us to dwell on, to poke at. What would it take and what would it mean to shape the evolutionary origins of new creatures and to set them loose in the wild? In doing so, we could bend the laws of life and we could add a new evolutionary force into the world, a force that I call genetic welding.

The fossil record and the living world show us exotic creatures. If we set our minds and muscles and money to it, the future could unveil the most exotic creatures of them all. You and I may give the Anthropocene a new hallmark: populations of new and extraordinary species. Such hyperexotics would set loose the dragons of the Anthropocene, both figuratively and literally.

1.2   Evolving tomorrow

As a biologist, I make trade in living physical systems. So, let's first talk about the idea of human-mediated evolution of extraordinary creatures in a corporeal sense. Despite the bounty of exotic creatures in nature, technically speaking, could we create something new? Something that more completely captures the physical essence of what it is to be entirely novel? Could we create something that is the epitome of the Anthropocene? To answer this question, we will need to dig into what we are asking.

First, what characteristics might we shape? The possibilities abound: size and color and locomotion, as well as features of physiology and development and behavior and the senses. These are the key characteristics of all animals, how they interact with each other and with the rest of the world. We can explore the diversity of options that currently exist or resurrect features that we know to have existed from extinct remains, to help define the so-called morphospace of what is possible. Dipping into this animal morphospace is also a dip into a melting pot of inspiration. We will explore in depth the body and soul of what it takes to be a creature in the next chapters as we prepare a mental map for what it takes to be an evolutionary engineer of extraordinary species.

Second, we need to set our tools on the table. Some of these tools are ancient tools, the tools of evolution. Humans are more adept in using evolutionary tools than you might think. Consider, for example, the magnificent variety of dog breeds that bring joy and comfort, the extraordinary productivity of maize and rice plants that deliver a bounty of carbs, the phenomenal milk yield and muscle density of docile cattle that find their way to supermarket shelves near you. Humans shaped genetic changes in these species over the course of domestication, guiding traits through a process called artificial selection, the incarnation of natural selection directed by people.

Evolution requires that we have a population of organisms with genetic heterogeneity within it. The more abundant and variable the population, the more raw material there is for selection to chisel and chip away. What you might not also realize is that evolution is happening all around us, all the time. That evolution can happen fast enough for us to watch it, and fast enough to mold it.

Other tools are new tools, the tools of biotechnology and genetic engineering. One class of such modern tools are blunt but effective. For example, we know how to induce genetic change with chemical mutagens as a crude but potent means to an end. Other tools act with great precision, including a spectrum of applications to gene editing using CRISPR-Cas9 techniques. If you are unfamiliar with the tangle of keyboard characters that spells "CRISPR-Cas9," then take heart, as a later chapter will cover what you need to know. This box of molecular tools, co-opted from the cellular machinery that bacteria evolved as a counter strike against viral attacks, can let us manipulate or remove individual DNA letters, or whole genes, as well as insert novel genes or gene regulators at precise locations in genomes.

Genome editing, on its own, whether with CRISPR-Cas9 or other techniques, introduces novel genetic variants into DNA. In its most basic sense, this is nearly identical to what mutation does or to what gene flow does when migrants from one place interbreed with residents of another locale. Human ingenuity, however, has led researchers to meld genome editing with our understanding of how the forces of evolution operate. This blending of evolutionary principles with biotechnological tools now gives a way to propagate throughout an entire wild species any given novel DNA variant constructed by genetic engineering. In the controlled confines of the lab, they are called gene drives. Deployed into nature, we shall see how gene drives are the foundation for genetic welding, a new force of evolution.

We will unpack this slate of tools that we have at our disposal, from the tools of evolution to the tools of biotechnology. By combining these tools, we have the opportunity to become more skillful evolutionary engineers with the capacity to direct the evolution of wild organisms outside the confines of the laboratory, or even to direct the evolution of new species altogether. We'll see how these tools operate, by creating and harvesting genetic differences and then altering the abundance of those genetic changes in the face of constraints and limitations. Is humanity in a position to create biologically hyperexotic entities, new species with unique characteristics, a living-and-breathing hallmark of the Anthropocene? More so than ever before.

★

The creation by humans of robust populations of extraordinary new species through evolutionary engineering would, as it turns out, create something else on the sly, beforehand, whether we like it or not. It creates figurative dragons, the monsters that inhabit every corner of our imagination, the concerns, suspicions, and worries.

The prospect of disrupting the natural order of living things, understandably, stirs unease in all of our hearts. For some, this worry stems from religious sensibility. It comes from the sense that genetic engineering or perturbing natural systems is akin to playing god and, consequently, blasphemous and immoral. For others, like me, the apprehension is secular. The concerns are rooted more in a sense that there is intrinsic value in letting nature go about its business without undue human intervention. For the Star Trek aficionados, this is the logic of the Prime Directive.[12] People feel these views in their gut, as well as in their mind, sometimes translating into raw emotion. Whatever the basis, the sense is that messing with DNA is just plain wrong—no matter that we've been messing with DNA in other ways for millennia. And, of course, humans are just another kind of animal. Genetic manipulation of our own genomes could be undertaken, in principle, in exactly the same way as for any other kind of animal. Any application of biotechnology must confront these figurative dragons soberly and with nuance, as we will do in later chapters.

We should also anticipate a second realm of apprehension in any attempt to create new species. It deals with the political and economic views that people hold. While some folks ascribe to anything-goes capitalism, others are leery of strictly financial profit-driven motives and their negative knock-on effects to society. For example, corporate patents over genetically modified organisms and monopolies over biological entities represent moral affronts to many people. When public funds are invested, they necessarily come with the opportunity cost of not investing in something else that might benefit society. This tension in society about what to devote tax dollars to, or what not to, forms a fundamental axis of strife between political views. We must consider in earnest such legitimate differences in perspective. Deep spending by governments on a focused issue often comes with formidable long-term benefits: think how the Apollo missions, to send spacecraft to the Moon, sparked new innovations and entrepreneurial spin-offs. Yet, it also echoes the rationalizations of colonialism and the ravages that it exacted on the parts of our global society that wield less power. The politics and economics of biological adventurism is fraught with competing perspectives such as these that demand thoughtful public debate.

Life is full of risks, but we all vary in how ill at ease we feel from any given risk. The COVID-19 pandemic taught us a fast lesson in how individual risk tolerance varies from person to person as we came to navigate new social norms with friends and family and coworkers.[13] A foray into the biological unknown also has risks. The key risk is that we lose control in a way that cascades with detrimental effect on the natural world or on society, tomorrow or for the next generation or for the next millennium. After all, biological systems are predisposed to taking on a life of their own, evolution driving their destiny in unexpected ways. An exotic organism may prove invasive, a new kudzu, fire ant, or cane toad writ large. They may disrupt ecosystems, drive other species to extinction, or introduce diseases that devastate wildlife or even human health.[14]

We humans have good experience, as an invasive species ourselves, with driving animals and plants to extinction. Moa, ground sloths, and aurochs come to mind, just to name a few large examples. Nonnative species may change the face of landscapes or impose novel selection pressures on other species, or they may exact financial impacts—on tourism, ecological remediation efforts, and medical costs. There also are risks, sometimes even worse, of doing nothing. We must, as a society, assess how tolerant or averse we are to judge the risk-reward trade-off in decision-making, to do something or not.

★

Nature is full of exotics. With all this pre-existing wonder in the world, it is fair to ask: Why bother to embark on a biological mission to create yet another exotic species, even if it is so unique that the world has never encountered it before?

We humans are a curious species. When shown a map of the realm of knowledge, the edge calls to us to resolve the boundary with understanding. In the scribble that notes "here be dragons," we see a conjecture, a hypothesis worth testing. Our history of exploration proves it, from geographic migration out of Africa to the rest of the world, from surface to submarine to outer space. Our history of ingenuity proves it, too. The goal, and the result, of the Apollo Moon missions was not simply to set human footprints on the Moon and to return. It was more ambitious and successful than that. The Apollo Moon missions sought scientific exploration in its own right and to establish new insights and technologies along the way to do so. Starting with NASA's Artemis Moon Mission V, plans are afoot to even begin colonizing other celestial bodies.[15]

Let's envision a mission that makes a grand biological twin to NASA's Apollo and Artemis Moon projects, an Artemis mission here on Earth. Artemis was, after all, the goddess of wild animals and wilderness, the twin sister of Apollo, born out of wedlock from the ancient union of the second-generation Titan, Leto, and her cousin Zeus. We would need to meld the ancient with the cutting-edge. We would take life's analogs of slide-rules and quantum computing: the classic forces of evolution and biotechnologies with genetic welding. A biological Artemis agenda on Earth might spawn heretofore undiscovered insights into genome editing, novel biotechnologies, biological control of invasive organisms, restoration of disrupted ecosystems, reclamation of pre-Anthropocene food webs, new medical inventions, fertility interventions, deciphering of heretofore unknown developmental and physiological mechanisms of animals and plants, and who knows what else. Scientific curiosity and exploration through basic research has produced an astounding foundation of knowledge.[16] A grand vision lets us test and expand the limits of our knowledge, to test whether our ideas about how the world works hold up in practice when we stretch them to the edge of the map.

To empower a grand biological vision of creation, to evolve new and extraordinary species to inhabit the world of tomorrow, infused with the tools of biotechnology, we must confront figurative dragons as well as technical obstacles. Introducing new species into a given spot in the world uncorks a test tube for ecosystems to respond to. Species in novel circumstances can impart unexpected and unintended consequences. We will observe how ecosystems matter alongside evolutionary change. We will deliberate the minefield of ethical tangles associated with evolutionary engineering of the natural world, and the prospect of genetic engineering of ourselves. We will consider how change begets more change, which we may perceive as good, bad, or indifferent. We will see how evolution never stops, except with extinction, and how extinction can, itself, leave unexpected and unintended consequences.

Notes

1. A cladode is that fleshy pad of a prickly pear cactus. Botanically speaking, it's a flattened stem and not a leaf, despite being the main location of photosynthesis.

2. Human activities have increased global average temperature by 1.0 °C (1.8 °F) since 1850, putting today's global temperature average above that from any time in the past 100,000 years. Masson-Delmotte, V., et al. 2018. Global warming of 1.5 °C: An IPCC special report on the impacts of global warming of 1.5 °C above pre-industrial levels and related global greenhouse

gas emission pathways, in the context of strengthening the global response to the threat of climate change, sustainable development, and efforts to eradicate poverty. https://www.ipcc.ch/sr15. Accessed April 6, 2021; Doney, S.C., et al. 2020. The impacts of ocean acidification on marine ecosystems and reliant human communities. *Annual Review of Environment and Resources*. 45: 83–112. https://doi.org/10.1146/annurev-environ-012320-083019; Snyder, C. 2016. Evolution of global temperature over the past two million years. *Nature*. 538: 226–228. https://doi.org/10.1038/nature19798.

3. De Vos, J.M., et al. 2015. Estimating the normal background rate of species extinction. *Conservation Biology*. 29: 452–462. https://doi.org/10.1111/cobi.12380; Ceballos, G., et al. 2020. Vertebrates on the brink as indicators of biological annihilation and the sixth mass extinction. *Proceedings of the National Academy of Sciences USA*. 117: 13596–13602. https://doi.org/10.1073/pnas.1922686117; Barnosky, A., et al. 2011. Has the Earth's sixth mass extinction already arrived? *Nature*. 471: 51–57. https://doi.org/10.1038/nature09678

4. Cowie, R.H., et al. 2022. The sixth mass extinction: fact, fiction or speculation? *Biological Reviews*. 97: 640–663. https://doi.org/10.1111/brv.12816; Leung, B., et al. 2020. Clustered versus catastrophic global vertebrate declines. *Nature*. 588: 267–271. https://doi.org/10.1038/s41586-020-2920-6

5. David, P., et al. 2017. Impacts of invasive species on food webs: a review of empirical data. *Advances in Ecological Research*. 56: 1–60. https://doi.org/10.1016/bs.aecr.2016.10.001

6. As of 2015, humans use about 41.2% of the 131,804,039 square kilometers of land on Earth for crops and grazing, plus another 8.8% of the land is devoted to urban and village settlements of varying degrees of human impact. That leaves 50% of the land surface as largely natural, although only about half of this natural land area is considered truly wildland. Ellis, E.C., et al. 2020. Anthropogenic biomes: 10,000 BCE to 2015 CE. *Land*. 9: 129. https://doi.org/10.3390/land9050129

7. Rochman, C.M. 2018. Microplastics research: from sink to source. *Science*. 360: 28–29. https://dx.doi.org/10.1126/science.aar7734; Haave, M., et al. 2021. Documentation of microplastics in tissues of wild coastal animals. *Frontiers in Environmental Science*. 9: 575058. https://doi.org/10.3389/fenvs.2021.575058

8. Elhacham, E., et al. 2020. Global human-made mass exceeds all living biomass. *Nature*. 588: 442–444. https://doi.org/10.1038/s41586-020-3010-5

9. What start date to use for the Anthropocene is still a point of discussion. The nuclear detonations starting in 1945 represent one clear demarcation (with peak plutonium deposition in 1964), but other contenders include the start of the Industrial Revolution (1760, or a bit before in 1610) or onset of farming agriculture (~11,000 years ago) or even the megafaunal extinctions dating to ~50,000 years ago. Waters, C.N., et al. 2016. The Anthropocene is functionally and stratigraphically distinct from the Holocene. *Science*. 351: aad2622. https://dx.doi.org/10.1126/science.aad2622; Lewis, S., and M. Maslin. 2015. Defining the Anthropocene. *Nature*. 519: 171–180. https://doi.org/10.1038/nature14258

10. Pearce, B.K.D., et al. 2018. Constraining the time interval for the origin of life on Earth. *Astrobiology*. 18: 343–364. https://doi.org/10.1089/ast.2017.1674

11. https://www.nobelprize.org/prizes/chemistry/2020 accessed October 12, 2021

12. For more on how Star Trek intertwines with real-world biology, Mohamed Noor goes where no one has gone before in his engaging 2018 book *Live Long and Evolve.*

13. As I write these words, humanity suffers the toll of the zoonotic SARS-Cov-2 virus that jumped a species barrier to kill over 6.2 million people so far, as of May 24, 2022, up nearly fourfold from the 1.6 million deaths tabulated when I first checked in writing this book: https://coronavirus.jhu.edu/map.html.

14. Mooney, H.A., and E.E. Cleland 2001. The evolutionary impact of invasive species. *Proceedings of the National Academy of Sciences USA.* 98: 5446–5451. https://dx.doi.org/10.1073/pnas.091093398; Mazza, G. et al. 2014. Biological invaders are threats to human health: an overview. *Ethology Ecology & Evolution.* 26: 112–129. https://dx.doi.org/10.1080/03949370.2013.863225

15. https://www.nasa.gov/mission_pages/apollo/missions/index.html and https://www.nasa.gov/specials/artemis accessed October 20, 2021

16. Pavitt, K. 1991. What makes basic research economically useful? *Research Policy.* 20: 109–119. https://doi.org/10.1016/0048-7333(91)90074-Z

2

Selection, naturally and otherwise

When you take a slow walk, jaw-dropped in awe, through the paleontology wing of a natural history museum, your eyes absorb the dusty minerals of fossil bones, bones trapped in cracked brown rock. You see them at a distance, resting in their old age, venerated behind the glass of the well-lit display. In my case, it is the Royal Ontario Museum (ROM) at the southwest corner of Bloor Street and Queen's Park in downtown Toronto. I am in the ROM's impressive dinosaur gallery, surrounded by its high angular ceiling in the portion of the museum built to resemble an enormous crystal, fabricated out of architectural glass and steel. The sign on the display in front of me declares that the fossil remains belonged to an ankylosaur named *Zuul crurivastator*, an intimidating extinct creature discovered by my colleague David Evans, that rummaged around what is now northern Montana, just over 75 million years ago.[1] Those old rocks and foreign skeletons might leave you with the false impression that evolution is a thing of the past, something glacial in speed, an abstract concept that does not impinge on the timeframe of daily experience. This impression is, I've got to tell you, entirely mistaken.

Just ask the *E. coli* bacteria in your gut that can jump from one generation to the next every 22 minutes.[2] That pace means a bacterial cell can become a great-grandmother in about an hour. In the time it takes you to live through the weekend, *E. coli* experience a generational span equivalent to the time since the dawn of the Roman empire, over 100 generations in human terms. If bacteria don't seem a fair reference for animal evolution, let's look at an animal. The little nematode roundworms that populate the Petri dishes in my laboratory at the University of Toronto are a good example. They have nerves, muscle, and skin, skin that is technically termed *cuticle*; they eat and they poop and they can have sex. They aren't as quick to propagate as the *E. coli* that they like to eat for breakfast, but in that same single weekend, a baby worm can grow up to make babies of its own. The figurative Roman empire started about six months ago for these *Caenorhabditis elegans* nematodes, as they are known in Latin. Do you prefer something with a backbone? With up to five generations per year, house mice had their four-pawed Caesar Augustus about two decades ago. Even your trusty mutt speeds its way from puppy to papa in the relative blink of an eye, meaning that, in dog years, a canine Rome would have packed the alleyways behind the taverns of the barking 1920s.

This contrast of timescales tells us that, for many creatures, time proceeds much faster than it does for humans. But they aren't breaking any laws of physics. Evolutionary *time*

Evolving Tomorrow. Asher D. Cutter, Oxford University Press. © Asher D. Cutter (2023). DOI: 10.1093/oso/9780198874522.003.0002

is measured not in maximum lifespan or in years but in generations, and generation times differ from one species to the next. Evolutionary *change* is measured across those generations from DNA differences and from the traits that are influenced by those genetic differences. Many organisms pass through generations in just a fraction of a year, not the nearly three decades, on average, that it takes us. Of course, our current average generation time of about 30 years is slower than it could be,[3] given that the onset of reproductive maturity takes less than half that time. As a consequence, the genetic changes that alter the development of what those other organisms look like can evolve at a much faster clip, when we use years as a yardstick. That faster clip can be fast enough to see as we live out our lives.

★

Time is important, but it is not the only way to calibrate evolutionary change. I am sure that you also think of evolution as the change in how creatures look. Paleontologists measure the change in the measurable traits of species over eons, millions of years. But plenty of other biologists measure evolutionary change in the traits of today's animals and plants. And they do it with an eye on their annual calendars. They measure evolution happen, as it happens.

Seasonal flu marks the calendar with its evolution as we watch it and fret. Flu is seasonal because it evolves, and because it evolves faster than we do. When we catch the flu virus and recover from it, we become immune from reinfection by that viral genotype forevermore. But the viral genome mutates: as it replicates in infected cells, its imperfect copying mechanism creates molecular changes in the flu's genetic material. Some of those mutations alter the shape of key viral proteins so that our immune system no longer recognizes them to defend itself. In particular, new mutations to the proteins that cloak the outside of influenza virus particles—the hemagglutinin and neuraminidase proteins—give flu a clever new disguise.[4]

This evolution is rapid enough that, each year, the flu virus evolves into a new form that evades our prior immunity, a fact that public health programs try to counteract by distributing a new flu vaccine each year. By sequencing the flu genomes across the world on an annual basis, we can track the rapid evolutionary changes as they occur (Figure 2.1). We can track the changes through time to see evolution as it happens and, to some degree, to predict the genotypes of future outbreaks to assist with vaccine design.[5] Influenza might even seem quaint in comparison with humanity's recent experience with SARS-CoV-2 (a.k.a. the cause of the COVID-19 coronavirus disease), which has evolved month over month to outpace even the exceptionally fast pace of COVID-19 vaccine deployment.[6]

Viruses aren't the only things that evolve rapidly. Moths might not kill us, but they can evolve fast. The peppered moth *Biston betularia* seems an unassuming species, as if it were determined to avoid notice. The reality is that natural selection picks off the conspicuous moths, leaving the camouflaged wallflowers to survive in its wake. As a result, *Biston* moths changed the color of their coats, twice, evolving to be less conspicuous to avian predators—all as a byproduct of human impacts on the color palette of the places that the moths like to sleep.

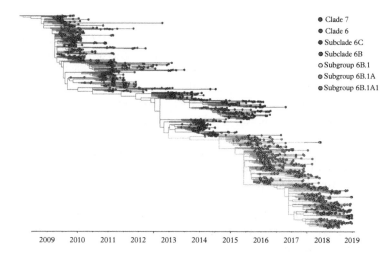

- ● Clade 7
- ● Clade 6
- ● Subclade 6C
- ● Subclade 6B
- ○ Subgroup 6B.1
- ◉ Subgroup 6B.1A
- ◉ Subgroup 6B.1A1

2009 2010 2011 2012 2013 2014 2015 2016 2017 2018 2019

Figure 2.1 *Gene tree relating isolates of A(H1N1)pdm09 influenza viruses in Kenya based on evolutionary divergence over time in their genome sequences. Each dot represents an individual viral genome sampled at a given point in time, with branches representing an evolutionary inference of their historical relationships to one another. Color coding of different subgroups ("clades") corresponds to distinct versions of the hemagglutinin gene (HA) that are important in infection, with different HA types evolving to become common at different points in time.*

Image credit: Owuor, D.C., et al. 2021. Characterizing the countrywide epidemic spread of influenza A(H1N1)pdm09 virus in Kenya between 2009 and 2018. *Viruses.* 13: 1956. https://doi.org/10.3390/v13101956; reproduced under the CC BY 4.0 license.

Back in Charles Darwin's day, peppered moths mostly had white wings speckled with black and gray flecks that allowed them to blend in with the mottled texture of the tree trunks of southern England. Then the Industrial Revolution came into full force. Bark became sooty and black around urban centers, and by the early 1900s, nearly all the peppered moths to be found were of a melanic color, this dark black form having evolved as a consequence of rapid increase in the abundance of a genetic variant of the *cortex* gene that influences wing pigment deposition.[7] The camouflage afforded by moth wing coloration when resting on dark versus light substrates was a decisive factor in bird predation: nine species of birds devour most efficiently those moths with a mismatch.[8] Pale moths sleeping on sooty tree trunks didn't stand a chance against hungry birds.

In the mid-1900s, however, pollution levels declined sharply. The pale form re-emerged from obscurity in the 1970s, along with the corresponding genetic variant of the *cortex* gene that made them whitish of wing. The peppered form re-evolved over the span of a few decades into the predominant coloration of *B. betularia* that you can find in the UK today. Nowadays, it's the melanic form that gets picked off by peckish birds.

A third icon of real-time evolution are the birds of the Galapagos Islands, the collection of species known as Darwin's finches. Charles Darwin snapped up 31 specimens of *Geospiza* during his stopover in the Galapagos in September to October of 1835, before his ship, the *HMS Beagle*, headed west across the Pacific Ocean. Those 31 birds, however, created more questions than answers as his colleague John Gould began picking them over back in England. Over 100 years later, in 1973, Peter and Rosemary Grant initiated what would become half a century dedicated to the study of those questions. The Grants and their team spent months of each year with Darwin's finches in the Galapagos to characterize the birds' trials and tribulations as they lived and died, generation after generation, on the dry and rocky volcano tops that pin-prick the salty Pacific 1000 kilometers (600 miles) offshore of the Ecuadorian mainland of South America.

What the Grants chronicled was a compendium of evolutionary change.[9] Among the many aspects of evolutionary change documented in Darwin's finches since the 1970s, evolution of the birds' beaks has proven especially profound. The finch beaks play a key role as a tool in acquiring energy from hard-to-handle food. Beak size and shape of a given species of *Geospiza* changed year over year—some annual intervals of evolutionary increase in size, and other intervals of decrease—with corresponding changes in abundance of genetic variants associated with beak morphology.[10] The rapid evolution of Darwin's finch beaks comes about due to the combination of two major factors: one, fluctuations in the availability of seeds that comprise their food in the dry season, and two, competition within their own species and with other species that have distinct specializations to hunting down and accessing seeds to eat. Gene variants that make beaks with the best match to the kind of food that is most available lead the corresponding birds to get enough food to survive and reproduce, those genes then being more likely to propagate into the next generation.

2.1 Evolution by natural selection

The hasty evolutionary change seen in influenza, peppered moths, and Darwin's finches owes its speed to natural selection. Natural selection, the most exalted of the ancient forces of evolution, is our name for the biological pressure on the characteristics of organisms that separate individuals by their ability to survive and reproduce. Evolutionary change will result when those characteristic differences can be inherited. What evolution by natural selection really does is to extinguish—kill off—those heritable features of organisms that make them worse at surviving or worse at successfully making offspring, leaving those alternative forms of the features to persist.

From this terse description, I hope you have gleaned three things about natural selection. First, a population, a group of organisms from the same species, must contain individuals that differ in some measurable characteristic in order for natural selection to have opportunity to sift among the alternatives.[11] That is, there must be *variation* in a trait. Second, the trait—beak size, wing color, immune evasiveness—will experience natural selection only when particular values of the trait are associated with the tendency of individuals to survive to reproduce successfully or unsuccessfully. In biologist's jargon, differences in the trait must influence *fitness*. These first two elements are enough to get natural selection, but not enough to get evolution as a result of natural selection.

Third, to get evolutionary change across generations, that is, to get a change in the average value of the trait that you can measure among individuals in the population, it must be the case that the trait differences can be *inherited* from parent to offspring.

When this trinity of circumstances comes to pass—inherited variation in fitness—then the population has adapted, and that new aspect of the trait has earned the name of adaptation. We do not mean adaptation in the sense of *acclimation*, which usually refers to a transient physiological state, as when you acclimate to the initial chill of a lake after jumping in. This word *adaptation* is a bit funny, it's both a noun and a verb. The word *adaptation* describes, on the one hand, a feature that has evolved as a result of natural selection that bestows a match to the creature's life circumstances and, on the other hand, the ongoing generation-to-generation change in response to natural selection. In both cases, adaptation requires natural selection.

The reciprocal, however, is not necessarily true: natural selection on a trait might *not* produce an adaptation if there are no genetic differences creating the source of variability between individuals. For example, survival in a war zone might favor the ability of people to see infrared light. However, differences between individuals in whether they can see infrared light depends entirely on differences in whether they have a pair of night-vision goggles,[12] rather than genetic differences. No evolutionarily adaptive response to infrared vision can result.

If you had a conception of evolution as operating on geologic timescales, you are not alone. Charles Darwin had a similar thought,[13] and he was a pretty bright guy who figured out from scratch quite a lot of the logic for how evolution works. One of his grand ideas was to apply the idea of geologic uniformitarianism to living systems. In geology, simple and slow but perpetual physical processes, like erosion and volcanism and tectonic uplift, can nonetheless create massive features on the landscape when carried out over millions of years. Those slow but perpetual processes gave us the Grand Canyon, the Hawaiian Islands, and the Chilean Andes.

Darwin conceived of natural selection as a similarly simple force that shaped the inherited characteristics of organisms. His tremendous legacy meant that most folks who learn something about evolution, even now, more than 160 years after the publication of Darwin's most famous book,[14] learn that it involves the slow accretion of minute changes across millennia that eventually add up to something big. This view of evolutionary change as slow and plodding is not wrong, but it is incomplete: it neglects the fact that evolution can proceed with rapid change.

The examples of viruses and insects and birds point out how evolution has cantered along at a pace that scientists can measure in real time in their own lifetimes. They show how we can quantify evolutionary change in the wild over timespans of decades or years. I will soon show you even faster evolution.

★

At its simplest, we can summarize what evolution is as the change across generations in the genetic composition of a population.[15] The traits that evolved in peppered moths and Darwin's finches were due to changes in genetic composition, changes to versions of genes that influenced camouflage coloration or feeding ability through beak shape. Not every change to a genome, however, alters the outward appearance of an organism in a way that you can perceive with your eyes or measure with calipers. Evolutionary changes to the immune system that confer resistance to pathogens, for example, may appear invisible to the naked eye under normal circumstances. In other words, evolution can happen despite no obvious alteration of the visible characteristics of a species. In fact, this is a very common aspect of evolutionary change, and these days it is easy to prove. To prove that genetic change has indeed taken place, we simply peer into the genome to see evolution recorded in its DNA.

Some species look so similar from one to the next that even experts on anatomy are hard-pressed to see any evolutionary differences between them. And yet, they may inhabit different places or be incapable of interbreeding, and so ought to be considered to be biologically distinct species. In such situations, the groups of organisms are sometimes called cryptic species. It takes analysis of something you can't see with the naked eye, DNA sequences, to sort out how cryptic species are biologically distinct from one another.[16]

You might think crocodiles are big enough that they'd be easy enough to tell apart. In western and central Africa, however, it turns out that the species question has long been an open question for *Mecistops*, slender-snouted crocodiles. Across the range, the crocs vary, but it's tough to get close to live animals for careful inspection to be certain of consistent differences. After measuring skull shape and DNA sequence for over 100 croc specimens, however, biologists found that those animals from Ghana, The Gambia, and Côte d'Ivoire are, in fact, biologically distinct from the crocs on the other side of the Cameroon Volcanic Line that find their home in Gabon and the Democratic Republic of the Congo. The fish-eating slender-snouted crocs comprise two distinct but cryptic species that split from one another about 7 million years ago.[17] This crocodilian separation arose, in part, as a geographical outcome of the region's volcanic activity that divides the landscape with a great physical barrier; that geological barrier influences other species groups as well. Comparisons of DNA, invisible to the naked eye, revealed the genetic changes of millions of years of evolution.[18]

The idea of so-called living fossils offers a variation on the cryptic species theme. In these cases, the characteristics of a modern-day species may look so similar to some ancestral form recorded in the fossil record that it seemingly appears like the creature has persisted in its present form and will continue to stay the same for all eternity. There are many such examples: horseshoe crabs, ginkgo trees, tree ferns, deep-sea soft sea urchins, *Lingula* lamp shell brachiopods, Australian lungfish, and coelacanths.

The coelacanth, a deep-water fish now known to inhabit the Indian Ocean, was recognized by Western scientists for 100 years only from fossil rocks dating to the Cretaceous period and earlier.[19] That is, until a fleshy specimen caught by fishermen off the coast of South Africa made headlines in 1938. It was a spitting image of the fossil forms. More recent decades have recovered many additional specimens for study, even leading to their genome getting sequenced in 2013, a feat that documented how it has accumulated substantial DNA sequence change despite the seeming stasis in many aspects of its outward appearance.[20] But such stasis is superficial: populations of "living fossils" are still subject to evolution on the inside. Their DNA inevitably acquires mutations each generation, some of which accumulate over time to cause the DNA of the present to differ from the past. All that genetic change marches on invisibly, sometimes involving extensive reconfiguration of their genome structure, despite little obvious outward consequence when constraints are sufficiently intense on the measurable traits of their body.

We will talk more in a bit about what can cause such constraints on evolution. A little later, too, we will explore other ways that we can use an understanding of DNA to our advantage for understanding evolution's past and for influencing its future.

2.2   The breeder's equation

Traits will evolve faster when selection more strongly favors extreme forms. That is, evolution is fastest when extreme forms confer an improved likelihood of survival and reproduction. One way to have stronger selection is for humans, rather than ecological circumstances, to do the selecting. In contrast to natural selection in the wild, this artificial selection operates through the selective breeding of a subset of individuals. A human chooses some individuals that have particular characteristics that they want to persevere or elaborate. Artificial selection, it turns out, is one of the first and most influential biotechnologies ever devised by humanity. It was only several millennia later that Charles Darwin coined the term *artificial selection* as a human hijacking of the rules for how natural selection works.[21]

Darwin's finches of the Galapagos, as we have seen, bear the weight of his name as the avian emblem of evolution by natural selection in the wild. But pigeons—also known by the more magniloquent moniker of *rock doves*—are the birds that Darwin knew best and that informed his ideas about artificial selection. The more than 10 pages devoted to the rock dove in the opening chapter to *On the Origin of Species*, apparently, simply wasn't enough to give the pigeon its due. In volume one of his 1868 book *The Variation of Animals and Plants Under Domestication*, Darwin devoted nearly 100 pages to pigeons between Chapter V "Domestic pigeons" and Chapter VI "Pigeons—*continued*," which reads as a sort of biological love story to selective breeding of these birds.

"Fancy" pigeons were bred by pigeon fanciers, a hobby and livelihood that long predates Darwin and even Shakespeare.[22] In the metropolises of the world, you may know them as rats-with-wings, but *Columba livia* rock doves were brought into the aviaries and rooftop coops of fanciers from cliffs in the wild of their wide native range from Spain to Sri Lanka, Burkina Faso to Kyrgyzstan. An early book on the subject, with the evocative title *Columbarium*, gave the state-of-the-art methods for pigeon breeding, as it stood in 1735.[23] Fancier clubs around the world now record over 1000 breeds of fancy pigeons, with some individual race events awarding over $1 million US dollars in prize money.

What makes a pigeon fancy? There is the style of the feathers, of course: the wide arc of the Fantail similar to a tom turkey, the poofy head crest of the Jacobin, the curlicue ringlets that coat the wing feathers of a Frillback, the long feathery slippers on the feet of the Trumpeter.[24] Feathers may be black or white or gray, blue-ish or reddish or iridescent, checked or barred or barless. Trumpeters also coo with a distinctive call reminiscent of the brass instrument that gives them their name. Tumbler pigeons are airborne acrobats, performing backflips in midflight, whereas Parlor Rollers are terrestrial acrobats, traversing across the ground head-over-heels in repeating somersaults. There is the Giant Runt, which is big and fat and good to eat. The various breeds of Pouters have a broad flap of skin beneath their chin, a crop, that they inflate into a balloon as big as the rest of their bodies, a mimic of the inflated throats of frigatebirds and frogs. There are pigeon breeds with long straight necks, or with recurved necks set so far back that their heads practically rest on their rumps; there are pigeon breeds with long and slender beaks, beaks thick like a parrot or a dodo, and there are short stumpy beaks that are barely there at all. This diverse spectrum of characteristics is all due to genetically distinguishable variants that human breeders selected upon for their novelty over the course of several thousand years.[25]

Throughout much of human history, artificial selection proceeded with whims and sharp eyes, and the wisdom wrought by a hungry stomach. This is how we got corn kernels the size of peas adhered to a cob nearly as long as your forearm from the grass seeds of its little, hard, brown ancestor teosinte. It's also how we got the droopy skin-folds of Neapolitan mastiffs and the scurrying legs of a Dachshund. These traits are, in essence, adaptations to being raised by humans. Rather than call them adaptations, as it was artificial selection not the wilds of nature imposing natural selection, we might call such features "domestications." Nowadays, we're more inclined to take a quantitative approach to producing domestications, meaning that we take numerical measurements.

By measuring the features of each member of the population, the subset used to breed, and the resulting offspring, we can calculate predictions for how quickly a feature can change across generations. We can even write down a short three-piece equation to describe it. The so-called breeder's equation in its simplest form is: $R = h^2 S$. This equation does a very good job at predicting the evolutionary response as the change from one generation to the next (R, sometimes also referred to as ΔZ). More specifically, it does a good job provided that we know two things.

First, we need to know the genetic heritability of the trait (abbreviated as h^2). We can calculate heritability by comparing measurements of both parents to measurements of all their offspring in a collection of individuals from the population. The heritability tells us how predictable the feature of offspring is, given that we know the value in the parents. A value of zero means that the parent traits give no extra information about what the offspring traits will be. A value of one means that you know with certainty what parents will produce in their kids. Most real traits have a heritability value between 0.1 and 0.8, depending on the kind of characteristic. For example, human caffeine consumption has a heritability of approximately 0.4, and height has a heritability of 0.79.

Second, to predict an evolutionary response accurately, we need to know the difference in the value of the traits that we measured for the subset of breeding individuals relative to all individuals (S, also known as the selection differential, sometimes referred to as the selection gradient β). With artificial selection, this info is easy to get hold of.

After all, with artificial selection, we are the ones who decide which individuals breed and which don't.

Say, for example, you loved guinea pigs and wanted more of the adorable little fuzzballs. On average, guinea pig sows produce 3.3 guinea piglets, but they can birth up to 8 in a litter. You might apply artificial selection to breed an especially fecund population of guinea pigs by only letting those parents that produce five or more in a litter propagate the next generation. With an average selected litter size of 6, the selection differential would be $S = 6-3.3 = 2.7$ pups. Because litter size in guinea pigs has a heritability of about 10% (in other words, $h^2 = 0.1$), we would expect this procedure to increase the average litter size by one pup after four generations of selective breeding ($R = h^2 S$, so $R = 2.7 \times 0.1 = 0.27$ pups in the first generation).[26]

This whole mathematical procedure was formalized in the mid-1900s in a branch of biology that is now known as quantitative genetics. Sophisticated versions of the breeder's equation are still used by animal and plant breeders to enhance food production, whether milk volume, poultry muscle, or wheat harvest date.

Just as temperature can be measured in units of kelvins and electric potential in units of volts, we have a unit of measure for evolutionary change: the "haldane."[27] The haldane was named after the famous-in-select-circles geneticist John Haldane (1892–1964). Known as Jack to some and JBS to others, Haldane capped off his politically active and colorful life by migrating from Britain to India in 1958. As Peter Medawar put it, "He could have made a success of any one of half a dozen careers—as mathematician, classical scholar, philosopher, scientist, journalist or imaginative writer. To unequal degrees he was in fact all of these things." During the First World War, Haldane also made successes of his fascination with explosives as a disturbingly fearless Trench Mortar Officer of the Black Watch on the front lines and while running a bomb training school for the British military.[28]

Let's look at the explosion of evolutionary change at the hand of artificial selection by horse breeders, as measured in haldanes. The racing speed of Swedish standardbred trotter horses changed by 0.37 haldanes over a period of about nine horse generations. What this means in racing terms is that this selective breeding achieved horses that take, on average, less than 78 seconds to run one kilometer (0.6 miles) from ancestors that averaged a one-kilometer race in about 81 seconds just nine generations prior. If you want to make a long bet, continued evolution along this trajectory predicts the best average race time to dip below 70 seconds around the year 2046.[29]

2.3   Selection in relation to sex

Survival defines one aspect of fitness, leading natural selection to distinguish among organisms by their ability to get the chance to reproduce. But in that postsurvival window before reproductive success actually gets to manifest as the next generation's offspring, there is further jockeying for who will mate with who and for who gets to decide. This is the stuff of sexual selection—a special incarnation of natural selection that gets its own name and its own purview—the competition between rivals, the choice of mates, and the potential antagonism between the sexes over courtship and copulation and fertilization.

Eggs and sperm—gametes—and the reproductive organs that make them define the primary sexual characteristics of an individual. These primary sexual characteristics arrive on the scene of adulthood with an entourage of secondary sexual characteristics that are the hallmarks of maturity. The secondary characteristics further distinguish the sexes from one another. In species with two sexes, it makes them sexually dimorphic.[30] Typically, the dimorphic characteristics have something to do with the lead up or aftermath of reproduction.

The teats and mammary glands to nurture young, for example, form the secondary sexual structure that give us mammals our name, despite typically developing fully only in females. Here are three other quick examples of sex difference, known as sexual dimorphism. The full-body scarlet plumage of an adult male cardinal, in contrast to the camouflaged and mostly dusty-brown plumage of female and juvenile birds. The 125-fold-more-massive body of the female golden orb-weaving spider *Nephila pilipes*.[31] The massive antlers of bull moose that these males grow in advance of the autumn rutting season, accounting for nearly 5% of their body weight. A cow moose doesn't make antlers at all. As testament to the seminal role of antlers as a weapon for sexual combat rather than some other kind of weapon, bulls will discard the bulky appendages at the end of each mating season, shed for the winter just to regrow a bigger rack starting the next spring. Scaled to human proportions, it would be like a man growing a 3.5 kg (eight-pound) disposable doodad on his noggin in preparation for his wedding day, and doing so again for every anniversary thereafter.

Two of these three example dimorphic traits arose through sexual selection and sexual conflict, that corollary mode of natural selection (by the end of the chapter, you'll know which example didn't). If ecological circumstances are the selecting agent with natural selection and humans are the selecting agent with artificial selection, then mates and rivals are the selecting agents with sexual selection. Sexual selection operates through interactions like competition between members of the same sex as they try to gain access to resources—nest sites, food-rich territories, and prime mating locations—or access to members of the other sex to allow them to reproduce. At 3200 kg, more than triple the mass of a female,[32] a bull elephant seal, *Mirounga leonina* uses his own sheer body weight as a weapon against rival males, underscoring the power of sexual selection to shape animal form when the benefits to survival and reproduction are high enough.

Sexual selection also can occur between members of different sexes, involving traits that influence things like the choice of mates and how frequently mating will occur. In particular, males and females may have distinct optima for such traits. In many species, the high resource investment in eggs and parental care by the female sex leads to selection favoring just a few choice matings, with matings to too many males placing excess demands on energy and risk of injury, predation, or disease. By contrast, the low per-gamete investment of males can lead selection to favor many mating partners as a means to contribute genetically to as many progeny as possible.

This situation yields so-called sexual conflict, which has the potential to drive an evolutionary arms race in the elaboration of sexually dimorphic traits and, as a result, for secondary sexual characteristics to diverge between species. You can see the evolutionary consequences of such sexual conflict if you look closely at the long legs and antennae of water striders, for example. Water striders are those long-legged black bugs that skim across the surface of still water in ponds and lakes and streams. They are small, but sexually bellicose. Consequently, females have evolved lever-like spines to facilitate mate-rejection to prevent unwanted copulation with undesirable or harassing males: female water striders don't seem to benefit from repeated mating (unlike males), and the distraction of mating reduces their ability to evade predators, making them more likely to get eaten. Males, in turn, have evolved sophisticated grasping structures to facilitate mate-retention to evade rejection, their side of an ongoing evolutionary antagonism of sexual adaptation and counter-adaptation.[33]

Sexual selection is not the only thing that can drive sexual dimorphism, however. For example, it was natural selection acting in a sex-biased way that favored the production of more offspring that drove the evolution of female gigantism in those 5-cm-long *Nephila pilipes* spiders (2-inch bodies with 8-inch leg spans) that are 125-times bigger than their mates. Larger females can produce disproportionately more offspring, regardless of how many male mates they have.

Animal weaponry can be one evolutionary outcome of sexual selection, and it is sometimes intended to be obvious. Horns and antlers and tusks advertise the might of their bearers, a warning of what comes from messing with something hard and backed by muscle. Their shape may be a simple curved cone, as for the broad black arcs of bison horns. More elaborate, they may coil or twist, the better for ramming, as in *Ovis canadensis* bighorn sheep, the males of which crash horn-to-horn at velocities over 80 km/h (50 miles/h).[34] The embellishments of bifurcating antlers into multi-pronged crowns among elk and deer are astonishingly intricate. Their chemical compositions vary by evolutionary origin: the keratin of a rhino horn made tougher than simply a glorified fingernail, the calcified bone of up to 37 kg (80 pounds) for those bull moose antlers, and the stony hydroxyapatite mineral of dentin in the exaggerated teeth that grow into tusks approaching 1 m (3 feet) in walruses and 2 m (7 feet) in elephants.[35] While these conspicuous displays and weapons can be used defensively against would-be predators, it is the pressures of mating that are most crucial in their origins and persistence.

Conspicuous displays of one species to another often serve as a scare tactic: they set off alarm bells of warning. But when presented to the opposite sex of one's own species, glitzy colors often are intended to chime wedding bells. These are the salacious displays, the behaviors expressed explicitly to solicit sex. Some are simple, the wilderness equivalent to swiping right; some are elaborate, the wilderness equivalent to a synchronized flash mob wedding dance routine.

Growing up as a Yankee New England kid, summertime family lobsterfests on the coast introduced me to the largest arthropods in the world, the Atlantic lobster *Homarus americanus*,[36] as well their small cousin the fiddler crab. The claws of *Uca* fiddler crabs are not all the same size. Every child knows this who has found themselves mucking about along the muddy banks of a salty Cape Cod estuary in summer and who can't help but get pinched when they try to catch the quick creatures. The left-right asymmetry of big and small claws of fiddler crabs is unusually extreme, however, and asymmetric between the sexes.

Female fiddler crabs actually have two normal-sized claws. It is only the males that have one hyper-exaggerated claw that accounts for up to two-thirds of their total body weight.[37] Those big claws, however, are almost all for show (Figure 2.2). Males wave it at one another, in the hopes of winning confrontations without need for physical contact. Males also wave their cumbersome claw at females, who prefer the superstimulation of larger claws waved faster and higher, as males do in the hopes of winning their hearts. By which I mean: winning the chance to lure her into his mud cave, plugging the exit for a few days to then copulate in the dark wet muck until she leaves to let the hatched zoea emerge from her egg sponge into the brackish water.

The seemingly preposterous size of a male *Uca*'s big fiddler claw seems out of sync with what natural selection would favor for survival. It is bulky and requires extra effort to make and to use; it is costly; and it is a squandering of resources. It takes a vigorous individual to wield it, a high-quality male in good condition to produce and use it effectively. In other words, it is a good signal of an especially good mate, an honest display of quality for females to discriminate among potential mates. In this way, evolution has led males to handicap themselves in terms of natural selection on survival for the sake of enhanced sexual selection on reproduction.

★

Evolutionary time is measured in generations. Evolutionary change across generations requires a population of individuals with heritable differences among them. Under some circumstances, evolution by natural selection can proceed with remarkable speed.

Figure 2.2 *Fiddler crabs, such as this* Uca leptodactyla, *display striking sexual dimorphism in claw morphology and behavior as an evolutionary consequence of sexual selection.*
Image credit: Wilfredor, reproduced under the CC0 1.0 license.

Some evolutionary change is cryptic to the naked eye, but easy to decipher from DNA sequences. Sexual selection is also responsible for spectacular evolution of characteristics and genes important for reproduction. Selection can be manipulated by people, with such artificial selection capable of producing predictable evolutionary change.

Notes

1. Arbour, V.M., and D.C. Evans. 2017. A new ankylosaurine dinosaur from the Judith River Formation of Montana, USA, based on an exceptional skeleton with soft-tissue preservation. *Royal Society Open Science*. 4: 161086. https://doi.org/10.1098/rsos.161086.

2. Chandler, M., et al. 1975. The replication time of the *Escherichia coli* K12 chromosome as a function of cell doubling time. *Journal of Molecular Biology*. 94: 127–132. https://doi.org/10.1016/0022-2836(75)90410-6.

3. According to 2019 data from about 30 countries collated by the international Organization for Economic Co-operation and Development, the average age of first childbirth was 29.2 years, and the average age at which women birthed a child overall was 30.5 years. https://www.oecd.org/els/family/database.htm Accessed April 19, 2022.

4. The epidemiological importance of hemagglutinin (H) and neuraminidase (N) is highlighted in the nomenclature that we use to reference distinct subtypes of flu, such as the H1N1 or H3N2 subtypes of influenza type A. Earn, D.J.D., et al. 2002. Ecology and evolution of the flu. *Trends in Ecology and Evolution*. 17: 334–340. https://doi.org/10.1016/S0169-5347(02)02502-8.

5. You can interact with the history and future of flu evolution, along with a number of other infectious diseases, through this publicly-available resource: https://nextstrain.org/flu/seasonal/h3n2/ha/12y?c=ep. Accessed April 28, 2021. Agor, J.K., and O.Y. Özaltın. 2018. Models for predicting the evolution of influenza to inform vaccine strain selection. *Human Vaccines and Immunotherapeutics*. 14: 678–683. https://dx.doi.org/10.1080/21645515.2017.1423152.

6. Flu only seems quaint, given Covid's echo of the 1918 flu pandemic and the mortality from seasonal flu accounting for nearly half a million lost lives each year. Paget, J., et al. 2019. Global mortality associated with seasonal influenza epidemics: new burden estimates and predictors from the GLaMOR Project. *Journal of Global Health*. 9: 020421. https://doi.org/10.7189/jogh.09.020421.

7. van't Hof, A.E., et al. 2016. The industrial melanism mutation in British peppered moths is a transposable element. *Nature*. 534: 102–105. https://doi.org/10.1038/nature17951.

8. Cook, L.M., et al. 2012. Selective bird predation on the peppered moth: the last experiment of Michael Majerus. *Biology Letters*. 8: 609–612. https://doi.org/10.1098/rsbl.2011.1136.

9. Jonathan Weiner regales the first 15 years or so of the prodigious research from the Grants' who's-who conveyor belt team of scientists in his 1994 book *The Beak of the Finch*. More recently, the Grants themselves summed up four decades worth of study, though even this book truncates the most recent analyses that delve into the genomes of Darwin's finches.

Grant, B.R., and P.R. Grant. 2014. *40 Years of Evolution: Darwin's Finches on Daphne Major Island*. Princeton, NJ: Princeton University Press.;Grant, P.R., et al. 2020. Darwin's finches, an iconic adaptive radiation. In *eLS*, John Wiley & Sons, Ltd (Ed.). https://doi.org/10.1002/9780470015902.a0029107.

10. Grant, P.R., and B.R. Grant. 2006. Evolution of character displacement in Darwin's finches. *Science*. 313: 224–226. https://dx.doi.org/10.1126/science.1128374.; Lamichhaney, S., et al. 2015. Evolution of Darwin's finches and their beaks revealed by genome sequencing. *Nature*. 518: 371–375. https://doi.org/10.1038/nature14181.

11. Biologists often refer to measurable traits as "phenotypes," but I will mostly avoid that jargon term in favor of synonymous words like *trait, feature*, or *characteristic*. The key aspect of all of these terms for our purpose is that they refer to biological features that one could distinguish or measure in some way.

12. Or, supposedly, application of the cancer treatment compound Chlorin e6. https://scienceforthemasses.org/2015/03/25/a-review-on-night-enhancement-eyedrops-using-chlorin-e6. Accessed December 15, 2021.

13. As Darwin stated eloquently in one famous passage from his magnum opus, "We see nothing of these slow changes in progress, until the hand of time has marked the long lapse of ages, and then so imperfect is our view into long past geological ages, that we only see that the forms of life are now different from what they formerly were," p. 84 in Darwin's 1859 book *On the Origin of Species by Means of Natural Selection*, http://darwin-online.org.uk. Accessed October 12, 2021.

14. Charles Darwin's seventh book, *On the Origin of Species by Means of Natural Selection*, first published in 1859 and now available publicly online along with all his other works at http://darwin-online.org.uk. Accessed October 12, 2021.

15. In Dobzhansky's words, "The most general definition of evolution is change in the genotype of a population."; Dobzhansky, T. 1951. *Genetics and the Origin of Species*. New York: Columbia University Press. p. 21.

16. Struck, T.H., et al. 2018. Finding evolutionary processes hidden in cryptic species. *Trends in Ecology & Evolution*. 33: 153–163. https://doi.org/10.1016/j.tree.2017.11.007.

17. Shirley, M.H., et al. 2014. Rigorous approaches to species delimitation have significant implications for African crocodilian systematics and conservation. *Proceedings of the Royal Society B*. 281: 20132483. http://dx.doi.org/10.1098/rspb.2013.2483.

18. My favorite cryptic species, I must admit, is *Caenorhabditis latens*, a nematode roundworm from East Asia that I study in the laboratory on Petri dishes. To the naked eye, they look like any other roundworm and even under the microscope, they are nearly indistinguishable from several other species. And yet, these different kinds of animals can't interbreed with one another in an entirely successful way. Using genome analysis and experiments, we discovered this evolutionary distinctiveness. My colleagues and I gave *C. latens* its name, meaning "previously hidden," after discerning that it was biologically distinct in terms of both DNA sequence and cross-fertility from another species known as *C. remanei*, which it originally had been lumped in with. Our ongoing experiments seek to discover the genetic secrets of how its

genome evolved to be incompatible with the genomes of its nearest-known relations. Felix, M.A., et al. 2014. A streamlined system for species diagnosis in *Caenorhabditis* (Nematoda: Rhabditidae) with name designations for 15 distinct biological species. *PLoS One*. 9: e94723. http://dx.doi.org/10.1371/journal.pone.0094723.

19. The Cretaceous geological period spanned 145 to 66 million years ago, the famed time that saw the diversification of flowering plants, the origin of mammals, and the extinction of dinosaurs.

20. Because evolution of DNA sequence proceeds relentlessly over time, both with and without evolution of outward characteristics, Darwin's "living fossil" term often is considered misleading. Amemiya, C., et al. 2013. The African coelacanth genome provides insights into tetrapod evolution. *Nature*. 496: 311–316. https://doi.org/10.1038/nature12027.; Casane, D., and P. Laurenti. 2013. Why coelacanths are not "living fossils." *Bioessays*. 35: 332–338. https://doi.org/10.1002/bies.201200145.

21. Darwin, C.R. 1859. *On the Origin of Species by Means of Natural Selection*. London: John Murray. http://darwin-online.org.uk.

22. In William Shakespeare's 1623 play *As You Like It*, the character Rosalind proclaims to her suitor, Orlando, "I will be more jealous of thee than a Barbary cock-pigeon over his hen," referring to the old fancy pigeon breed now typically referred to as the English Barb. The face of the Barb is notable for the pronounced circular orange wattles that circumscribe the eyes and its short beak beneath a rather protruding forehead.

23. This treatise documents the details for dozens of breeds and even contains sage advice regarding provision of a pigeon loft with a "salt cat," apparently a source of micronutrients and rubble for the birds' gizzards, "call'd by the Fanciers a Salt Cat, so nam'd, I suppose, from a certain fabulous oral, Tradition of baking a Cat in the Time of her Salaciousness, with Cummin-Seed, and some other Ingredients as a Decoy for your Neighbour's Pigeons" (p. 14). Moore, J. 1735. *Columbarium: or, the Pigeon-House. Being an Introduction to a Natural History of Tame Pigeons*. London: J. Wilford.

24. The American Pigeon Museum and Library of Oklahoma features an especially delightful gallery of pigeon mugshots: https://www.theamericanpigeonmuseum.org/pigeon-breed-gallery. Accessed October 12, 2021.

25. Many of the gene variants that cause pigeon trait differences have now been characterized at the level of DNA. There's even a pigeon genetics video game: https://learn.genetics.utah.edu/content/pigeons/pigeonetics. Accessed October 7, 2021; Domyan, E.T., and M.D. Shapiro. 2017. Pigeonetics takes flight: evolution, development, and genetics of intraspecific variation. *Developmental Biology*. 427: 241–250. https://doi.org/10.1016/j.ydbio.2016.11.008.

26. Note that R would get slightly smaller with each generation of breeding if the average selected litter size stayed steady at six pups, because the average preselected litter size increases each generation to result in a smaller selection differential. Cedano-Castro, J.I., et al. 2021. Estimation of genetic parameters for four Peruvian guinea pig lines. *Tropical Animal Health and Production*. 53: 34. https://doi.org/10.1007/s11250-020-02473-6.

27. One Haldane of change corresponds to a trait changing at a rate of one "phenotypic standard deviation" per generation. This statistical language of "standard deviation" may sound peculiar. The logic of it is that it lets us compare rates of change for different traits. Different traits may have different units of measure—say, length versus hardness—so, if we were to compare their average change from one generation to the next, we would be hard-pressed to say which feature experienced more evolution. The solution is to standardize the values by how much variability there is among the measured individuals, and this variability is captured with the statistical concept of the standard deviation. About 68% of all individuals have a value within one standard deviation of the average value. For more on Haldanes as a unit of evolutionary change, see Hendry, A.P., and M.T. Kinnison. 1999. The pace of modern life: measuring rates of contemporary microevolution. *Evolution.* 53: 1637–1653. https://www.jstor.org/stable/2640428.

28. Medawar's quote comes from the preface to Ronald W. Clark's 1968 biography, *J.B.S.: The Life and Work of J.B.S. Haldane.*

29. Árnason, T. 2001. Trends and asymptotic limits for racing speed in standardbred trotters. *Livestock Production Science.* 72: 135–145. https://doi.org/10.1016/S0301-6226(01)00274-3; Geiger, M., and M.R. Sánchez-Villagra. 2018. Similar rates of morphological evolution in domesticated and wild pigs and dogs. *Frontiers in Zoology.* 15: 23. https://doi.org/10.1186/s12983-018-0265-x.

30. *Biological sex* refers to the propensity of an individual to produce gametes of a given type, based on their reproductive structures, as well as secondary sexual characteristics. Females typically have organs that can produce eggs, whereas males typically have organs that can produce sperm; hermaphrodites have organs that can permit them to make eggs and sperm. In mammals and birds, chromosomal sex determination is widespread. In many fish and reptiles, among other organisms, however, genetic factors play little to no role in the development of biological sex. These terms are distinct from gender identity, which is how someone self-determines their own sexual representation, and distinct from *sexual orientation*, which refers to the direction of erotic interest. https://teachmag.com/archives/11261. Accessed October 12, 2021.

31. Kuntner, M., et al. 2012. *Nephila* female gigantism attained through post-maturity molting. *Journal of Arachnology.* 40: 345–347. https://doi.org/10.1636/B12-03.1.

32. Weckerly, F.W. 1998. Sexual-size dimorphism: Influence of mass and mating systems in the most dimorphic mammals. *Journal of Mammalogy.* 79: 33–52. https://doi.org/10.2307/1382840.

33. Khila, A., et al. 2012. Function, developmental genetics, and fitness consequences of a sexually antagonistic trait. *Science.* 336: 585–589. https://doi.org/10.1126/science.1217258.

34. Schaffer, W. 1968. Intraspecific combat and the evolution of the Caprini. *Evolution.* 22: 817–825. https://dx.doi.org/10.2307/2406906.

35. Huxley, J.S. 1931. The relative size of antlers in deer. *Proceedings of the Zoological Society of London.* 101: 819–864. https://doi.org/10.1111/j.1096-3642.1931.tb01047.x; Fay, F.H. 1982. Ecology and biology of the Pacific walrus, Odobenus rosmarus divergens *Illiger. North American Fauna.* 74(74): 1–279. https://doi.org/10.3996/nafa.74.0001; Elder, W.H. 1970.

Morphometry of elephant tusks. *Zoologica Africana.* 5: 143–159. https://dx.doi.org/10.1080/00445096.1970.11447388.

36. Atlantic lobster bodies can grow over two feet long (60 cm) and weigh over 40 pounds (20 kg), with the crushing claw being a bit more than half the length of the animal. The typical fare in seafood shacks on the coast of Cape Cod, Maine, and Nova Scotia, however, make meal-sized portions of one-to-two-pound lobsters (0.5–1 kg). Out at sea with my uncle one summer as a teenager, while fishing for scallops off the Cape Cod coast, we pulled up a monster lobster in the trawl the size of a small dog, the biggest I've ever encountered. Wolff, T. 1978. Maximum size of lobsters (*Homarus*) (Decapoda, Nephropidae). *Crustaceana.* 34: 1–14. https://doi.org/10.1163/156854078X00510.

37. Sadly for male fiddler crabs with larger claws, they have lower endurance, poorer locomotor activity, and suffer heavier metabolic costs. Swanson, B.O., et al. 2013. Evolutionary variation in the mechanics of fiddler crab claws. *BMC Evolutionary Biology.* 13: 137. https://doi.org/10.1186/1471-2148-13-137; Bywater, C.L., et al. 2018. Legs of male fiddler crabs evolved to compensate for claw exaggeration and enhance claw functionality during waving displays. *Evolution.* 72: 2491–2502. https://doi.org/10.1111/evo.13617.

3

Ancient forces

Evolution isn't the result of selection alone, whether natural or sexual or artificial. Selection may be the star player, but it is only one member of the team. To understand the whole game of life, we must appreciate the roles played in nature by each evolutionary force. That is how we will understand how the genetic composition of a population of individuals changes over time. It is those changes to genetic composition that cause a species to evolve bigger or smaller claws, longer or shorter legs, brighter or darker coloration. And appreciating all the ancient forces of evolution is how we will become adept in knowing how to guide that change.

3.1   Getting from A to T

If you have ever visited eastern North America, then you know the eastern gray squirrel. Their penchant for city life leads some folks to call them tree rats, despite their fluffy tails and daytime habits. Once upon a time in Washington, D.C., all the gray squirrels were gray. That fact changed in 1902. In that year, under the presidency of Theodore Roosevelt, the National Zoo imported a scurry of eight squirrels from Rondeau Provincial Park, located on the north shore of Lake Erie in Ontario, Canada.[1] The curious thing about gray squirrels in Ontario, however, is that two-thirds of them are not gray, but black.[2]

This migration of squirrels from one population to another introduced the genetic variant responsible for black, melanic fur. That genetic variant is due to a shorter version of the gene that encodes a protein called *melanocortin-1 receptor*, often abbreviated as MC1R.[3] Nowadays, it's not uncommon to spot melanic squirrels throughout the greater Washington, D.C., area, just as I can in my Toronto backyard (Figure 3.1).

In the genetics of nature, we see stories of migration like this as the basis of "gene flow." Gene flow is the exchange of novel genetic variants from one population to another.[4] Gene flow is one of the fundamental ancient forces of evolution that can change the genetic composition of populations.

On occasion, gene flow happens not between disjunct populations of a species but between two wholly different species altogether. This idea might seem far-fetched, as usually we think of separate species as those groups of organisms that can't interbreed

Evolving Tomorrow. Asher D. Cutter, Oxford University Press. © Asher D. Cutter (2023). DOI: 10.1093/oso/9780198874522.003.0003

Figure 3.1 *The dark melanic form of the eastern gray squirrel* (Sciurus carolinensis) *is common in Ontario, Canada, with migration having spread it around eastern North America. The black versus gray fur of these squirrels in Toronto's Queen's Park develops as a result of distinct genetic variants controlling the MC1R protein.*
Photo by the author.

with one another. That is true, as far as it goes. It turns out, however, that species boundaries are not always as impermeable as they might seem—just as sometimes the seams of your raincoat aren't perfectly impermeable to the wet. Males and females, while exploring far fields in their fervor to procreate, may hop the proverbial species fence and make the mistake of commingling with an overly distant relative.

Most of the time, we expect such interspecies breeding to result in abject failure. If any progeny result at all, the hybrid offspring often suffer developmental defects or a mismatch to the environment in such a way that it renders them inviable or sterile. Mules and hinnies are the sterile hybrid offspring of donkeys and horses, the more benign outcome; oftentimes, the interspecies mating of donkeys with horses leads to fertilization failure or to spontaneous abortion. Nonetheless, just the right circumstances may allow hybrid animals to then breed back with members of one of the parental species. Such back-breeding can allow genes from that *other* species to take up residence in the population.

This leaky-seam kind of gene flow is called introgression, where the genetic material from one species manages to make its way into another, usually when it confers some adaptive benefit. It is rare, but detectable. If you've taken a DNA test to explore your ancestry, you might have been shocked to see that your genome may contain up to 4% of its DNA from another species of human, a byproduct of prehistoric interbreeding between different species of *Homo*.[5] Depending on your ancestry, DNA from prehistoric interbreeding of *Homo sapiens* with *Homo neanderthalensis* and *Homo denisova* could comprise up to 104 million DNA base pairs of your genome. The introgressed DNA fragments from these other, now extinct, species of humans are common in the genomes of people with European, Asian, and Indigenous North American and Australian ancestry. Neanderthals and Denisovans inhabited Europe and Asia before *Homo sapiens*, and so *Homo sapiens'* expansion out of Africa brought those subgroups of humans into geographic and sexual proximity. Even though any one present-day person may only have 1% or 2% Neanderthal DNA, modern human genomes around the world have nonetheless archived a lot of Neanderthal and Denisovan DNA. If you were to combine it all together, it could tile across 1600 million DNA base pairs. This amount of DNA squirreled inside of our species represents about half of the length of our genome, if all that archaic DNA were concentrated into a single genomic copy of one person.

Interestingly, the genetic variant of the MC1R gene in gray squirrels that confers black fur also seems to have originated in just this way. The original mutation appears to have arisen in fox squirrels, with subsequent interspecies interbreeding leading the genetic variant for black fur to introgress into gray squirrel populations at some point in their squirrely history.[6]

★

Once novel versions of genes percolate into one population from another, a second evolutionary force can start to do its thing. The genomes of the newcomers, when they interbreed with the locals, intermingle, and, one generation after the next, their genomic contents scramble into new genetic combinations among the descendants. This genetic scrambling is called recombination.

Recombination is the genetic process that can produce novel combinations of distinct versions of genes when those genes are located in different spots in the genome. It changes the relative abundance of *combinations* of genetic variants but does not in itself make any single gene variant any more or less common in the species. Recombination is how you can pass along to one of your children the genetic variant for the dashing cleft chin that you yourself inherited from Dad along with the genetically separate variant that you received from Mom that leads to your elegant detached earlobe. Without recombination, you'd only be able to pass along one or the other of these genetic features—cleft chin or detached earlobe, but not both—in any one of your sperm or egg cells.

This power of recombination to create genetic novelty has its limits, however. It only produces new combinations of pre-existing things. As a result, the power of recombination lies at the mercy of whatever pre-existing genetic differences happen to be present within a species. This means that recombination gives rise to genetic and evolutionary innovation only insofar as the raw materials are already sitting there but sitting in a suboptimal arrangement. This inability of recombination to change gene frequencies from one generation to the next on its own means that sometimes recombination isn't considered to be a fully fledged force of evolution. But recombination is an important aspect of evolutionary change, a seminal sidekick, Robin to natural selection's Batman.

★

Where *do* the raw materials of genetic uniqueness originate? In nature, if you trace back the ultimate source of a distinctive genetic version of a bit of DNA then you will find that its origin lies with a mistake. A mistake in DNA gets a special name: *mutation*. Mutation gives us another member of the team as an ancient force of evolution. There are myriad ways that mutation can alter DNA, but they all come down to a biochemical error in the process by which the DNA double-helix manipulates, copies, and repairs itself as the cell that it resides in follows its developmental path to becoming a sperm cell or an egg cell.[7]

Sometimes the error means that a segment of DNA inserts an extra letter from the chemical alphabet of DNA, one of the nucleotides so often abbreviated as A, T, G, or C, the chemical shorthand for adenine, thymine, guanine, and cytosine, respectively— DNA, of course, being the chemical shorthand for deoxyribonucleic acid. Other times, mutation mixes up letters, say, swapping a G for a C to turn a GATTACA into a CAT-TACA. Sometimes the string of DNA loses a letter, or a whole mouthful of letters, or a segment of DNA inadvertently moves from one spot in the genome to another.

Cells are exceptionally good at avoiding this panoply of blunders. The fidelity of the molecular machines that copy DNA is astoundingly refined, typically making fewer than one mistake for every 30 million nucleotide letters that they copy. But genomes are very long books. Your genome is written with over three billion DNA letters, in duplicate copies, in each of your cells. That means you are bound to have inherited a solid 100 mutations, on average, from your parents.[8] Not only is every individual unique because of the distinct combination of DNA that they inherit from each parent's ancestors, but every individual is unique because of the distinct new mutations cradled in the nuclei of the sperm cell and egg cell that fused to make the single zygote cell from which they then proceeded to develop.

An inheritance of 100 mutations from Mom and Dad might not sound like the bounty of riches you would have hoped for. Part of the sinking feeling in your belly has to do with the pejorative sense of what a mutant means in our cultural imagination. *Mutant* conjures the popular storytelling since the Manhattan project came to its earth-shaking fruition, having spawned movie and comic book takes on mutant-filled postnuclear catastrophe.

But in a genome that is three billion nucleotides long, what are the chances that any one of those 100 mutational events will be detrimental and monstrous? Or, thinking glass half full, how many will confer a special advantage? It turns out that changes to 15% or less of the DNA positions in our genomes could potentially affect our ability to survive and reproduce.[9] Key to how natural mutations arise, however, because they are molecular mistakes, is that mutations arise independently of whether the change might confer a detrimental effect, a beneficial effect, or no detectable effect at all. As a consequence, at least 85% of new mutations to our genomes will be without consequence for our ability to grow to adulthood or to be fertile. Every single one of them, however, does have consequences as a source of evolutionary change and a record of evolution in our DNA.

★

Mutations create genetic uniqueness. This rarity of being unique, however, means that distinctive genetic variants arise in an evolutionarily precarious position. A new muta-tion does not by itself immediately transform the species as a whole. Initially, that new mutation is present in only a single individual. There is some chance that the single individual that was bequeathed the new mutation may not produce offspring in turn. Statistically speaking, the probability that such a brand-new mutation will fail to persist even into just the next generation, because of such chance events alone, is about 37%.[10] Rare things—good, bad, or indifferent—like new mutations, are especially susceptible to chance events. This kind of chance event, unrelated to whether the genetic variant will lead a particular individual to end up surviving to maturity and then reproducing or not, will cause the abundance of a particular genetic variant to fluctuate up or down unpredictably from one generation to the next.

The insecure prospects of a new mutation are all the more acute because mutations often confer detrimental effects, jeopardizing the survival or reproductive capacity of individuals who happen to hold them. This means that individuals that have a new muta-tion that happens to be detrimental would, in fact, be less likely to have offspring of their own. As a consequence, in the wild, natural selection will tend to eliminate those delete-rious genetic variants, driving them extinct. Many other mutations, however, aren't so bad. That is to say that, for many new mutations, for that 85% figure that I mentioned a moment ago, selection neither favors nor disfavors them.

What happens to these not-so-bad a.k.a. neutral mutations? What happens, it turns out, is up to another ancient force of evolution, a subtle and often-forgotten force that works in the shadows of nature: genetic drift.

Genetic drift is easiest to think about for the situation where the effects of one genetic variant or another are neither good nor bad. Regardless of whether your suspenders are brown or blue, they both will hold up your pants. Consequently, the identity of the genetic variant (the color of your suspenders) has no say in the survival and reproduc-tion of its host organism (whether or not your pants stay up). It is a passenger simply along for the ride; it does not create a genetic basis for natural selection to discriminate between different versions of the gene. From one generation to the next, any given ver-sion of such a gene might increase or decrease within the species. The changes to how common gene variants are in the species arise by chance. These chance fluctuations to the genetic composition of a population by genetic drift actually represent a key form of evolutionary change, an evolutionary force distinct from selection, distinct from gene flow and recombination and mutation.

★

Genetic drift leads genetic variants to ebb and flow across generations within the gene pool of a species, like genetic driftwood floating on the sea. The driftwood genetic variant has a life destined for one of two possible eventual fates: to wash up on shore or to list out into the deep to sink to the bottom. We can think of the shore-bound genetic variants as having become a permanent part of the coastline: they have become a fixed feature of the genome for all members of that species. Those genetic variants that sink have gone extinct, lost forever from the gene pool of a species. How fast these fates come to pass depends crucially on how big the population—the gene pool—is.

In this watery analogy, we can think of chance events as the effects on floating flotsam when an ambling toddler stomps their rubber boots in the water. Smaller populations are like a puddle, which will experience bigger splashes relative to the size of the gene pool.

That means a given genetic variant will get tossed about to higher or lower abundance in bigger jerks each generation, more rapidly landing on the shore or drowned to the muddy floor. The gene pool of a very big population, say the size of Lake Erie, will experience only small ripples as a result of that stomping toddler. Genetic drift, therefore, is a weaker force in big populations and a more powerful force when population gene pools are small. Genetic drift will cause faster changes to the genetic composition of a species when that species is made up of fewer individuals than when the species occurs in greater abundance.

Back in 1956, Peter Buri illustrated in an elegant experiment how smaller populations lead to bigger shifts in gene frequency each generation. He painstakingly peered at little flies, *Drosophila melanogaster* fruit flies like those you might find hovering around your kitchen counter bananas in summer. He paid special attention to their eyes, counting those flies with white eyes or with brown eyes. Their eye color depends on which version of the *bw* gene that they happened to inherit from their parents.[11] When he tracked the gene frequency changes from one generation to the next in 107 vial populations, each containing just 16 breeding flies, the relative abundances of the *bw* gene variants shifted up or down by about 6% each generation. By the sixth generation, 4 of those 107 vials contained flies with *only* white eyes or *only* brown eyes: one version of the *bw* gene or the other had become fixed in a vial, making up 100% of the gene copies in each of those four miniature populations. After 19 fly generations, genetic drift had fixed one gene variant or the other in *more than half*—55 out of 107—of the vial populations. In large containers that held hundreds of breeding flies, by contrast, none of the populations had fixed a single version of the *bw* gene during his experiment. Buri showed, with the evolution that he documented in these experiments, how genetic drift is more potent in smaller populations.

Do you think I have wasted your time telling you about how mutations can get buffeted about in abundance from one generation to the next due to genetic drift? After all, neutral mutations exert no influence on a species for good or ill, no benefit or detriment to an animal's survival or fertility, and, therefore, don't deserve our notice? But they do.

Those big, erratic jumps and drops caused by genetic drift in small populations have another important consequence. It means that genetic variants that are beneficial have a harder time swimming their way to shore: natural selection is less effective at differentiating good from bad from inconsequential pieces of genetic flotsam in the gene pool. As a result, sometimes the genetic variant that actually confers a worse ability for organisms to survive and reproduce can end up splashing onto the sand to become a fixed feature of the genome, simply because the size of the population is so small. In this way, genetic drift can influence not only mutations that are neutral in effect, but also the likelihood that natural selection will be able to mete out the all-else-equal fate of beneficial or detrimental mutations. Sometimes, the best mutation can lose, and a worse mutation can win, and that happens because of genetic drift.

This kind of outcome has proved to be a serious problem for the wolves of Isle Royale. An American island isolated in Lake Superior about 20 km (13 miles) from the Canadian shore and due south of the mainland Ontario city of Thunder Bay, Isle Royale was home to just two wolves in 2018 from a peak of only 50 in 1980.[12] Such a tiny breeding population led to severe inbreeding, and to the persistence of detrimental mutations in many genes. These detrimental gene variants are rare overall in wolves but happened to be present in the founders of Isle Royale. Genetic drift led some of them to become the predominant genetic variants. The accumulation of detrimental gene variants by genetic drift in the genomes of these wolves created a deleterious genetic load, depressing their survival and reproductive potential below what it could be otherwise. The genetic load from some of these deleterious gene variants cause measurable defects. Their skeletons often developed in deformed ways, including extra vertebrae and ribs or malformations of vertebrae.

Small breeding groups also afflict man's best friend. Small breeding groups for artificial selection during dog domestication contributed to genetic risks of health problems in modern-day purebred dogs, in addition to the intended effect of fostering their distinctive sizes and shapes. For example, Greyhounds and other long-limbed breeds are notoriously susceptible to bone cancers; Newfoundlands are predisposed to coronary disease because of obstruction of their heart ventricles; and Shar-Peis are subject to recurrent bouts of inflammation and fever, among many other afflictions.[13] Such consequences of small populations gives one of many factors to consider when embarking on a program of selective breeding.

3.2   A history of mutational events

Even in big populations that don't suffer the severe deleterious consequences of genetic load, the genome still accumulates DNA sequence changes as a result of genetic drift. These changes give us a powerful tool. They give us a tool to measure time, a genetic stopwatch, so to speak, to measure the span of history as it gets recorded in DNA. You've probably heard of radiocarbon dating. After death, organic material accumulates carbon-12 and carbon-13 isotopes in place of carbon-14. The rate of decay of one isotope of carbon to another over time lets an archaeologist calculate how long ago it was that an organism had been alive. Analogous to the accumulation of alternative isotopes of carbon, genomes accumulate neutral mutations from one generation to the next to give us a molecular clock based on DNA sequence change.

Let me walk with you through a short chain of logic. It will lead us to understand how we can count up the number of DNA sequence differences in the genomes of distinct species and use that sum to calculate how much time has elapsed since those species shared a common ancestor. We know that we can sequence DNA from each of two species and compare those sequences, counting up the differences in As, Ts, Gs, and Cs. What we'd like to know is, what is it that this count of differences, this sequence divergence, depends on?

The eminent mathematical biologist Motoo Kimura answered this question in 1968 with a deceptively simple solution.[14] The sequence divergence—a quantity that we call K—that accumulates in a species over time is the product of two things. The sequence divergence is made up of 1) how many new mutations a gene will receive in any given generation multiplied by 2) the chance that any given mutation will drift to fixation and become present in 100% of the DNA copies at that spot in the genome for the species.

That first piece, the total number of new mutations to the gene in the species, is simply the mutation rate to the gene multiplied by the number of gene copies that could get mutated. If there are 5000 individuals in the species, each of which has two copies of each gene that it inherited from its parents, and the mutation rate is one error in every million, then the cumulative input of new mutations for any given spot in the genome is two times 5000 times 1/1,000,000. This would mean that the species gets an average of one new mutation every 200,000 generations at that particular spot in the genome. Because species differ in abundance and mutation rate, to generalize, we can talk about there being N individuals and that the mutation rate is μ ("mew," where μ is the Greek letter for "m"). Consequently, the mutational input into the species is $2N\mu$.[15]

The second piece of Kimura's multiplication is about the evolutionary fate of those new mutations. A new mutation could get lost. Or, eventually, it could become ubiquitous, what evolutionary geneticists term *fixed*. For us to see DNA sequence differences between species, the only fate we care about is fixation: those new mutations that happen to drift upward in abundance to eventually represent the version of DNA that occurs in 100% of the copies of DNA at that spot in the genome for the species. For a new mutation, the chance that this happens is very small. For a population of 5000, there are 10,000 copies of the gene and just that one copy with the new mutation (0.01% of copies). It turns out that the probability that it will, eventually, get fixed in such a species is one in 10,000, and, on average, it will take 20,000 generations for that to happen. In the language of mathematics, the chance that any given new mutation will eventually drift to attain 100% frequency is $\frac{1}{2N}$.

Now, let's put these two parts together. Together, it will tell us about how predictably DNA sequence divergence accumulates when we compare two species. It is the combined outcome of mutations getting added to a population and the subsequent likelihood that genetic drift leads those mutations to come to represent the new version of that gene for the species. That is, the number of DNA sequence differences between species that we call K, is equal to $2N\mu$ times $\frac{1}{2N}$. We can write this down as $K = 2N\mu \times \frac{1}{2N}$. But note that the "$2N$" cancels out in this arithmetic, leaving us with the simplest of equations: $K = \mu$.

In words, the average accumulation of DNA sequence change in each generation depends only on the average rate of input of new neutral mutations each generation. This forms the logic behind the idea of a molecular clock. If we have some means of determining the rate of input of new mutations, then we can directly calculate how much time has elapsed for a given observed measurement of DNA sequence divergence.[16] Experiments and fossils both give us ways to calculate those mutation rates to be able to convert observed DNA sequence differences into an estimate of elapsed time.

These last few paragraphs might have seemed a long road to navigate to get to a signpost that simply reads, "$K = \mu$." I've belabored this point, but for good reason.

This simple result to explain the accumulation of DNA sequence change is, I would argue, the single most elegant mathematical result in all of evolutionary biology.

★

The molecular clock helps us to date the timing of when different species emerged even when the fossil record is sparse or missing. Mutation rates often are similar among related species, which lets us extrapolate from species with good information to those with sparser information. We can also use it to help date events in our own human history for which there are gaps or conflicts in archaeological evidence. For example, this same technique helps us to date how long ago *Homo sapiens* hybridized with *Homo neanderthalensis*.

One of the first popularized applications of molecular clocks within humans led to the ignominious notion of Mitochondrial Eve. What is special about mitochondria and mitochondrial DNA is that, for the most part, it gets passed down only through eggs, making such DNA sequences a pure record of maternal descent. Sperm cells have lots of mitochondria but are dead ends as far as mitochondrial inheritance is concerned. As a consequence, humans transmit mitochondrial DNA in an asexual way, in a manner of speaking. It doesn't recombine with genetically distinct versions that occur in other members of a species, unlike the genes encoded on, say, human chromosome 17. This lack of recombination means that mitochondrial DNA accumulates differences between members of the same species in much the same way that different species accumulate DNA sequence differences. By comparing mitochondrial DNA between different people, we can learn how long ago they shared a common female ancestor.

As a result, the idea for using a molecular clock was simple enough: identify DNA sequence differences among human mitochondrial genomes to calculate how long ago these genetically variable bits of DNA coalesce into a single common ancestral sequence of DNA. It is a truism that there was such a sequence of mitochondrial DNA and that there was some woman in prehistory whose mitochondria encoded that sequence. It is true only for that stretch of mitochondrial DNA, however, with different coalescences in different individuals for other pieces of DNA in the rest of our genome. At that distant time, eons ago, this fact of a mitochondrial common ancestor to all present-day peoples of the Earth—a so-called Mitochondrial Eve—did not demarcate anything especially special about that individual person who lived in a world of thousands of other people. That calculation did, however, help to underscore the African origins for all of humanity's ancestry.

★

Tracing the molecular history of mitochondrial DNA was the easy low-hanging fruit at the dawn of the era of DNA sequencing. Now, however, we sequence the entirety of genomes, not just the part that transmits asexually from generation to generation. The ancestry of the recombining portion of our genome, the 98.2% of our DNA on chromosomes 1 through 22 plus the X chromosome, tells a more nuanced story of relatedness.

Take out a pencil and paper. When you draw a genealogical tree that connects you to your ancestors, you start with a pair of lines pointing from you to your parents. The lines then double into four lines to *their* parents, your grandparents. Because of biparental reproduction in humans, each generation deeper into your pedigree doubles it in size. You could continue this doubling and doubling again for as far back as you choose to go. But once you hit about 30 generations back, around the year 1300 CE, the number of lines you'd have to draw would be too many. Not just too many to fit on the piece of paper, but more lines than the number of people that were alive in the entire world at that time—roughly 350 million.

How could that be? After all, it is impossible to have had more ancestors than the number of people in existence. The doubling of lines, as it turns out, hides an assumption. It assumes that each parental pair in a given generation are entirely unrelated. The further back in time you go, however, the less likely that assumption is to be true. It becomes more and more likely that some great-great-to-the-n^{th}-degree-grandparent got together with a cousin of a cousin, or an even closer relation. It creates loops in your genealogy, much the way Charles Darwin and his wife Emma Wedgwood formed a genetic loop through the two grandparents that they had in common (Josiah and Sarah Wedgwood).

Those genealogical loops of distant-but-not-totally-unrelated pairings work to rein in the expansion of the genealogical tree. They rein it in so much that, at some point, rather than it being an impossibly large number of ancestors, it's just *all* of them. Your set of ancestors includes everyone who was alive at that point in time who left any descendants at all that are found in the present day. Another way of saying what this means, when you get that far back, as Joseph Chang put it, "*all individuals who have* any *descendants among the present-day individuals are actually ancestors of* all *present-day individuals.*"[17]

The weight of the genetic link that you have to each person in this spider web of ancestral relations, however, changes the deeper back in time you go. You got one copy of chromosome 19 from Mom, and one from Dad: 50% of your DNA from each parent. But the sequence of DNA in that copy you got from Mom was a mosaic of DNA from each of her parents. That mosaic of your grandparents' DNA came about because of recombination during meiosis in the production of the egg that led to you; the same goes for the copy of chromosome 19 that you inherited from Dad, which was a meiotic mosaic of his parents' DNA. On average, you'd expect that the contribution of DNA from someone's mother's mother to their own genome would be 25%. The mosaic of DNA made by recombination, however, is not precise: the true percentage could be as low as 17% or as high as 33%.[18]

This same logic applies to every egg and sperm cell made in meiosis, tracing back through the entire pedigree of your genealogical history. As a result, your genome shares more DNA with some ancestors than others that otherwise come from the same generation (Figure 3.2).

Look back at the scrawl of pencil lines on your paper. When you trace back more than seven generations, there is a very high likelihood that some of the names pointed to by the scribble of lines in the diagram of your genealogy will point to ancestors who contributed a grand total of zero DNA to your genome. In this peculiar way, you are not actually genetically related to some of the people who, by pedigree, truly are your

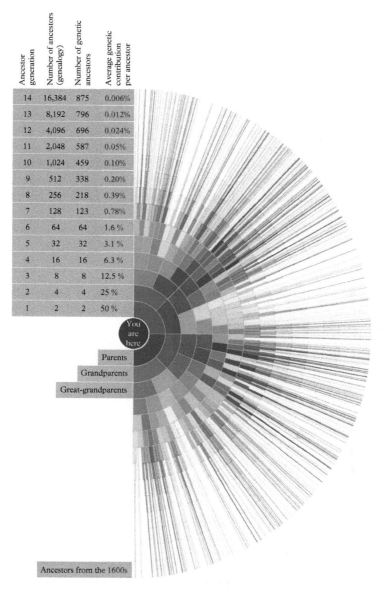

Ancestor generation	Number of ancestors (genealogy)	Number of genetic ancestors	Average genetic contribution per ancestor
14	16,384	875	0.006%
13	8,192	796	0.012%
12	4,096	696	0.024%
11	2,048	587	0.05%
10	1,024	459	0.10%
9	512	338	0.20%
8	256	218	0.39%
7	128	123	0.78%
6	64	64	1.6 %
5	32	32	3.1 %
4	16	16	6.3 %
3	8	8	12.5 %
2	4	4	25 %
1	2	2	50 %

Figure 3.2 *Our genetic relatedness to our direct genealogical ancestors decreases with each generation back in time. We are very likely to have inherited zero DNA sequence from some of our direct genealogical ancestors that are seven or more generations removed from us. This diagram illustrates the outcome of a genetic simulation of human ancestry for autosomal chromosomes (excluding sex chromosomes and the mitochondrial genome) going back one generation for each concentric arc radiating out from "you" in the middle, for a total of 14 generations back in time. Paler shading indicates lower percentage of DNA inherited from that ancestor relative to the highest amount inherited in that generation, with blue for male ancestors and red for female ancestors. White cells in the diagram indicate*

continued

ancestors through a direct line of parent-child descent.[19] Of those roughly 350 million people alive back in the year 1300, you probably only inherited DNA sequence from about 1900 of them.

★

Evolution, at its simplest, is changes to the genetic composition of a population across generations. Natural selection contributes to evolution with a team of ancient forces that influence change in the genetic composition of populations: gene flow, mutation, genetic drift, and recombination. All of these forces contribute to the pattern of DNA sequences in genomes and their changes over time. We can quantify evolutionary change and we can influence evolution by influencing these forces. Some aspects of evolutionary change are predictable. Under some circumstances, evolution can proceed with remarkable speed.

Notes

1. Feinstein, J. 2011. *Field Guide to Urban Wildlife*. Mechanicsburg, PA: Stackpole Books. 42–43. https://www.cbc.ca/news/world/black-squirrels-washington-dc-canada-history-smithsonian -ontario-1.4744392 accessed January 7, 2021.

2. Lehtinen, R.M., et al. 2019. Dispatches from the neighborhood watch: using citizen science and field survey data to document color morph frequency in space and time. *Ecology & Evolution*. 10: 1526–1538. https://doi.org/10.1002/ece3.6006.

3. Melanism in eastern gray squirrels also has come to be more prevalent among urban squirrels than those in nearby countryside. Cosentino, B.J., and Gibbs, J.P. 2022. Parallel evolution of urban–rural clines in melanism in a widespread mammal. *Scientific Reports*. 12: 1752. https://doi.org/10.1038/s41598-022-05746-2; McRobie, H.R., et al. 2019. Multiple origins of melanism in two species of North American tree squirrel (*Sciurus*). *BMC Evolutionary Biology*. 19: 140. https://doi.org/10.1186/s12862-019-1471-7.

Figure 3.2 *continued*

ancestors who contributed zero DNA to the genome of "you" in the present day. In this genetic outcome, "you" inherited some amount of DNA from just 875 of the 16,384 direct ancestors (5.3%) in the pedigree from 14 generations ago (dating to the 1600s, presuming a historical span of about 25 years between generations). In my case, even though I can trace multiple ancestral paths to people who arrived in North America on the Mayflower in 1620, it is very unlikely that I inherited any DNA from them through the intervening generations. Percentages will vary from person to person, because of the variability in recombination along chromosomes and the chance transmission of genetic information during fertilization. Simulation and visualization code based on that developed by Graham Coop (https://github.com/cooplab/Genetic_ancestors).

4. Geneticists typically use the term *allele* to refer to a given variant form of a piece of DNA, but I will generally avoid this piece of jargon in favor of the phrase *genetic variant* as a synonym. I hope this phrasing makes for a more accessible read.

5. As it turns out, however, at most 7% of our genome contains genetic variants that are truly unique to *Homo sapiens*, in part because of the genetic variants that were present in the species that existed as the common ancestor of us and Neanderthals and thus that we shared with them. Vernot, B., et al. 2016. Excavating Neandertal and Denisovan DNA from the genomes of Melanesian individuals. *Science.* 352: 235–239. https://dx.doi.org/10.1126/science.aad9416; Schaefer, N.K., et al. 2021. An ancestral recombination graph of human, Neanderthal, and Denisovan genomes. *Science Advances.* 7: eabc0776. https://doi.org/10.1126/sciadv.abc0776.

6. McRobie, H.R., et al. 2019. Multiple origins of melanism in two species of North American tree squirrel (*Sciurus*). *BMC Evolutionary Biology.* 19: 140. https://doi.org/10.1186/s12862-019-1471-7.

7. Mutations can occur when any cell divides, not just in producing sperm and eggs. These other mutations are termed *somatic* changes to the genome, and they will not be passed on to the next generation. They can produce striking features, however, such as the variegation in leaf color in some species of plants, in the moles and birthmarks on our skin—and, of course, in the origin of cancers.

8. Xue, Y., et al. 2009. Human Y chromosome base-substitution mutation rate measured by direct sequencing in a deep-rooting pedigree. *Current Biology.* 19: 1453–1457. https://doi.org/10.1016/j.cub.2009.07.032.

9. Grauer, D. 2017. An upper limit on the functional fraction of the human genome. *Genome Biology & Evolution.* 11: 3158. https://doi.org/10.1093/gbe/evx121.

10. Templeton, A.R. 2018. *Human Population Genetics and Genomics.* Cambridge, MA: Academic Press, p. 104. https://doi.org/10.1016/B978-0-12-386025-5.00004-X.

11. The *bw* gene is now known to encode a membrane transporter protein that works together with another transporter protein called *w* to import guanine into cells as a precursor for making red pigments in fly eye cells. http://flybase.org/reports/FBgn0000241.html. Accessed October 12, 2021; Buri, P. 1956. Gene frequency in small populations of mutant *Drosophila*. *Evolution.* 10: 367–402. https://doi.org/10.2307/2406998.

12. Despite its closer proximity to the Canadian mainland, Isle Royale is part of Michigan and forms a United States National Park subject to regular monitoring of wolf and moose populations. https://www.nps.gov/isro/learn/nature/wolf-moose-populations.htm. Accessed April 21, 2021; Robinson, J.A., et al. 2019. Genomic signatures of extensive inbreeding in Isle Royale wolves, a population on the threshold of extinction. *Science Advances.* 5: eaau0757. https://dx.doi.org/10.1126/sciadv.aau0757.

13. Schoenebeck, J.J., and E.A. Ostrander. 2014. Insights into morphology and disease from the dog genome project. *Annual Review of Cell and Developmental Biology.* 30: 535–560. https://doi.org/10.1146/annurev-cellbio-100913-012927.

14. Kimura, M. 1968. Evolutionary rate at molecular level. *Nature*. 217: 624–626. https://doi.org/10.1038/217624a0.

15. Here, we are only talking about those mutations that are selectively neutral—like the 85% in the human genome.

16. In practice, the computation and interpretation of calculations of this sort involve careful attention to some assumptions and technical considerations that I have neglected in this summary, but the basic principles remain the same.

17. The timing, such that there is an identical set of genealogical ancestors among all present-day people, was about 4000 to 5000 years ago. Rohde, D., et al. 2004. Modelling the recent common ancestry of all living humans. *Nature*. 431: 562–566. https://doi.org/10.1038/nature02842; Quotation from p. 1002 in Chang, J.T. 1999. Recent common ancestors of all present-day individuals. *Advances in Applied Probability*. 31: 1002–1026. http://www.jstor.org/stable/1428340.

18. A nice explanation of this idea is presented in the following blog post: https://dna-explained.com/2020/01/14/dna-inherited-from-grandparents-and-great-grandparents. Accessed August 30, 2021.

19. Similarly, your child's genome and your second-cousin's genome also are seven generations apart, quite possibly sharing no segments of DNA from the genetic inheritance of your great-grandparent that they have in common. Graham Coop provides an excellent deeper explanation of these ideas in a series of posts to his blog https://gcbias.org/2017/12/19/1628. Accessed October 12, 2021; Ralph, P., and G. Coop. 2013. The geography of recent genetic ancestry across Europe. *PLoS Biology*. 11: e1001555. https://doi.org/10.1371/journal.pbio.1001555.

4

Evolution's racetracks

When she was four years old, my daughter invented the Go-Snap™. It was a teleportation device designed to instantaneously transport its occupant to her cousin's home in San Francisco or, alternately, to a grandparent's living room. She constructed it out of a large cardboard box, laid on its side, with the flaps folding down to enclose a small person inside. Emblazoned on one interior wall was a large round button, adhered via glue stick, colored red from a magic marker's scribblings. The user interface, it seems, was modeled on the user interface of a hotel elevator, which also transports a button-presser to a brand-new place in the blink of an eye. Upon what was to be the Go-Snap's maiden voyage, after pressing that red button for the first time and opening up the flaps, Oona encountered deep disappointment when the laws of physics taught her a cold lesson about the hidden mechanisms by which things work.

How do you prove to yourself that you truly understand something, a new toy, a new gadget, a new idea? You play with it. You poke it and squeeze it and juggle it between your hands. You bounce it off the floor to check whether it reacts in the way that you expect. If you had a spare, you might even cut it open or unscrew its assembled parts, dissecting and tinkering with the inner workings of your sample, perhaps even measuring and cataloging the properties of each component piece before putting it back together. You might construct your own version of it, out of clay or spare parts lying around the house. Curious play, just like a child with a new toy, is the spirit that biologists take in experiments with evolution to figure out how it works.

4.1 Everyday evolution

Pig farmers and chicken breeders and rice horticulturists make focused use of evolution. They accelerate evolution with artificial selection, selectively breeding the leanest and highest yielding stock to produce lasting change in the next generation. This technique of speeding up evolutionary change through selective breeding not only works in the farmyard. It also works in the laboratory. Artificial selection is, in fact, a key tool in the toolkit of molecular biology labs around the world. In some applications, artificial selection is so routine that scientists may not even fully appreciate that they are using

Evolving Tomorrow. Asher D. Cutter, Oxford University Press. © Asher D. Cutter (2023). DOI: 10.1093/oso/9780198874522.003.0004

evolutionary principles on their benchtop. In other situations, they knowingly and intentionally wield the forces of evolution to help in their scientific quest to solve problems entirely unrelated to evolution.

Do you have your shots? If so, then you are the direct beneficiary of evolution running in real time. The classic approach for creating live attenuated vaccines depends on evolution in the laboratory.[1] But it all started with chickens.

In the late 1800s, Louis Pasteur experimented with a devastating bacterial pathogen that infects poultry, a bug now known as *Pasteurella multocida* that leads to chicken cholera.[2] He and his lab assistant let a culture flask of the bacteria sit in its broth on the lab bench for a month while on holiday, an accident so the story goes. Back from vacation and eager to jump into work, they used the forlorn flask to inoculate a batch of chickens to study. An odd thing happened: the chickens didn't die of cholera. The chickens developed only mild symptoms. Even more peculiar, when they followed up by injecting the same chickens with a fresh batch of *P. multocida*, rather than suffering premature death by diarrhea, their health persisted.

The batch of lab bench-neglected bacteria had come to be attenuated: they were capable of inducing an immune response but incapable of inducing disease. By inducing an immune response, the chickens could fight off subsequent infections with ease. As Pasteur wrote in 1881 about his method of attenuating bacteria for vaccination, after refining the technique, "we pass from one culture to that which follows it, from the hundredth to, say, the hundred and first, at intervals of a fortnight. . ."[3] What had happened was that the bacteria evolved over the course of all those transfers between culture flasks: they had adapted to life in a jar and this came at the expense of being unable to grow well in a chicken.[4] At the time, Pasteur did not understand that the mechanism was evolution that led to attenuated pathogens as an effective vaccine. Nevertheless, his 1879 discovery of attenuation set medicine on the path to exploiting evolution to aid human health, for more than a century and counting.

By 1927, Albert Calmette and Camille Guérin had successfully expanded on this idea. They created an attenuated vaccine for tuberculosis by evolving the *Mycobacterium bovis* bacterium. It took them 14 years by transferring the bacteria 230 times, an approach now termed *serial passage*, which is code for controlled laboratory evolution. In the case of *M. bovis*, they passaged the bacteria between batches of culture medium that sound like a bad soup, its essential ingredients being glycerinated beef bile and potato.[5]

But things went back to chickens for the next major milestone. In the late 1930s, Max Theiler established the use of chicken eggs as a form of tissue culture to passage viruses. The egg provides a convenient self-contained microcosm of animal cells for the viruses to infect and grow, an environment easier to mass produce than synthetic culturing of animal cells.[6] This method of using laboratory evolution to develop an attenuated vaccine for the yellow fever virus proved instrumental in creating standard operating procedures for subsequent industrial development of attenuated vaccines. For pioneering this eggy approach to exploiting the forces of evolution in the laboratory to create attenuated vaccines, Theiler earned a Nobel Prize in 1951.

This technique is still in use today. If your doctor has ever asked whether you are allergic to eggs before giving a shot, it may be due to the use of chicken eggs in the evolutionary culturing of an attenuated live vaccine. During the 1950s, scientists devised methods that use cultures of cells rather than chicken eggs. If you received the oral polio vaccine as a child, then you ingested an attenuated version of the polio virus derived from evolution. Similarly, vaccines for anthrax and rabies were evolved in the lab. Your childhood MMR shot depended on the evolution of viral populations of measles, mumps, and rubella through their serial passage that led to attenuation, letting your body produce antibodies to fight potential infection by nasty unattenuated genetic forms of those viruses lurking out there in the world. In this way, we can thank human-mediated evolution for saving millions of lives that otherwise would have suffered disease-driven morbidity and mortality.

★

Laboratory evolution, however, sometimes works against us. The hallways and gurneys of hospitals also are laboratories, of sorts, and cauldrons of evolution. With every new antibiotic drug that we come up with, bacteria sooner or later evolve resistance to it. This happens whether the drug was synthesized from scratch like the fluoroquinolone antibiotic ciprofloxacin or extracted from a natural compound like penicillin, the very first mass-produced antibiotic, discovered serendipitously in 1928 by Alexander Fleming.[7]

You may not have heard of the antibiotic methicillin, however, unless someone dictates to you the full words behind the MRSA acronym. Methicillin quickly became medically useless because of the rapid evolution of resistance emerging in clinical settings within just one year of its introduction for hospital use around 1960. This rapid resistance evolution occurred despite the fact that methicillin, as a semisynthetic compound, was something that nature had never encountered previously. In fact, the genetic variants capable of resistance appear to have originated in nature long before methicillin as a chemical compound was even developed.[8]

Worse still, strains of bacteria like *Staphylococcus aureus*, known more ominously as the superbug MRSA, frequently have evolved multidrug resistance, meaning that there is virtually nothing left in our antibiotic arsenal to stave off their proliferation in staph infections. Current estimates put the number of genes encoding resistant variants at 55,994, which come from 82 different bacterial pathogens.[9]

The problem is that antibiotic agents are a double-edged sword: the fact that they act to kill microbes means that they are a superb agent of natural selection on those microbes. The antibiotic can readily kill off most of the susceptible bacteria in an infection. Some bacterial cells may remain alive, however, especially if the administered dose was too weak or was stopped too soon. If any of those bacterial survivors have genetically encoded resistance, then they can continue to proliferate and outcompete their susceptible siblings.[10] Such selection will favor the increased abundance of any genetic variants that allow the bacteria to resist, evade, detoxify, or tolerate the toxicity of the antibiotic compounds.

Resistant genetic variants in a population of bacterial cells can come about in multiple ways. The variant might have been pre-existing in the bacterial population, perhaps having arisen long ago and preserved as a defense against chemical attacks by other microbes; after all, most of our antibiotics are derived from chemicals produced by living things in nature. Or a new mutation in the genome might confer the resistance. Given the enormous numbers of bacterial cells in a colony, the chances of such new mutations arising from scratch is higher than you might think.

One of the most important ways that bacteria acquire the right genetic stuff to resist antibiotics, however, is from that ancient force of evolution called gene flow. Even though bacteria mostly reproduce asexually, budding off one from another, they sometimes have a funny sort of sex called conjugation that allows the genetic material from different cells to intermingle and recombine. Even kinkier, bacteria are capable, once in a blue moon, of conjugating successfully with cells of totally different species. The genetic fruit of that labor is the transfer of genetic material from one evolutionary lineage to another, often referred to as horizontal gene transfer (HGT).

Such HGT is responsible for the intensely baroque web of evolutionary relationships among the genomes of bacterial species.[11] Importantly for the evolution of antibiotic resistance, HGT often leads to the shuttling of tidy genetic packages of multiple genes conferring resistance to antibiotics. It is in this way that laboratory evolution in the form of hospital intervention has worked against us, our own actions letting cassettes of genes evolve to high prevalence throughout the microbial world to produce the multidrug resistant superbugs like MRSA.[12]

The evolution of drug resistance even happens in our houses. And not just with bacterial bugs. Insects, like lice, evolve, too. The perennial threat of head lice among elementary school kids makes parents scratch their own heads with anxiety, infested or not. Lice have been with us for eternity, found even in the scalps of Egyptian mummies and other mummified human remains from around the world,[13] their close affinity to people emblazoned on their Latin name: *Pediculus humanus* (Figure 4.1). Once upon

Figure 4.1 *Human head lice rapidly evolved resistance to the pyrethrin-based insecticides since the 1990s.*

a time, starting in the mid-1980s, it was the case that commercially available insecticidal shampoo was a simple and effective treatment to cure a child of cooties. But the genomes of some individual lice were especially lousy, from the perspective of human comfort. Those bugs carried a gene variant that let them escape death when exposed to the insect neurotoxic effects of pyrethrin and chemically related insecticides.[14] The causal culprit was the so-called *kdr* allele that affects three amino acids in a protein known as the voltage-sensitive sodium channel (VSSC) α-subunit.[15] This gene variant spread like a crown fire from head to head across the classrooms of the world, with resistance becoming common by the mid-1990s. More than 97% of head lice had evolved resistance to pyrethrin by 2010, including in the city I call home.[16] Fortunately, head lice do not transmit disease, unlike their body lice brethren that nibble on your nether regions. Unfortunately, the best cure for a louse infestation, these days, is manual labor: a fine-toothed louse comb and a fine dose of patience.

★

Evolution is inevitable. Evolution by natural selection is inevitable. These statements are true because of the base simplicity of the conditions that give rise to them. DNA replication is imperfect; this inevitably creates new genetic variants through mutation. A population of organisms is made up of a finite number of individuals; this inevitably creates chance fluctuations in the incidence of any given gene variant through genetic drift. In other words, gene frequencies will change across generations. Evolution is inevitable. Resources required for survival and reproduction are finite in abundance and vary over space and time; this inevitably creates competition for the access and use of resources. And, inevitably, some genetic variants will confer on organisms a greater or lesser ability to acquire and make use of those finite and varying resources. Such genetic variants will increase or decrease in their relative abundance across generations. Evolution by natural selection is inevitable.

These properties arise in nature with no special intervention. They depend on such basic facts of life as mistakes and noninfinite numbers of things and some things being more capable than others. There is no value judgment in the existence of these features of the world.

Evolution does not depend on intervention, but it is not immune from it. Humans have devised many ways to intentionally shape the course of evolution in the species around us. We can easily modulate rates of mutational input through chemical means, and we can create genetic novelty through biotechnological manipulation. We can skew the influence of genetic drift by nurturing growth of large populations or by culling populations to small numbers. Selective breeding and artificial selection give us say over which genotypes will survive and reproduce to propagate the next generation, whether directly or indirectly, by altering the environmental circumstances experienced by a species in a particular way. If there is intelligent, or not-so-intelligent, design in the world, it is designed by us. If there is creation to be directed in the world, it is directed by us.

4.2   Experimental evolution

Vaccines and drug resistance are important for public health, but laboratories apply evolution in other ways, too. One of the big challenges for biologists is to know how the measurable features and characteristics of an organism actually map to specific locations in their DNA to drive the cellular development of those characteristics. That is, what genes make different individuals have different values of a given trait? We can measure how long an animal is, or how well it can digest sugar, or what color it is, but often there is a fat black box of ignorance about the genetic causes. Research scientists try to shine light into the black box with genetic mapping, and genetic mapping sometimes makes use of artificial selection in the laboratory.

Take the example of the simple single-celled creature known as *Saccharomyces cerevisiae*. As you might recognize from the similarity of the second half of its name to *cerveza*,[17] this creature's nickname is brewer's yeast, harkening to the fact that it is the heralded workhorse in winemaking and beer brewing.[18] This same species is the yeast that goes by another *nom de plume*, baker's yeast, when used to leaven bread. Scientists on friendly terms with this solitary fungus call it budding yeast. Whatever name you whisper into its vat or Petri dish as you ponder the 5195 genes in its genome,[19] yeast's agroeconomic importance puts a premium on figuring out which genes influence which traits. The traits we care about are life's fundamentals in a fungus like this one with its individuals made up of just a single cell: how it takes in and digests different types of nutrients, what metabolic byproducts it expels, how it handles noxious conditions and compounds.

Temperature tolerance is a key factor for the growth and survival of all organisms, and is important for yeast. To figure out what genetic differences between different strains of budding yeast made them more or less able to withstand high temperatures, geneticists Leopold Parts, Gianni Liti and their team exploited artificial selection.[20] Because yeast are tiny, because they are single-celled, you can grow vast numbers of them on a Petri dish. Parts and Liti made populations with up to 100 million individual cells, representing similarly enormous numbers of unique genetic combinations attributable to recombination across their genomes. They then placed the Petri dishes at a hot 40 °C or at a balmy 23 °C (104 °F vs 73 °F), the lab equivalent of spending an eternity of midday summer in Tucson versus San Diego.

The next step was to simply let selection sift among all those unique genomes over the course of 12 days, which corresponds to about 30 generations for the yeast.[21] This experimental evolution allows cells with gene combinations best suited to their thermal environment to proliferate fastest, meaning that the genetic variants that confer the ability to multiply most quickly will increase disproportionately to higher abundance.

At the end of this trial by fire, the final experimental step is to learn which parts of the genome evolved. The trick is to figure out which genetic variants became more common or less common in the dishes that experienced 40 °C compared to the dishes that experienced 23 °C. The researchers used modern high-throughput DNA sequencing to scan the genomes of the populations of yeast. With this technique, they could look at the individual letters in the 12 million letter-long book of the yeast genome to see whether it was the G versus the T nucleotide at position 522,887 on Chromosome II that reached greater abundance in the high temperature environment (it was the G). Altogether, this approach pinpointed 21 places in the genome that responded to artificial selection, revealing novel genetic factors that influence heat stress.

This example told the story of a microbe, the single-celled brewer's yeast. But the same method also works in animals.[22] To do it in short order, it helps if they are small and can grow fast so that you can produce lots of them. The *C. elegans* nematode round-worms that I've mentioned previously certainly fit the bill: you can grow literally millions of these 1-mm-long animals on Petri dishes, with the 300 or so babies they make growing up to make their own babies in as little as two days, giving you 10 generations of evolutionary change in less than a month.

★

Watching evolution in the laboratory is more inventive than using evolution simply as a standard operating procedure in molecular biology protocols and vaccine design. Another approach is to design experiments to understand the evolutionary process itself by measuring evolution as it unfolds.[23] The scientific craft of experimental evolution lets us test the depth of our understanding and our certainty about how the different forces of evolution operate; it lets us explore the bounds and constraints on evolutionary change. It also proves just how fast the speed of evolution can be.

In experimental evolution, distinct from how artificial selection works, the experimenter does not directly pick the individuals that will breed. The experimenter sets the conditions and controls the environment. But we let survival and reproduction unfold in a natural way within the confines of those imposed circumstances. This kind of experimenter is like a set designer who arranges the stage for a no-auditions-necessary improv troupe, a contrast to the writer-director who hand-picks the lead actors in a scene of artificial selection. The cast is usually made up of a collection of people off of the street: the starting, base population for experimental evolution uses a large random sampling of genetic variants found in nature. The set designer then puts on a different hat to transform their job title to that of critic. As critic they evaluate what was said, how the plot developed, and how the show resolved when the curtain dropped, if it resolved at all. To learn more about the outcome, the critic experimenter might then conduct interviews, so to speak, and arrange to see an encore of subsequent vignettes. In this way, you can measure the changes in genetic composition of the populations in response to the details of how the experiment was set up.

Evolution experiments happen both in the laboratory and in the field, where the field situation is seminatural to integrate more of the vagaries of the real world at the expense of being less finely controlled. First, the lab. Your typical pint of beer was made through the work of roughly 50 billion yeast cells.[24] As fine as a good pint may be, they take up a lot of room. Michael Desai and his team grow yeast in so-called microtiter plates, a molded 12.8 cm × 8.6 cm piece of plastic (5 inches × 3.4 inches) patterned with an 8-by-12 array of deep dimples. Each of the 96 dimples is, in effect, a miniature bottle (Figure 4.2). Each miniature "bottle" gets filled with 0.128 milliliters of yeast broth, about the volume of three raindrops. The ultimate

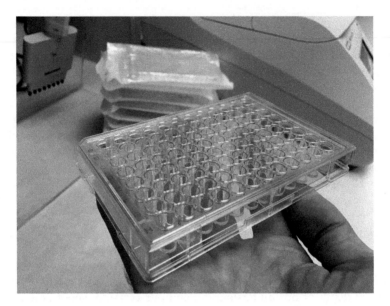

Figure 4.2 *Experimental evolution studies with microbes, such as the brewer's yeast* Saccharomyces cerevisiae, *can be conducted rapidly through time and with a minimum of space using physically tiny populations. "Microtitre plates," such as this one, can house populations of millions of cells in each of its 96 wells.*

Photo by the author.

microbrew. Each of Desai's tiny bottles is its own evolving population of up to 7 million cells.

This sub-pint-sized setup allowed a robot arm to transfer the yeast to fresh broth every day, culling the population down by a factor of nearly 10,000 to grow up again, letting them propagate at a rate of 70 generations per week. After 145 weeks—nearly three years—the team had witnessed over 10,000 yeast generations of experimental evolution,[25] about half the number of generations that has elapsed since *Homo sapiens* split as a separate lineage from our common ancestor with *Homo neanderthalensis*.

By freezing representatives over the course of the experiment, Desai created a frozen fossil record, of sorts. His team cryopreserved the raindrop-sized bottles to preserve the history of those microbrew populations of yeast, making historical records to trace the evolution of 205 such populations in all. At the end of this kind of experiment, you become an evolutionary archaeologist with a time machine: thaw out the frozen fossils to compare, simultaneously, the properties of early generations to what evolved in later generations. Because each of the 205 micro-bottles started out exactly the same, watching what happens is the evolutionary equivalent of 204 alternate answers to the question: what if I lived my life all over again?

Over the course of evolution, the yeast adapted: mutations arose and increased in abundance that enabled faster reproduction in the carefully controlled conditions of their simple universe in a raindrop-sized bottle. By sequencing their genomes, the research team could pick out which mutations evolved to become fixed features of the populations, documenting recurrent adaptive changes to the same genetic pathways independently in different populations. The exact course of evolution, however, depends on what happened before; as Desai's team put it, "mutations in certain genes make mutations in others more or less likely" to result in adaptive evolution.[26]

This real-time characterization of 10,000 generations of evolution was a tour de force. And then there is Richard Lenski's long-term experimental evolution with *E. coli*. As of April 2021, Lenski's team has been tracking continuously the evolution of 12 isolated populations of these bacteria as they adapt to the same, constant environment since 1988, for what sums to 74,000 generations so far.[27] To distill that incredible duration of evolution down to one thing: evolution is interminable. The rate of fitness increase has slowed across the bacterial eons, but the populations nevertheless experience repeated bouts of rapid adaptation and gene frequency change.[28] Even after all this time, new mutations continue to arise that confer a benefit to cell survival and reproduction, making the populations better and better adapted to the exact same environment that their ancestors experienced thousands and tens of thousands of generations before.

★

Now, what happens when experimental evolution happens out of doors? The trick is in defining self-contained experimental areas that are otherwise exposed to the whims of the environment. The streams and rivers of the Northern Range Mountains of Trinidad are both opportune for experimental purposes and picturesque. With their intervals of sharp waterfalls and rapids, the upstream length of the Aripo and El Cedro rivers get divided up into convenient isolated pools of populations called home by the guppy *Poecilia reticulata*. The guppies share the river with other small fish, of course, but the river segments upstream of a waterfall are devoid of the major predators that eat adult guppies.

David Reznick and his team used this natural landscape to conduct experimental evolution. Their goal was this: measure how guppy evolution proceeds when a genetically variable population gets transplanted from a high-predation stream segment to a segment with low predation risk. After letting the guppies live and die on their own for up to 19 generations (11 years), checking on them every few years in between, they found that the absence of adult predation led to the evolution of guppies that take longer to reach adulthood and are bigger as adults.[29] Without predators picking off the biggest guppies, natural selection favored genetic variants that allocated resource investment in growth to larger size; larger guppies devote reproductive resources toward fewer, larger offspring. This profound change in circumstances led to incredibly fast evolution, as fast as artificial selection can do, even when measured in units of haldanes.

Islands also make for convenient open-air experimental units. Especially in the Caribbean, so thought Jonathan Losos and Thomas Schoener, as they introduced batches of *Anolis sagrei* lizards to each of 14 tiny Bahamian islands starting in 1977, the smallest of which was about the size of a two-bedroom apartment in Toronto.[30] This procedure mimicked natural colonization of previously uninhabited environments.

These new island habitats also had a novel ecological structure: the lizards' source habitat had lots of complex vegetation with scrub and forest, whereas the colonized islands had few if any trees, mostly occupied by low vegetation with narrow stems.

Analyzing the new lizard populations 14 years later, amounting to about 30 anole generations, led to the discovery that they had evolved shorter legs as well as wider and grippier toe-pads. Moreover, the extent of evolutionary divergence was greater on those islands that differed most substantially from the source island that they came from prior to colonization. The lengths of their limbs had evolved toward a match to the diameter of available plants to perch on. Following up on the little islands through 2016 showed no further consistent trend in evolutionary change in anole legs, though intermittent hurricanes appear to impose natural selection favoring smaller limbs.[31] Part of the changes may actually reflect developmental acclimation to the novel environmental circumstances, what biologists term phenotypic plasticity. This plastic component of the change is non-evolutionary. Only the part of the change that is encoded by genetic differences is what contributes to an evolutionary response to the novel natural selection encountered by the lizard populations in the newly colonized islands.

★

Experimental evolution teaches us that the forces of evolution operate in practice, not just in theory. By setting up multi-generation studies, experimental evolution lets us tease apart the roles of each force of evolution to understand it individually and in combination. These experiments demonstrate that mutational input operates as a source of genetic novelty; we can assess mutation by sequencing DNA and by measuring the effects of mutations on how organisms develop, survive, and reproduce. They show us how genetic drift influences gene frequencies more strongly in small populations and how gene flow can shift the genetic material available to influence subsequent evolutionary change. They illustrate how natural selection also influences gene frequencies to eliminate detrimental variants and lead beneficial variants to make up a greater and greater fraction of the population over time, and how recombination can facilitate this adaptive response. Experimental evolution demonstrates how evolution never stands still. Evolution experiments show how evolution can repeat itself with similar trajectories of change in characteristic traits and even genetic pathways. The trajectories of evolutionary changes that are most likely also depend, in part, on the mutational steps taken previously. Experiments of evolution prove how evolution is true.

★

Natural selection and evolutionary change are inevitable, both in nature and in human-controlled settings. Evolution impacts our daily lives, in the evolution of pathogen infectiousness, the development of medical interventions like vaccines, and in the features that we see in the organisms all around us. This evolution can be fast, fast enough for us to document and monitor and shape, with the natural timescale of its speed measured in generations of the organism of interest. Experimental evolution proves how evolution is true. Experimental evolution has allowed scientists to design biological microcosms to scrutinize our understanding of the forces of evolution.

Notes

1. Plotkin, S.A., and S.L. Plotkin. 2011. The development of vaccines: how the past led to the future. *Nature Reviews Microbiology*. 9: 889–893. https://doi.org/10.1038/nrmicro2668; Ebert, D. 1998. Experimental evolution of parasites. *Science*. 282:1432–1435. https://dx.doi.org/10.1126/science.282.5393.1432; Hanley, K.A. 2011. The double-edged sword: how evolution can make or break a live-attenuated virus vaccine. *Evolution Education Outreach*. 4: 635–643. https://doi.org/10.1007/s12052-011-0365-y.

2. https://www.nature.com/articles/d42859-020-00008-5. Accessed January 13, 2021.

3. Pasteur, L. 1881. An address on vaccination in relation to chicken cholera and splenic fever. *British Medical Journal*. 2: 283–284. https://dx.doi.org/10.1136/bmj.2.1076.283.

4. Adaptation (by the bacteria) to a novel environment (the jar) can lead to maladaptation to the previous environment (the chicken) in two main ways. First, there can simply be mutational degradation of the genetic factors that promote growth in chicken because there is no longer selection to retain them. This is called relaxed selection. Relaxed selection usually degrades genes and measurable traits relatively slowly because it depends on mutations getting fixed randomly by genetic drift. Second, there may be a physiological trade-off between the mutations that are beneficial for growing well in a jar versus a chicken. Under these circumstances, a mutation that is good for growing in a jar is necessarily worse for growing in a chicken. This mechanism represents one instance of so-called antagonistic pleiotropy, and it can lead to rapid evolution of maladaptation to the previous environment.

5. Gheorgiu, M. 2011. Antituberculosis BCG vaccine: lessons from the past. In S.A. Plotkin (ed.), *History of Vaccine Development*. New York: Springer. pp. 47–55. https://dx.doi.org/10.1007/978-1-4419-1339-5_7.

6. The emerging synthetic meat industry, however, may soon give eggs a run for their money.

7. Ciprofloxacin became available in 1987, with resistance to it now above 10%. Penicillin was first used clinically in 1942, with resistance following later that decade. Mulder, M., et al. 2017. Risk factors for resistance to ciprofloxacin in community-acquired urinary tract infections due to *Escherichia coli* in an elderly population. *Journal of Antimicrobial Chemotherapy*. 72: 281–289. https://doi.org/10.1093/jac/dkw399; Davies, J., and D. Davies. 2010. Origins and evolution of antibiotic resistance. *Microbiology and Molecular Biology Reviews*. 74: 417–433. https://dx.doi.org/10.1128/MMBR.00016-10; Fleming, A. 1929. On the antibacterial action of cultures of a *Penicillium*, with special reference to their use in the isolation of *B. influenzæ*. *British Journal of Experimental Pathology*. 10: 226–236. https://www.ncbi.nlm.nih.gov/pmc/articles/PMC2048009/.

8. *MRSA* stands for methicillin-resistant *Staphylococcus aureus*, the bane of clinical wards because it is a harbinger of bacterial staph infections that are resistant to treatment by multiple antibiotic drugs. Harkins, C. P., et al. 2017. Methicillin-resistant *Staphylococcus aureus* emerged long before the introduction of methicillin into clinical practice. *Genome Biology*. 18: 130. https://doi.org/10.1186/s13059-017-1252-9.

9. Alcock, B.P., et al. 2020. CARD 2020: antibiotic resistome surveillance with the comprehensive antibiotic resistance database. *Nucleic Acids Research*. 48: D517–D525. https://doi.org/10.1093/nar/gkz935.

10. Stephen Palumbi delivers one of my favorite expositions on how bacterial antibiotic resistance can evolve rapidly in his 2001 book *The Evolution Explosion*.

11. For an engaging story about the science and scientists that led to full appreciation of the importance of HGT, I recommend David Quammen's 2018 book, *The Tangled Tree*. Puigbò, P., et al. Search for a "Tree of Life" in the thicket of the phylogenetic forest. *Journal of Biology*. 8: 59. https://doi.org/10.1186/jbiol159.

12. For a sobering overview of the evolution of antibiotic resistance and its implications, that is still relevant despite its age, see Levy, S.B. 1998. The challenge of antibiotic resistance. *Scientific American*. 278: 46–53. https://www.scientificamerican.com/article/the-challenge-of-antibiotic-resista/.

13. Mumcuoglu, K.Y. 2008. Human lice: *Pediculus* and *Pthirus*. In D. Raoult and Drancourt (eds), Paleomicrobiology. Berlin: Springer. pp. 215-222. https://doi.org/10.1007/978-3-540-75855-6_13.

14. Interestingly, pyrethrin itself is the product of evolution: ancestors of the Dalmatian daisy *Tanacetum cinerariifolium*, a plant sometimes called chrysanthemum or pyrethrum, evolved the ability to produce this set of compounds as a defense against insect herbivores. The synthetic version is called permethrin. Yamashiro, T., et al. 2019. Draft genome of *Tanacetum cinerariifolium*, the natural source of mosquito coil. *Scientific Reports*. 9: 18249. https://doi.org/10.1038/s41598-019-54815-6; Lybrand, D.B., et al.2020, How plants synthesize pyrethrins: safe and biodegradable insecticides. *Trends in Plant Science*. 25: 1240–1251. https://doi.org/10.1016/j.tplants.2020.06.012

15. Marshall Clark, J., et al. 2013. Human lice: past, present and future control. *Pesticide Biochemistry and Physiology*. 106: 162–171. https://doi.org/10.1016/j.pestbp.2013.03.008.

16. Meinking, T.L. 1999. Infestations. *Current Problems in Dermatology*. 11: 73–118. https://doi.org/10.1016/S1040-0486(99)90005-4; Marcoux, D., et al. 2010. Pyrethroid pediculicide resistance of head lice in Canada evaluated by serial invasive signal amplification reaction. *Journal of Cutaneous Medicine and Surgery*. 14: 115–118. https://dx.doi.org/10.2310/7750.2010.09032; https://www.scientificamerican.com/article/revenge-of-the-super-lice/. Accessed January 13, 2020.

17. *Cerveza* is, of course, the word for *beer* in Spanish.

18. Liti, G. 2015. The fascinating and secret wild life of the budding yeast *S. cerevisiae*. *eLife*. 4: e05835. https://doi.org/10.7554/eLife.05835.001.

19. https://yeastmine.yeastgenome.org/yeastmine/bagDetails.do?scope=all&bagName=Verified_ORFs. Accessed January 8, 2021.

20. Parts, L., et al. 2011. Revealing the genetic structure of a trait by sequencing a population under selection. *Genome Research*. 21: 1131–1138. https://dx.doi.org/10.1101/gr.116731.110.

21. For humans, 30 generations would take 750 years, presuming that one generation follows the next every 25 years.

22. Burga, A., et al. 2019. Fast genetic mapping of complex traits in *C. elegans* using millions of individuals in bulk. *Nature Communications.* 10: 2680. https://doi.org/10.1038/s41467-019-10636-9.

23. Kawecki, T.J., et al. 2012. Experimental evolution. *Trends in Ecology & Evolution.* 27: 547–560. http://dx.doi.org/10.1016/j.tree.2012.06.001.

24. Given an inoculum of 40 million cells per milliliter, what brewers call a pitching rate, the yeast cells will roughly triple in concentration during fermentation, the metabolic byproducts of which produce the special properties appreciated in all 473 milliliters that comprises one pint of beer. Verbelen, P.J., et al. 2009. Impact of pitching rate on yeast fermentation performance and beer flavour. *Applied Microbiology and Biotechnology.* 82: 155–167. https://doi.org/10.1007/s00253-008-1779-5.

25. Johnson, M.S., et al. 2021. Phenotypic and molecular evolution across 10,000 generations in laboratory budding yeast populations. *eLife.* 10: e63910. https://doi.org/10.7554/eLife.63910.

26. Quotation from p. 11 in Johnson, M.S., et al. 2021. Phenotypic and molecular evolution across 10,000 generations in laboratory budding yeast populations. *eLife.* 10: e63910. https://doi.org/10.7554/eLife.63910.

27. This corresponds to nearly 2 million years, if scaled to human generation times. https://telliamedrevisited.wordpress.com. Accessed May 4, 2021.

28. Good, B., et al. 2017. The dynamics of molecular evolution over 60,000 generations. *Nature.* 551, 45–50. https://doi.org/10.1038/nature24287.

29. The guppy traits evolved at a rate up to 0.15 haldanes. Reznick, D.N. 2011. Guppies and the empirical study of adaptation. In J.B. Losos (ed.), *In the Light of Evolution: Essays from the Laboratory and Field.* Greenwood Village, CO: Roberts and Company. pp. 205–232;Reznick, D.N., et al. 1997. Evaluation of the rate of evolution in natural populations of guppies *Poecilia reticulata. Science.* 275: 1934–1937. https://dx.doi.org/10.1126/science.275.5308.1934; Hendry, A.P., and M.T. Kinnison. 1999.The pace of modern life: measuring rates of contemporary microevolution. *Evolution.* 53: 1637–1653. https://www.jstor.org/stable/2640428.

30. Each island ranged from 89 to 5790 square meters (960 to 62,000 square feet). Losos, J., et al. 1997. Adaptive differentiation following experimental island colonization in *Anolis* lizards. *Nature.* 387: 70–73. https://doi.org/10.1038/387070a0.

31. Schoener, T.W., et al. 2017. A multigenerational field experiment on eco-evolutionary dynamics of the influential lizard *Anolis sagrei*: a mid-term report. *Copeia.* 105: 543–549. https://doi.org/10.1643/CE-16-549.

5

Evolutionary accelerants and speed bumps

Now that we see that we can watch evolution happen before our eyes, whether in the lab or in the wild, what makes it easiest for experimental evolution—or evolution in any setting—to go fast?

We've already talked about how we measure evolutionary time in generations. So, the simplest ingredient to add to experimental evolution is an organism with a short genera- tion time. Viruses are king in this regard: the T4 phage virus that infects *E. coli* bacteria can practically break the sound barrier by exploding its population numbers to more than 100,000,000,000 (100 billion) in just two hours.[1] I'll also remind you of the fast- growing *E. coli* itself and the *E. coli*-eater *C. elegans* roundworms from Chapter 2. For vertebrates, we shift further into lower gears. Even the fastest developing vertebrate, the 5.4-cm-long (2-inch-long) African killifish *Nothobranchius furzeri*, takes 14 days to reach reproductive maturity.[2] There are lots of vertebrate species with generation times around two months, however, including mammals like house mice and birds like chickens.

One trend you may notice in this blitz across the tree of life is that generations are quicker for smaller creatures. So, all else equal, smaller organisms will be able to speed experimental evolution along more easily. Small organisms also often come with the benefit of simpler rearing, tolerating with stoicism a life lived with an austere food supply in a Petri dish, a glass flask, a shoebox cage, or a backyard coop.

Our next criterion for encouraging fast evolution dips its toe into abstraction, as it steps directly into the forces of evolution. This criterion harkens back to a venerable idea known, in genetics circles, as Fisher's fundamental theorem of natural selection. Fisher's fundamental theorem describes the idea that the rate of evolutionary adaptation that happens in a population depends on the quantity of genetic variants that are present in the population.[3] What this means is that having more genetic diversity in a population will allow it to evolve faster. In turn, the aggregate amount of genetic variability will be greatest in those populations with the biggest collection of individuals. As a result, we should expect natural selection to drive evolution most effectively when the size of the population is large.

The reason for large populations having more genetic variability is twofold. First, more individuals means that there are more gene copies circulating in the gene pool that

Evolving Tomorrow. Asher D. Cutter, Oxford University Press. © Asher D. Cutter (2023). DOI: 10.1093/oso/9780198874522.003.0005

could get hit by mutation to introduce genetic novelty. Second, genetic drift has less of an influence in large populations, as we discussed in Chapter 3. A weaker influence of genetic drift makes those rare new mutations less likely to go extinct in any given generation simply because of the vicissitudes of random events. The mutations will float around for longer in a big gene pool, riding only gentle ripples of genetic drift.

On balance, these two factors mean that bigger populations will contain a greater diversity and abundance of genetically distinct forms than will small populations. This gives yet another practical advantage of small organisms: those critters that are small in size and have small resource demands are most feasible to propagate in abundant quantities.

There is one more key factor that will help evolution speed along: sex. I don't mean sex in the sexy sense portrayed in blockbuster Hollywood romantic dramas, but sex in the clinical way that it expedites genetic rearrangement.[4] Sex is what allows that sidekick force of evolution, recombination, to be effective in scrambling alternate versions of different genes into different combinations. Recombination can only be effective in producing unique combinations of genetic variants along a chromosome when different individuals can regularly exchange genetic material. Recombination allows the genome to match different beneficial mutations together in the same individuals, even when they arose in separate individuals. By the same token, recombination can disentangle detrimental mutations at one gene from nearby beneficial mutations at another. All of this rearranging helps to create the optimal combinations to speed evolution along.

But not all organisms that get made by different individuals reproduce by fusing sperm and egg. Some organisms don't even make sperm or egg cells at all. In such cases, where sex and recombination are rare or absent, evolution won't speed along as quickly as it could otherwise. Fast experimental evolution can get away without sexual reproduction when generation times are short enough and mutation rates are high enough, as in viruses and bacteria. But the longer generation times of animals and plants make sex and recombination essential features for their rapid evolution. The ideal scenario for fast evolution is to have large populations of rapidly maturing organisms with lots of genetic variability and lots of sex.

5.1   Putting on the brakes

Just as some factors can rev evolution's engine, others set out speed bumps. The speed bumps that we care about most are those that interfere with the ability of selection and selective breeding to give a genetically stable response to a change in a trait. Selection can get overwhelmed by the noise that environmental variability infuses into the development of a trait, the environment masking genetic sources of variability in the trait. We've already mentioned another constraint, the demographic constraint of population size: the fewer individuals we have, the less easily selection can differentiate among the different genetic variants that are present in that population. When population size is too small, selection can get overwhelmed by noise. That noise, in an evolutionary sense, can come from the random fluctuations of gene frequencies caused by genetic drift.

One way to think about population size is as the total number of individuals that live in a place. This census size is the crucial count to tell us how much food and space are required to host a species, as well as how many new mutations will arise into it. Selection, however, is a kind of genetic bottleneck on that total number of individuals. Only a subset of the total population actually ends up transmitting their genetic material to the next generation. This number of breeding individuals defines the size of the active gene pool of the population, what geneticists call the genetically *effective* population size.

This distinction in what *population size* means is analogous to different ways that we can count people in a country. For example, we can distinguish between the number of people counted in the U.S. census versus the number of registered voters. While all residents of the United States contribute to the economy in one way or another, only the subset of people who are registered to vote can contribute to the election of politicians to set laws for future generations.

Another kind of constraint on evolution, one owing to physical limitations, is simple to think about: evolution cannot break laws of physics.[5] Nonetheless, history shows how biological systems have proved extremely inventive in exploiting the laws of physics, skirting the edges of physical boundaries, as if life itself had a crack team of lawyers. Let me give you just one quick example.

In addition to all the incredible animal characteristics that we've already encountered, there is the ever-mind-boggling tardigrade. You can find them on moss, though you'll need a microscope to view them properly. Affectionately known as *water bears,* these miniscule creatures that define their own phylum of animals are capable of tolerating the vacuum of open space.[6] This feat does not result from adaptation to interstellar travel but is simply a byproduct of adaptation to the stress of being parched. Tardigrades often are faced with drought, their little world drying up completely. When confronted with a lack of water, their bodies reconfigure themselves anatomically as they enter what is called an anhydrobiotic state: they shrink their bodies down like a folded accordion to minimize surface area to the outside and, inside, produce special intrinsically disordered proteins that help stabilize their cellular structures as the animals become metabolically quiescent. This quiescent state also makes them resistant to extreme temperatures, radiation, and some noxious chemicals. These adaptations for a tardigrade's ability to survive extreme drought also just happen to prove effective in surviving low Earth orbit.

★

Genetics itself introduces a washboard road of potential speed bumps to evolutionary change. Genes have to work together to produce a given characteristic of an organism, forming the "genetic architecture" of a trait. Several braces of the scaffolding of this architecture, so to speak, can limit evolutionary responses.

The most fundamental genetic constraint is a lack of genetic variants. Individuals might differ, but those differences could result from environmental influences and *not* from genetic differences. After all, the environment contributes enormously to the characteristics that an individual expresses. But there can be no evolution without genetic differences. This absolute genetic constraint would mean that selective breeding or natural selection on the trait would have no genetic baton to pass from one generation to the next: no heritable transmission of differences. We will explore detours around this limitation shortly, but first let's consider a few other ways that genetics can constrain the speed of evolution.

Mutations will be least evolutionarily successful when they cause the most collateral damage.[7] Some types of mutation, and mutations to some types of genes, will be more likely to cause collateral damage. A mutation might improve survival, but also cause a detriment to reproduction that trades off to make it a wash overall, or even a net detriment. This trade-off reflects the multiple impacts that mutations can have. For example, a genetic variant of the aryl hydrocarbon-interacting protein (AIP) causes human gigantism. That gene variant of AIP can lead a person to grow to an impressive height of 2.3 m (7 feet, 7 inches) or more, as for Charles Byrne in 1700s Ireland. Unfortunately, this gene variant of AIP also causes the collateral damage of a 30% reduction in life expectancy attributable to cardiovascular and other health complications.[8] More subtle genetic trade-offs could involve effects on distinct anatomical features, or varying performance in one habitat versus another, or even the consequences of expression by one sex versus the other. In the lingo of geneticists, some genetic changes have stronger pleiotropic effects than others, and these pleiotropic effects that alter multiple features usually slow down evolution.

To minimize this kind of evolutionary speed bump, it helps when a given genetic variant influences as few features of an organism as possible. The kinds of changes that fit this idea tend to be mutations that alter when or where or how much of a protein gets expressed. These kinds of regulatory changes will tend to have effects in only a subset of tissues or of a subtle magnitude. In contrast, mutations that alter the protein structure itself (as for AIP), or that make it impossible to make the protein at all, will exert their effects everywhere that the gene would ordinarily get expressed, leading to a greater likelihood of multiple impacts. Mutations that induce a big change to one feature also are more likely to impact others, making it harder for evolution to balance the benefit with the potential costs of the side effects.

★

Collateral damage of mutations sometimes is indirect and *not* caused directly by the mutation itself. The linear nature of DNA on chromosomes means that genes can be linked close together on a chromosome. When this happens, the genetic variant for one gene may have detrimental effects that partially counteract the benefits of a linked genetic variant of a different gene nearby. The happenstance of mutation made them genetic neighbors on that same linked string of DNA.

This circumstance of antagonistic genetic neighbors will inhibit selection, making selection unable to work separately on each genetic component. This interference is due to the speed bump of linkage. When genes are far apart, the evolutionary force of recombination can more easily rearrange distinct gene variants into optimal combinations. When genes are stuck close together, however, it may take many generations for sex and recombination to create the ideal genetic combinations. The evolutionary consequences can be drastic in the interim when, for example, we impose strong artificial selection that favors one gene variant. The variant of the *other* gene will get dragged along for the ride.

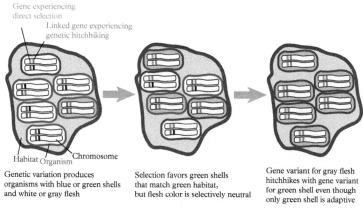

Genetic variation produces organisms with blue or green shells and white or gray flesh

Selection favors green shells that match green habitat, but flesh color is selectively neutral

Gene variant for gray flesh hitchhikes with gene variant for green shell even though only green shell is adaptive

Figure 5.1 *Recombination can fail to create distinct combinations of gene variants for genes that are closely linked to one another. Natural selection and artificial selection that directly targets one of the genes, because of how it affects survival and reproduction, can cause a genetic variant of a different gene to evolve higher abundance in the species, termed* genetic hitchhiking. *Genetic hitchhiking led to the evolution of the gene variant of SLC2A9 that causes hyperuricosuria in Dalmatians, as a correlated byproduct of artificial selection directed at a genetic variant affecting the spot size on their fur.*
Photo credit: Dalmatiner24.eu reproduced from public domain.

Getting dragged along like this is termed genetic hitchhiking (Figure 5.1). Genetic hitchhiking can drive the unintended evolution of one characteristic of an organism simply as an incidental byproduct of selection on another—even though different gene variants affect each trait separately. Genetic hitchhiking can even let *detrimental* variants get dragged along. This phenomenon is especially conspicuous in dogs. Dog genomes are notorious for having long stretches of DNA that were passed along intact in the formation of a given dog breed. Inside those long stretches of DNA lurk genetic variants that define both the breed's myriad outward characteristics and its predispositions to disease.[9]

One such example has plagued Dalmatians, a striking dog breed dating to 1300s Croatia.[10] The stereotypical feature of Dalmatians is their erratic pattern of black spots on a white coat, known as ticking. A genetic region affecting the size of the pigmented spots, however, is adjacent to the genetic region responsible for another unusual feature of Dalmatians. They excrete uric acid in their urine as humans do, and unlike other dogs and other mammals, a feature known as hyperuricosuria, or huu for short. Hyperuricosuria predisposes Dalmatians to bladder stones that require surgical removal and is caused by a distinctive genetic set of changes to the SLC2A9 gene. Artificial selection for ticking led to the unintended hitchhiking of the genetic variant with disrupted SLC2A9 function, simply because of the close proximity of these two features on canine chromosome 3. Breeders have since used careful selective breeding to recombine away the defective copy of the SLC2A9 gene to produce so-called low uric acid (LUA) Dalmatians. LUA Dalmatians have genomes just like other Dalmatians except for a fully functional version of the SLC2A9 gene.[11]

★

These kinds of unintended effects of correlated evolutionary change aren't restricted to cases of just one or two genes with large effects on a characteristic. They can involve tens or hundreds or thousands of genes. Each of those genes can have genetic variants that exert only a tiny effect. Most characteristics of organisms, in fact, including humans, depend on the influence of many, many genes. These kinds of traits are polygenic or, in the extreme, omnigenic: the trait value that we can see and measure will depend on the particular variants a creature has for many genes, up to essentially *all* of the genes in the genome. Such traits are the epitome of genetic complexity, altered by the minute influence of each genetic variant on the overall measurable characteristic that an organism develops.

Take height. How tall someone is depends a lot on how tall their parents were.[12] This fact simply means that height is highly heritable. But no one single genetic variant affects height very much—excepting anomalies like the C to T mutation at position 910 of the AIP gene that causes gigantism, of course. On average, any one genetic variant only alters a person's height up or down by 0.14 mm (1/180th of an inch), but there are thought to be over 100,000 of such genetic variants sprinkled along human chromosomes.[13]

In evolutionary engineering, we may actually be interested in selectively favoring many characteristics, with each characteristic influenced by many genes. Each characteristic may be genetically correlated with one another to different degrees because of genetic factors they have in common. The cumulative effects of both direct (pleiotropic multiple impacts) and indirect (linkage) genetic influences on the thousands of contributing genes can explain how selection directly on one feature can slow down evolution or lead to unintentional side-effect evolutionary responses on features. Hyperuricosuria in Dalmatians was a simple single gene case, but the same idea applies to traits that are influenced by many genes.

If you've ever eaten a 1980s commercial tomato raw, you know what I mean. As a kid, I never much liked the taste of tomatoes. I've since come around, having discovered the delicious flavor of heirloom varieties that contrast so starkly with the watery and grainy and mucilaginous supermarket products that pervaded the salad bars of my youth (Figure 5.2). Horticulturists had selected tomato plants for high yield, disease resistance, and firm fruits.[14] This breeding proved successful, with many, many genes contributing to these characteristics. Farms could ship lots of blemish-free orbs that were robust to long-distance transport in the back of a truck. But lousy flavor came along for the ride. In particular, the bigger the tomato, the less sugar content it is likely to have. This trade-off comes about, in large part, because of gene variants of the *Lin5* gene, which encodes an enzyme found in plant cell walls that affects the amount of soluble sugar in the fruit cells—the tomato is, after all, botanically speaking, a fruit. The unintentional consequence of evolving tomato varieties with desirable agricultural characteristics like large size was the correlated evolution of undesirable flavor characteristics like low sweetness.

The web of interactions among genes controls how they go about their business of making cells to divide and move and differentiate into specialized types. This structure of this genetic web makes changes to some genes much less prone to causing collateral

Figure 5.2 *Selective breeding on the high genetic diversity of heirloom tomato varieties led genotypes with undesirable taste features to hitchhike along with features desirable for commercial production. Some forms of genetic editing can act to prevent some of the negative unintended consequences of artificial selection.*
Photo by the author of heirloom tomatoes at a farmer's market in Newmarket, Ontario.

damage than others. Some genes in that genetic web can channel genetic changes to produce novel traits without perturbing other traits. These genes are what biologist David Stern refers to as input-output genes. In the long term, this goldilocks kind of input-output gene is primed to be a hotspot of repeated evolutionary change by presenting little if any bump in the path of evolution.

There may be no simple mutational path along the genetic web, however, that could result in the organism changing in the direction that we desire, or that nature would favor. The translation of this idea into one of my grandfather's idioms would be, "*you can't get there from here.*" This constraint will be stronger when there are few genes and, therefore, few mutational targets. More genes make for more mutational targets and more genetic pathways, more genetic backdoors capable of tweaking how an organism develops. The idiosyncrasies of connections among partners in that genetic web also may lead some developmental outcomes to be more likely. For example, the effects of random mutations may be more likely to make a trait smaller than to make it larger, a phenomenon sometimes referred to as a developmental bias.[15]

5.2   Skirting the limits

Traits affected by variants at thousands of genes present a challenge. We can't just think about increasing one version or another of a given gene to shape an evolutionary response. And, often, we are interested in shaping the evolution of many traits, and most traits have such complex genetic causes. To deal with this challenge, breeders can create a selection index. This index is a composite metric that integrates across all of the different characteristics of interest. Rather than using artificial selection on any one gene or any one trait, we select for the composite index. As Steven Arnold put it,[16] a selection index lets us "devise a selection program that will give the best results across the board." Each characteristic gets weighted by how important it is to us; for agricultural breeding, from popcorn to pigs, that importance usually is related to economic value. The characteristics to prioritize, however, are limited only by one's imagination. The traits included in a selection index can even include seemingly ethereal features like the gut microbial community that defines an organism's microbiome or the resilience of animals to variable environmental stresses.[17]

Traits, however, often are slow or logistically difficult to measure. These days, sequencing DNA is quite quick and easy. As a consequence, the selection index, in actual fact, could be based on DNA directly.

Agricultural breeders have created statistical links between measurable characteristics and the array of genetic variants that a given species has in its genomes. The DNA variants of individuals, measured at thousands of genomic locations, associate with valuable or interesting characteristics and let a breeder avoid dealing with noisy and tedious measures of the characteristics themselves every generation. This genomic information means that breeders can make a selection index based on DNA that is predictive of what traits individuals will express, and what traits will evolve. This technique creates a "genomic selection index," rather than a selection index based just on trait measurements. By using a genomic selection index, horticulturists can identify individuals for breeding before they even manifest measurable traits or they can incorporate and propagate traits that you can only measure after death. In pigs, this approach gives a 25% boost to how fast the generations of artificial selection can unfold to drive their evolution.[18]

With these techniques, we can apply artificial selection to nudge the genetic composition of a population. It can let us shift the relative abundances of genetic variants in such a way that it profoundly alters the average features seen among individuals in just the way that the breeder's equation predicts.[19] The rate of change from one generation to the next, however, may get tempered by the genetic details, as we explored earlier in this chapter. The evolutionary response will be sensitive to population size, physical constraints, and the existence and abundance of distinct genetic variants, as well as the degree to which genetic variants exert multiple effects and are linked to one another along chromosomes.

★

Should you or I choose to intervene in the evolutionary trajectory of a population of organisms, to push its buttons so to speak? Artificial selection is a button that is big and red and begging to be pressed. But if the population contains no distinct genetic variants that influence the trait of the critter we hope to nudge, then the red button will be nothing but a red herring: there can be no evolutionary response without genetic differences to mediate the response. And yet, all is not lost. We have tools at our disposal to speed up the creation of genetic novelty, to once again let us try to prod evolutionary fate.

The classic way that geneticists will add genetic novelty to a lab population is to mutagenize them. In its essence, this method increases, in a controlled manner, the mutation rate that cells normally experience. There are lots of ways to introduce new genetic variability with mutagens, depending in part on the kind of organism or the kinds of mutations to induce. For some critters, you can give them a chemical bath. Others might receive these compounds in their food or by injection, depending on the requisite animal care guidelines. The chemicals have rightfully ominous names like ethyl methanesulfonate and *N*-ethyl-*N*-nitrosourea. After all, they produce changes to DNA in stem cells that make gametes—and gamete cells fuse to make offspring—in organisms as diverse as plants, insects, and mammals. These chemicals have proved exceptionally good at adding new genetic variants that produce new characteristics, but they do so in an untargeted, random way across the genomes of the affected organisms. Many of the genetic changes are detrimental, and many of the new characteristics are monstrous,[20] despite their incalculable value in helping researchers to understand the mechanisms of life as it has been encoded in DNA.

Another way to ramp up the mutation rate so that novel genetic variants arise more readily is to disrupt the molecular machinery that repairs errors in DNA. As cells copy DNA each time they divide, they inevitably make lots of mistakes. They also have backup systems to correct those mistakes. One backup system depends on so-called mismatch repair proteins like MutS. By disrupting the activity of MutS and not letting it do its job, one may let those DNA mistakes remain uncorrected. Presto, a higher mutation rate and faster introduction of novel and entirely natural genetic variants. One could disrupt MutS permanently, by knocking out the gene and deleting it from the genome altogether. You could also disrupt it transiently by using molecular biology techniques like RNA interference (RNAi). RNAi gives experimentalists a way of tricking cells into degrading their own gene's expression by injecting a funky version of an intermediate form of that same gene.[21] By tricking the cells into degrading MutS, a higher mutation rate will result.

All these ways of boosting mutation rates do the job of creating genetic novelty, but in a scattershot way, blind to what result they might have. An alternative is to guide the formation of new genetic variants in a concerted way, directing changes in the genome based on prior knowledge of how to influence the expression of traits. This alternative is to genetically engineer evolutionary novelty.

★

Genetic engineering usually starts with a transgene. A transgene is a piece of genetic material that comes from somewhere else. That "foreign" DNA could be from a template sequence seen in another species or synthesized from the mind of a scientist. To begin genetic engineering with a transgene, you use the tools of the molecular biologist's trade to make a customized sentence of DNA that you synthesize in a little plastic tube. Because you get to write that sentence, you get to decide what it will say. It might call for a particular gene to get turned on or turned off. It might even add a gene to the genome of a species that ordinarily would only be found in another species.

A classic example of such a cross-species transgene encodes a bright molecule called green fluorescent protein (GFP). Whenever a cell or tissue makes GFP, it will glow a beautiful fluorescent green color when you shine a UV light onto it. Consequently, GFP gets lots of use, for logistical purposes, in the laboratory. Since prehistory, GFP found its home inside the crystal jellyfish *Aequorea victoria*, which inhabits much of the eastern edge of the Pacific Ocean.[22] Nowadays, you can find GFP expressed in just about any organism you can think of that biologists use in the laboratory. The exceptional utility of GFP in laboratory experiments explains why *C. elegans* biologist Marty Chalfie was awarded the Nobel Prize in 2008 for establishing it as a tool, along with Osamu Shimomura and Roger Tsien.

Biologists have devised a legion of recipes to insert one of these customized sentences of DNA into an organism's genome. One approach, which is surprisingly effective for the little roundworms that researchers in my laboratory study, uses brute force. First, you coat tiny gold beads with the engineered DNA. Then, you load this cargo of DNA-coated gold balls into a microscopic shotgun shell, and finally you shoot them out with a high-pressure helium gun to blast the gold particles into a pile of worms on a Petri dish.[23] Amazingly enough, a few of the DNA-slicked golden BBs lodge themselves in the gonad tissue and break a chromosome, and as the cell repairs the damage from the chromosomal breakage of this bombardment, the engineered DNA gets incorporated into the genome. In an alternate technique, you would inject a minuscule potion of engineered

DNA directly into gonad tissue using a needle, expose it to conditions favorable to integration of the transgene into the genome, such as UV or gamma-ray light exposure, and hope for the best.

Biologists have been using these methods for decades, but they leave a lot to chance: the location in the genome that the transgene gets inserted is more-or-less random. Since 2012, there's been a new game in town. This new game, this momentous step-change in how accurate and precise one may manipulate genomes, goes by the peculiar name of CRISPR-Cas9 genome editing.

I want to devote dedicated attention to CRISPR-Cas9 in its own chapter, but let me make just two points now. First, genetic editing with CRISPR-Cas9 gives us a well-stocked machine shop of tools for engineering genetic novelty. It allows the finesse of flipping one DNA nucleotide letter to another at a specific spot in a genome, or instead, the wholesale elimination of genes, or even the insertion of custom-designed sets of genes. It is fast and cheap and biologically elegant. CRISPR-Cas9 puts the entire genome within our jurisdiction to tweak according to our needs for new genetic variants in a population.

Second, one class of techniques that use the CRISPR-Cas9 molecular machinery, once applied to wild organisms, has the capacity to take on a life of its own. It has the capacity to alter the genetic composition of an entire population over the course of a few generations, not just the critters that first got injected with it. These engineered elements are known as gene drives. Gene drives reconfigure our notion of simple genetic engineering as a special, human-assisted kind of mutation into an entirely new evolutionary force. CRISPR-Cas9 gene drives put the world on the cusp of creation. Human-mediated genetic engineering with gene drives released into the wild, as we will soon explore, would instantiate a new force of evolution: genetic welding.

★

Evolutionary change can be facilitated or constrained. Among the evolutionary accelerants and speed bumps are population size, sexual reproduction, mutational availability, and the genetic architecture that underpins the cellular development of traits. A single mutation may impact multiple traits (pleiotropy) and different mutations in nearby genes may impact the same trait in opposing ways (interference by linkage), both of which can impede evolutionary responses to selection. Traits may have few genes with variants responsible for differences between individuals, or very many genes. Evolution of traits subject to polygenic genomic architectures can be guided with the aid of a genomic selection index. Humans can increase the stocks of genetic variation in a species by mutagenesis and by transgenesis. Genetic engineering biotechnology now offers unprecedented capabilities for inserting specific genetic variants into organisms.

Notes

1. Laemmli, U.K., and F.A. Eiserling. 1968. Studies on the morphopoiesis of the head of phage T-even. *Molecular and General Genetics*. 101: 333–345. https://doi.org/10.1007/BF00436231.

2. These small fish inhabit ephemeral pools of water in Mozambique's savannahs, spending most of the year in diapause as embryos buried underneath dry mud until the rains come. The rain-filled depressions can last as little as 20 days and still support *N. furzeri* populations. Vrtílek, M., et al. 2018. Extremely rapid maturation of a wild African annual fish. *Current Biology*. 28: R822–R824. https://doi.org/10.1016/j.cub.2018.06.031.

3. In Fisher's own words, it is stated as, "The rate of increase of fitness of any species is equal to the genetic variance in fitness," Fisher, R.A. 1930. *The Genetical Theory of Natural Selection*. London: Oxford University Press. p. 46.

4. As a graduate student, my wife's friend ended up on a date with a single professor in my department. She confessed to being unusually intimidated, as he was a world expert on sex. Little did she know, until later, that his sexual expertise, from an academic standpoint, was entirely theoretical and most finely honed in terms of algebraic equations.

5. Alexander, R. M. 1985. The ideal and the feasible: physical constraints on evolution. *Biological Journal of the Linnean Society*. 26: 345–358. https://doi.org/10.1111/j.1095-8312.1985. tb02046.x.

6. Weronika, E., and K. Łukasz. 2017. Tardigrades in space research: past and future. *Origins of Life and Evolution of Biospheres*. 47: 545–553. https://doi.org/10.1007/s11084-016-9522-1; Boothby, T.C., et al. 2017. Tardigrades use intrinsically disordered proteins to survive desiccation. *Molecular Cell*. 65: 975–984.e975. https://doi.org/10.1016/j.molcel.2017.02.018.

7. Stern, D.L., and V. Orgogozo. 2009. Is genetic evolution predictable? *Science*. 323: 746–751. https://dx.doi.org/10.1126/science.1158997; Connallon, T., and M.D. Hall. 2018. Genetic constraints on adaptation: a theoretical primer for the genomics era. *Annals of the New York Academy of Sciences*. 1422: 65–87. https://doi.org/10.1111/nyas.13536.

8. Chahal, H.S., et al. 2011. AIP mutation in pituitary adenomas in the 18th century and today. *New England Journal of Medicine*. 364: 43–50. https://dx.doi.org/10.1056/NEJMoa1008020.

9. Marsden, C.D., et al. 2016. Bottlenecks and selective sweeps during domestication have increased deleterious genetic variation in dogs. *Proceedings of the National Academy of Sciences USA*. 113: 152–157. https://doi.org/10.1073/pnas.1512501113.

10. Bauer, M. and N. Lemo. 2008. The origin and evolution of Dalmatian and relation with other Croatian native breeds of dog. *Revue de Médecine Vétérinaire*. 159: 618–623. https://www.revmedvet.com/artdes-us.php?id=1674; Bannasch, D., et al. 2008. Mutations in the SLC2A9 gene cause hyperuricosuria and hyperuricemia in the dog. *PLoS Genetics*. 4: e1000246. https://doi.org/10.1371/journal.pgen.1000246.

11. SLC2A9 stands for Solute Carrier Family 2 Member 9, with "solute carrier" pointing to the role that this protein plays in metabolism within the liver and kidney to process and transport compounds in blood for secretion in urine. The selective breeding to eliminate the genetically induced health problems caused by SLC2A9 in Dalmatians, however, has encountered

resistance by kennel clubs for LUA Dalmatians to be accepted in competition. https://ckcusa.com/blog/2019/november/the-dalmatian-back-cross-project.

12. As long as parents and offspring were well-fed and well-cared for, nutritional and environmental effects generally will be outweighed by the genetic contribution to differences in height. Nonetheless, do not discount the influence of nongenetic effects: the average height of humans has increased by 10 to 20 cm (4 to 8 inches) over the past few centuries because of improved nutrition. https://ourworldindata.org/human-height. Accessed June 16, 2022.

13. Boyle, E.A., et al. 2017. An expanded view of complex traits: from polygenic to omnigenic. *Cell*. 169: 1177–1186. http://dx.doi.org/10.1016/j.cell.2017.05.038.

14. According to panelists assessing tomato varieties, we most like varieties with intensely sweet and umami flavor and good texture. The Maglia Rosa Cherry comes out on top, an incredibly sweet fruit. But there must be a lot of individual preference, as my personal favorite when Mom grows them, brandywine, wasn't ranked all that highly by panelists, perhaps because of its chemistry being the sourest and among the most salty and bitter. A commercial variety, with the unappealing name of FL8735, came out on bottom. There are 28 molecular characteristics, however, that together significantly predict both the likability and texture of a tomato. Modern commercial varieties have reductions in 13 flavor-associated volatile compounds compared to old heirloom varieties, in addition to sugar content. Refrigeration destroys flavor, too. Tieman, D., et al. 2017. A chemical genetic roadmap to improved tomato flavor. *Science*. 355: 391–394. https://dx.doi.org/10.1126/science.aal1556.

15. Despite their already diminutive size, being just 1 mm long (1/25th of an inch), new random mutations that influence the length of *C. elegans* nematode roundworms will, on average, make them shorter rather than longer. Azevedo, R.B.R., et al. 2002. Spontaneous mutational variation for body size in *Caenorhabditis elegans*. *Genetics*. 162: 755–765. https://doi.org/10.1093/genetics/162.2.755; Svensson, E.I., and D. Berger. 2019. The role of mutation bias in adaptive evolution. *Trends in Ecology & Evolution*. 34: 422–434. https://doi.org/10.1016/j.tree.2019.01.015.

16. In practice, the weighted average of traits that goes into a selection index takes advantage of variability and correlations among the traits due to both genetic and environmental influences. Arnold, S.J. 1992. Constraints on phenotypic evolution. *American Naturalist*. 140: S85–S107. https://doi.org/10.1086/285398.

17. Weishaar, R., et al. 2020. Selecting the hologenome to breed for an improved feed efficiency in pigs: a novel selection index. *Journal of Animal Breeding Genetics*. 137: 14–22. https://doi.org/10.1111/jbg.12447.; Berghof, T.V.L., et al. 2019. Opportunities to improve resilience in animal breeding programs. *Frontiers in Genetics*. 9: 692. https://doi.org/10.3389/fgene.2018.00692.

18. Vieira, R.A., et al. 2016. Selection index based on the relative importance of traits and possibilities in breeding popcorn. *Genetics and Molecular Research*. 15: gmr.15027719. https://dx.doi.org/10.4238/gmr.15027719; Meuwissen, T., et al. 2016. Genomic selection: a paradigm shift in animal breeding. *Animal Frontiers*. 6: 6–14. https://doi.org/10.2527/af.2016-0002.

19. Recall from Chapter 2 how heritability and selection together let us predict evolutionary responses with the breeder's equation.

20. For a lively and consuming exposition of the effects of mutations on human traits, with both science and storytelling, I highly recommend Armand Leroi's 2005 book *Mutants*.

21. "Funky intermediate" is my choice non-technical term for double-stranded RNA, dsRNA. The NIH has a nice primer about how RNAi works with more technical details: https://www.ncbi.nlm.nih.gov/probe/docs/techrnai/. Accessed October 13, 2021.

22. Purcell, J.E. 2018. Successes and challenges in jellyfish ecology: examples from *Aequorea* spp. *Marine Ecology Progress Series*. 591:7–27. https://doi.org/10.3354/meps12213; Prasher, D.C. 1995. Using GFP to see the light. *Trends in Genetics*. 11: 320–323. https://doi.org/10.1016/S0168-9525(00)89090-3.

23. You can learn all of the intricate details of the protocol for integrating transgenic DNA via microparticle bombardment here: http://www.wormbook.org/chapters/www_microbombard/microbombard.html.

6

Hijacking genetics

On the fifth floor of the Ramsay-Wright Zoological Laboratories in the heart of the University of Toronto campus, nestled below the brutalist architecture of the iconic Robarts Library across the street, there is an animal whose brain sparkles. Literally, the neurons of this creature shine with light (Figure 6.1). Thanks to its transparent skin, you can watch the brain cells gleam as the animal winds in sinuous arcs throughout its day, making its quiet way through life. This animal has a foreign name to most ears, JAC1293, but now that we're on familiar terms, we'll call it Jackie.[1] Jackie is the brainchild of researcher John Calarco, and Jackie's glimmering mind owes its twinkle to CRISPR-Cas9 genome editing.

Genome editing with the CRISPR-Cas9 technique took the research world by storm in 2012, and for good reason. It is cheap, fast, precise, and flexible. It is a triumph of what basic scientific research can discover about evolution's inventions in nature.[2] It is a triumph of what women in science have achieved, with its co-discoverers Jennifer Doudna and Emmanuelle Charpentier quickly earning a Nobel Prize just eight years later. And, to top it off, CRISPR-Cas9 genome editing is beautiful: an elegant piece of biological artwork and an elegant tool for creating biological artwork all rolled into one.

6.1   What becomes of an unwelcome guest

It all begins with an unwelcome guest. Life inside a bacterial cell is a rambunctious party, the DNA of its genome dancing along in its circular conga line to the tune of proteins pumping out of the ribosomes. If the cell is ciliated, then its cilia wriggle, radiating from the cell surface like miniature millipede feet to make the cell dance its whole life long; if the bacterium is flagellated, then its tail of a flagellum gyrates in a microscopic emulation of a single-celled Elvis Presley.[3] Life is a blast, until a phage crashes the party. Phages are viruses that infect bacteria, injecting their genomes and hijacking the cell's machinery to crank out more copies of the virus before bursting the cell to release a new horde of phage that will seek out new parties to blow out. To avoid this balloon popping end to a bacterial bash, defense mechanisms have evolved. Bouncers. CRISPR-Cas9 is a bacterial bouncer whose job is to detect alien DNA, to identify the invading viral genome, and to rip it to shreds.

Evolving Tomorrow. Asher D. Cutter, Oxford University Press. © Asher D. Cutter (2023). DOI: 10.1093/oso/9780198874522.003.0006

Figure 6.1 *These* C. elegans *nematode roundworms have been genetically modified to express the jellyfish green fluorescent protein (GFP) in some of their 302 neurons.*
Image credit: Heiti Paves, reproduced under the CC BY-SA 3.0 license.

Really, though, the set of molecules that make up CRISPR-Cas9 serve as a bacterial immune system. They work against viruses by constantly checking a wall of wanted-poster mug shots of past viral genomes that the bacterial ancestors encountered. Different types of viruses have uniquely distinct sequences that make up their genomes. The bacterial version of a wall of mug shots is a record in DNA listing a unique entry for each would-be invader in a special part of the bacterial chromosome. This spot in the bacterial chromosome is called the "cluster of regularly interspaced short palindromic repeats," which in the interest of conversational brevity, is nicknamed CRISPR.

The CRISPR spot in a bacterial chromosome acts like a dictionary to keep track of fragments of viral genomes, writing them down in DNA as words to remember, venerated for their potential to wreak terror. To keep these words in living memory, the cell makes innocuous versions of them out of its own RNA in a series of steps that produce a pair of molecules that operate as "guide RNA."[4] The guide RNA teams up with a posse of proteins, the chief among them being a protein known as Cas9.[5] Cas9 is a humble name for such a ruthless sniper of a molecule. When teamed up with the guide RNA, Cas9 homes in on the foreign viral genome, perfectly identifying it by matching the mantra of the dictionary word spoken by the guide RNA. After that, Cas9 slices the viral DNA right down the middle to inactivate the nefarious potential of the phage.

In molecular terms, the CRISPR-Cas9 RNA-protein complex achieves this precision matching by using the inherent nucleotide complementarity of RNA and DNA. RNA and DNA might be different languages, but there is a one-to-one mapping to each of the four letters in their respective alphabets that allows for perfect translation between them. Once it finds exactly the right location where the guide RNA perfectly matches a string of DNA, Cas9, as an endonuclease protein, can cut the DNA. It is this teaming up of a special kind of RNA with a special kind of protein that provides the bacterial memory of a viral enemy and a means to recognize and destroy that enemy.

The perks of CRISPR-Cas9 go beyond such an immune response, however. This ability to cut precisely at a specific location in a string of DNA, it turns out, is a holy grail in molecular biology. Molecular biology labs already had lots of tools to repair broken DNA in customizable ways. What seemed intractable, however, was to easily create a DNA break in just the right spot. In a genome millions of nucleotides long, how do you create a snip in the DNA exactly where you want it to be? A bacterial cell's CRISPR-Cas9 machinery, as it turns out, knows how to do just that.

It is hard to overstate the importance of the discovery that the bacterial immune system had created, over the course of evolution, a way to specifically target an arbitrarily defined sequence to slice. The importance of this idea was not lost on Doudna and Charpentier, however, and they made the most of it.

★

What if we created our own guide RNA that matched a particularly enticing DNA sequence motif? What if we added it along with Cas9 to non-bacterial DNA as a target? Could we repurpose the bacterial immune system to our own needs and desires? These are exactly the questions that Jennifer Doudna, Emmanuelle Charpentier, and their teams answered in their ground-breaking study in 2012.[6] The answers were a resounding "yes." "[T]he Cas9 endonuclease can be programmed with guide RNA engineered as a single transcript," they wrote, "to target and cleave any dsDNA sequence of interest."[7] The ability of CRISPR-Cas9 to slice DNA down the middle, it turned out, was not limited to just seeking out phage DNA. Within a year, studies around the world proved that researchers could indeed apply the CRISPR-Cas9 toolkit to all sorts of animals to alter their genomes by precisely targeting changes to specific locations on chromosomes.

Genome editing with the CRISPR-Cas9 technique works like this, what I'll call the core protocol (Figure 6.2). First, you decide what spot in a genome you want to target, a particular stretch of DNA letters along a chromosome. You then design a string of RNA about 20 nucleotide letters long to match that spot. This is your engineered guide RNA. The target in the genome needs to be unique, with only one string of letters with that sequence. There are computer programs to help with this step, and commercial companies can then synthesize the custom string of RNA nucleotides for about $10.

Second, an optional step, you design a string of DNA to use as a template. This custom template sequence serves as the molecular splint for repairing in a specific way the break that you will soon make to the genome. We'll come back to the panoply of things that you could do with such a custom template. These first two steps are the key "engineering" phases where you design the feature to alter and where to alter it within the genome.

The next step is to get hold of Cas9 protein. There are a few ways to do this. One could isolate the messenger RNA precursor to the protein (a.k.a. mRNA) or engineer a

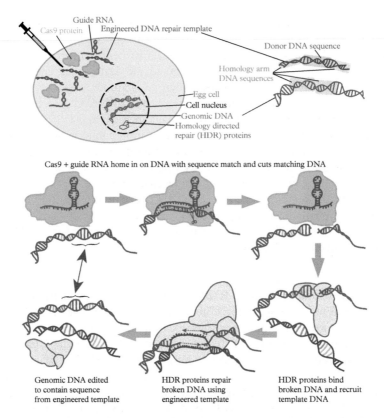

Figure 6.2 *To perform genome editing with CRISPR-Cas9, you must first introduce the reagents into appropriate cells so that Cas9 protein, guide RNA, and the engineered DNA repair template will be able to interact with genomic DNA within a cell's nucleus (top). The DNA sequence regions termed* homology arms *are engineered in the repair template to match the position in the genome to be targeted for gene editing, with an optional "donor" sequence in between them in the engineered DNA repair template. Then (bottom), the Cas9 homes in on the sequence of genomic DNA that matches the guide RNA sequence and cuts the genomic DNA to create a double-strand DNA break. The cell detects the DNA break and sends DNA repair molecules to fix the break. When homology-directed repair (HDR) takes place, the cell will often use the engineered DNA repair template instead of the true homologous chromosome to direct repair of the DNA break. HDR with the engineered DNA repair template leads the repaired genomic sequence to encode the new sequence of DNA that differs from the original genomic DNA sequence.*

circle of DNA known as a plasmid vector to encode both the Cas9 gene and the ability to entice cells to express it, which cells do by using their own usual molecular machinery to transcribe the Cas9 DNA into mRNA and then translate the mRNA into Cas9 protein. Alternately, with some laboratory elbow grease, you could biochemically purify your own protein from microbes that you engineer to be bioreactors. Doudna, Charpentier, and their collaborators originally isolated Cas9 protein from *Streptococcus pyogenes*, a sometimes-pathogenic species of bacteria. Or you could just order a little tube of the protein itself from a commercial vendor for about $100.

In the penultimate step, you inject a mixture of all these ingredients into a recipient tissue that makes sperm or egg cells. There are other ways to "inject" the CRISPR-Cas9 reagents than simply with a needle, however, including use of nanoparticles or even delivering them to cells inside of otherwise-innocuous viruses, such as lentiviruses and adenoviruses.[8] For many research or clinical purposes, however, you don't actually want to edit the genomes of gametes, so you can use other cell biology techniques to get the ingredients into other types of cells. Usually, this mixture also includes a signal that simply lets you easily determine whether the procedure worked.

Finally, you verify that the genetic editing worked as expected and test for potential off-target effects in unexpected locations of the genome. The essence of manipulating DNA by CRISPR-Cas9 genome editing, however, is what happens, exactly, when all the ingredients find themselves all together inside that novel cellular environment.

★

Let's look inside a cell that has been delivered a potion of CRISPR-Cas9 reagents. When the Cas9 protein finds its DNA target, directed and confirmed by the guide RNA, the Cas9 protein snips the DNA. This snip creates a break in the string of DNA. Breaks in DNA set off alarm bells within the cell, resulting in a rush of molecular emergency workers to the scene to repair the broken section of the genome. Just as there are different types of front-line workers with overlapping but distinct expertise—firefighters, police, ambulance paramedics, and ER doctors—there are distinct DNA repair mechanisms for reattaching the cut ends of broken DNA. Two are most important for gene editing.

First, there is a crude but effective repair pathway that simply takes the two loose ends of DNA sequence, sands them down a bit, and then glues them back together. It solves the problem of broken DNA, but it is a bit slipshod. That sanding of the loose tips at the break point, in molecular terms, involves chewing back some of the nucleotide letters that were adjacent to the site that was originally cut by Cas9. Once the two loose ends get glued, or ligated, back together to produce the repaired continuous string of DNA, there will be a few nucleotide letters missing. In other words, deletion mutations arise as a by-product of this mechanism of repair, a mechanism that goes by the name nonhomologous end joining, or NHEJ for short.

NHEJ is the cell's quick n' dirty means of fixing the potentially catastrophic effect of a physically broken genome. From the vantage of genome editing, it means that a biologist can induce a deletion into any gene, using the laser-targeting ability of CRISPR-Cas9. Deletions to a gene generally disrupt the gene's function or expression, which let a scientist manipulate the characteristics of the organism that result from the gene's activity.

The second key DNA repair pathway is responsible for how CRISPR-Cas9 genome editing earned its exciting reputation for versatility. This repair mechanism is termed homology-directed repair, a.k.a. HDR. The molecules responsible for HDR are clever: they say, hey, the genome actually has two copies of each chromosome, so why don't we use the other, unbroken, copy as a template for fixing this bit of broken

DNA? Remember that animal cells come from a zygote that received one set of chromosomes from Mom and a second set of chromosomes from Dad. These two sets of chromosomes are said to be homologous to one another. The unbroken homologous chromosome, therefore, can act as a molecular splint for the broken section of the other chromosome. HDR takes a bit more time to track down the homologous unbroken chromosome, to line it up with the loose ends around the DNA break, and to copy over the DNA sequences from that homologous chromosome as a natural template. But it makes for near-perfect repair. If NHEJ is a field medic in a war zone, then HDR is a crack surgeon in an operating room of the Mayo Clinic.

HDR is, however, susceptible to human hacking. We can provide an alternative template that superficially looks like an appropriate bit of DNA to repair a break, a mimic of what would ordinarily be that homologous unbroken chromosome. In actual fact, however, the alternate template is engineered to have distinctive features that differ from the true homologous unbroken chromosome. When this custom repair template is engineered and included in the injection mix along with Cas9 and guide RNA, then the cell will often grab it to use as the molecular splint instead of the homologous chromosome. When that happens, the cell will copy over the string of engineered DNA to repair the broken section of genome.

This engineered repair template is the "optional step 2" in the core CRISPR-Cas9 workflow summarized above. You can customize it and synthesize it to encode virtually any gene you might choose. Designing what this repair template looks like and what it can do is where creativity and ingenuity can shine.

★

All custom repair templates are variants on the following basic design: a string of DNA with a central "donor sequence" that is sandwiched between a pair of so-called homology arms. The homology arms are DNA sequences that match perfectly to the DNA on either side of the spot in the genome that was cut by Cas9. Usually, each homology arm is a few hundred nucleotide letters long.[9] The homology arms are the crucial strings of DNA that trick the HDR molecules into using the engineered repair template for DNA repair rather than using the cell's other unbroken homologous chromosome.

Depending on how you create the donor sequence of the custom repair template, the genome edit can let you insert a novel stretch of DNA, delete a specific sequence of DNA around the break, or recode the original sequence with particular nucleotide replacements to the DNA. As an example insertion, the donor sequence could encode a jellyfish green fluorescent protein so that the repaired cell could be made to glow green. This is what John Calarco did to make Jackie, the worm with the glittery mind. This would be a "transgenic" alteration to the genome, inserting DNA sourced from another species. Another example: you could recode the regulatory control of a pre-existing gene such that the gene gets expressed only under certain circumstances, such as when exposed to high temperatures. For medical or conservation of endangered species purposes, you might choose to recode a defective version of a gene that causes disease with a functional version of that gene modeled after DNA in another member of the species. This use of

DNA information from the creature's own species would represent a "cisgenic" rather than a "transgenic" alteration of the genome by CRISPR-Cas9 genome editing.

It just isn't possible to list all the possibilities. There are only two basic barriers to using CRISPR-Cas9 genome editing to create novel genetic variants: your knowledge of what kinds of changes would cause a desired outcome and your imagination.

6.2   What to do

We depend on agricultural crops, having benefited from a long history of human artificial selection and breeding. As we have seen, this process sometimes gives suboptimal characteristics to the foods we eat. Improving crops also may be limited by what genetic diversity exists within that species. Horticulturalists depend on that natural genetic diversity to use as the basis for breeding fruits, vegetables, and grains with greater nutritional content, higher yield, lower resource demands, or better aesthetics and shelf life.

It is also possible to create novel genetic variants using CRISPR-Cas9 genome editing. One such application targets the genetic elements that control when and where and for how long a given gene gets turned on. Such regulatory region engineering does not tinker with the sequence of the protein itself that could be ingested, instead altering the timing or tissue of when and where the plant makes that protein.[10] It could, of course, instead be engineered to disrupt the protein itself so that it can't perform any function at all. The alteration to the genome of the plant would look like just about any new mutation that got created spontaneously in meiosis through natural causes. In other words, the piece of the genome edited by CRISPR-Cas9 in this way wouldn't correspond to a transgene. A genome edited by CRISPR-Cas9 in this way wouldn't have DNA from another species. The presence of a transgene in a genome represents the oft-maligned bogeyman of anti-GMO opinions. Later in this book, we'll discuss GMO politics and ethics in more detail.

This kind of genome editing has been implemented to modulate pathogen resistance in rice and citrus, branching and fruit size in tomato, as well as salt tolerance of rice plants. In livestock like pigs, sheep, and goats, CRISPR-Cas9 disruption of the gene that encodes the myostatin protein leads animals to pack on greater muscle mass for meat production.

Of course, genetically more-invasive applications of CRISPR-Cas9 also can be used for the purposes of crop and livestock improvement. You could, for example, use CRISPR-Cas9 to insert DNA from one species into another. This application *does* leave a transgene in the edited genome. To make leaner pigs that are healthier animals and healthier for human consumption, pig breeders used CRISPR-Cas9 to introduce a copy of the UCP1 gene from the mouse genome into the pig genome.[11] Pigs, unlike most other mammals, ordinarily do not have a working copy of UCP1 encoded in their genomes. This fact makes it hard for them to effectively regulate their body temperature, especially when exposed to cold conditions as piglets. By editing the porcine genome to encode and express the UCP1 gene, inserting a version that was based on the DNA sequence of UCP1 in mice, the resulting pigs proved more cold-tolerant and had nearly 5% less body fat. In principle, commercial use of such pigs would improve animal welfare, reduce energy use and costs in production, and yield leaner meat.

How well does gene editing with CRISPR-Cas9 do what it is supposed to do? The guide RNA used in CRISPR-Cas9 genome editing has incredibly high fidelity to target just a specific sequence of DNA. Nonetheless, living systems are notoriously fiddly, and so any given experiment typically requires weeks, at minimum, of preparation, optimization, and redesigning. Living systems also are imperfect, and so there remains the potential for unintentional editing of nontarget locations in a genome, so-called off-target effects. The worry, which is most acute in a clinical setting, is that off-target mutations could unintentionally disrupt normal gene expression. One especially worrisome detrimental consequence of disrupted gene expression could be activation of cancer-promoting oncogenes within an individual, for example. Off-target effects are now exceedingly rare, thanks to biotechnological advances and guides to best-practice since CRISPR-Cas9 genome editing was first introduced, at least when guide RNAs and homology arms are well-designed and when you use high-fidelity Cas9 proteins in the set of reagents needed to make editing work.[12]

★

It may seem absurd that breaking a piece of DNA was such a stumbling block to genome editing. In truth, breaking DNA isn't hard; the problem is that it is hard to control *where* the break will take place. Most methods break the DNA randomly, like gamma-rays do. But there are other games in town besides CRISPR-Cas9.

Before CRISPR-Cas9 came along, two other molecular strategies gave an in-road to targeted genome editing. They might not be cheap and easy like CRISPR-Cas9, but these clever biotechnology tweaks to natural cell machinery do have splendid names: zinc-finger nucleases and TALENs.[13] Both of these techniques involve creating a protein chimera, fusing one piece that can bind to DNA (a zinc finger peptide domain or a TAL effector) with a second piece that can cut DNA (usually the cleavage domain from a FokI endonuclease protein[14]). In the case of zinc fingers, however, it is tricky to know what DNA sequence motif will be recognized by a given set of molecular zinc fingers. TAL effectors have a convenient matching to individual strings of nucleotides but are tricky to make because they depend on a repetitive sequence that is technically challenging to synthesize in a customized way. As a result, despite the proven capability of these tools to cut specific DNA sequences, their finicky attributes limit their scope in ways that CRISPR-Cas9 easily circumvents.

It turns out that Cas9 is just one of a big family of nearly one hundred types of CRISPR-associated proteins.[15] Cas9 is the best understood and its ability to make a so-called blunt cut of the double-helix of DNA is convenient for many purposes. Cas12a is another good choice, as well, differing from Cas9 in that when it snips DNA, it makes a staggered cut. Staggered cuts to DNA leave behind chemically sticky ends on the double-stranded DNA, in the case of Cas12a, such that there are five nucleotides left single-stranded. Yet another family member, Cas13, also has been co-opted for biotechnological applications, valued for its peculiar ability to cut RNA instead of DNA. Both

Cas12a and Cas13 are smaller proteins than Cas9, which also confers a convenient technical feature.

I don't want to get bogged down in too many of the minutiae of Cas proteins, but I must point out that molecular biologists have now devised a slew of fascinating tweaks to the Cas protein itself. They've genetically engineered the engine of genetic engineering itself. For example, modifying the domain within the protein that actually cuts DNA has let researchers make versions of Cas9 that simply nick one strand of the DNA or that adhere to the DNA without cutting it at all. This insightful modification lets a researcher experimentally and temporarily inhibit the expression of the gene it binds to in the genome, rather than making a perpetually inherited change to the DNA. Other alterations make Cas9 get activated only when you shine light on the cells that it resides in. Some researchers have split the Cas9 protein into two pieces that allow an experimenter to control more finely when and where in a tissue Cas9 will be activated. There are new tricks getting developed all the time, none of them magic—so long as you know how to interpret a barrage of acronyms and scientific jargon.

★

As impressive as the exploits of these molecular machines may be, to do their thing, they first need to get into the right cells at the right time in the right way. For a CRISPR-Cas9 genome edit to be transmissible from one generation to the next, the DNA alteration must be present in those cells that give rise to gametes (i.e., sperm or eggs). In animals like the nematode roundworm *C. elegans*, this procedure simply requires steady hands and a microscope. These tiny animals also require a tiny needle, made by carefully pulling a heated capillary tube to make a hair-thin tip. With this ultra-fine glass needle, you inject the cocktail of CRISPR-Cas9 ingredients into an animal's gonad. In particular, you inject the cocktail at a spot in the worm's tissue that is structured as a syncytium of cells, germ cells undergoing meiosis that happen to not be completely wrapped up individually in their own cell membranes. As a result, each member of this syncytium of cells can freely exchange the cytoplasmic fluid within those cells. The liquid reagents diffuse into every germ cell as it matures into sperm or egg, letting Cas9 cut DNA and then letting HDR repair the gap with the engineered custom repair template. Presto change-o, worm zygotes that develop into offspring of the next generation will have their genomes modified.[16]

In many animals, however, including mammals, the corresponding tissue is hidden away, tucked safely out of view and out of easy reach of a needle. For mice, the protocol is more baroque than for *C. elegans*, but researchers have established reliable standard operating procedures. One CRISPR-Cas9 technique modifies the DNA of embryonic stem cells grown in Petri dishes. These edited stem cells then get injected into embryos at the very earliest stages of development—as zygotes or blastocysts—that were harvested from a donor female. By microinjecting the stem cells into these early stage embryos, then transferring them into the uterus of a recipient mother, the genetic modification can get incorporated into the germline of the developing embryo to propagate the genetic variant to the next generation. This embryonic transfer step is exactly what happens in human fertility clinics for women who undergo in vitro fertilization (IVF) treatment, albeit without the prior genome editing. An alternate technique in mice lets researchers sidestep some of this complexity to perform heritable genome editing. Instead, you can directly inject CRISPR-Cas9 reagents into the oviduct of pregnant females shortly after conception.[17]

Still another option would be to perform a version of somatic cell nuclear transfer (SCNT). This daunting term refers to the protocol most typically associated with animal cloning. SCNT is the cloning technique made famous by Dolly the sheep.

SCNT starts with an unfertilized egg cell, an oocyte, and first involves the removal of its nucleus and all of its DNA. This "empty" enucleated oocyte then gets a new nucleus: its replacement comes from the nucleus taken from some other cell, a non-gamete somatic cell such as a fibroblast, a kind of cell ordinarily found in skin and tendons. After this egg cell with its new nucleus gets a special molecular bath in a Petri dish, it starts dividing as if it had been fertilized. You then transfer this nascent embryo into a recipient female's oviduct, as with IVF, so that it can continue to develop in her womb. For SCNT applied to genome editing, the idea is to first conduct the CRISPR-Cas9 genome editing in a set of cells in a Petri dish, using cells like those fibroblast cells. The DNA and nucleus from these genome-edited cells provide the source of the replacement nucleus that gets microinjected to the enucleated oocyte. Rather than cloning a genetic replica of whatever individual animal provided the fibroblast cells, you will "clone" an individual that has an entirely novel genetic composition due to CRISPR-Cas9 genome editing. This approach seems especially popular in livestock, with examples in pigs, goats, cows, and horses.[18]

Things are more complicated still in those animals like birds that lay eggs with an eggshell. The shell itself is not the problem, it can be gently peeled away. The trouble is that it is difficult to get access to the zygote at the point of fertilization when it is up inside the reproductive tract of a female bird, the point when it is just a single cell or a small bundle of cells. By the time a bird lays an egg, the embryo has up to 60,000 cells.

But dedicated poulterers have devised solutions: a three-step process and a one-step process.[19] In the original three-step approach, one must first isolate and grow in Petri dishes a collection of so-called primordial germ cells, a.k.a. PGCs. These PGCs may be extracted, for example, from testis tissue before their differentiation into sperm cells. In the second step, you use CRISPR-Cas9 genome editing to modify the DNA in the genomes of those PGCs that you've managed to grow in the laboratory, similar to what I just described for mammalian fibroblasts. You can also freeze the PGCs, before or after editing, to keep a cryopreserved frozen library of unique bird genomes in the freezer. Finally, the edited PGCs get carefully injected into embryos of the egg of, say, a chicken. This strategy creates an embryo with a mosaic of cells, some of which were engineered (from the laboratory PGCs) and some of which were not (those present in the egg that was injected into). When those cells with modified genomes get lucky and get incorporated into the collection of embryonic cells that will end up making gametes in the adult bird, then the genome edits will get transmitted to a fraction of the progeny of the next generation.[20]

The one-step technique for birds, however, doesn't depend on transferring PGCs. Instead, you inject the CRISPR-Cas9 ingredients along with a replication-defective virus, an adenovirus, into the blastoderm of the bird embryo when the egg gets laid. This procedure hijacks the capabilities of the adenovirus to infect cells. As the weak viral infection spreads from cell to cell, it delivers the ingredients for guide RNA and Cas9 protein expression into each cell of the developing embryo. The virus eventually gets cleared by the immune system, but not before the genome editing ingredients get expressed briefly within the cells of the embryo. The expression of the transgenic CRISPR machinery then edits the DNA within the cells of the embryo, including the primordial germ cells that will later give rise to gametes in the adult.

★

CRISPR-Cas9 originated in bacterial evolution as a defense mechanism, an immune system to target and degrade viral infections. Humanity has since co-opted the versatility of the CRISPR-Cas9 molecular machinery as a biotechnology tool to manipulate specific locations in genomes in creative ways. Such genome editing allows us to insert, delete, or change virtually any DNA sequence motif at any location in a genome of our choosing. Depending on the organism, genome editing can be combined with cloning or in vitro fertilization techniques.

Notes

1. JAC1293 refers to the unique genotype of a laboratory strain of the nematode roundworm *Caenorhabditis elegans*. John Calarco and his team used CRISPR-Cas9 to insert into its genome a DNA sequence that encodes a protein that can fluoresce when exposed to UV light. They engineered this transgene to be expressed in neurons so that they could study the molecular mechanisms that control gene regulation. There is, in fact, not just one individual worm named Jackie, but a legion of thousands of genetically identical Jackies. They reside in Petri dishes and, cryopreserved, in the deep freezers of the Calarco lab. To learn how Jackie can refer to more than one thing, I encourage you to read about all the Jacquelines in the surreal 2012 children's storybook *Stories 1,2,3,4* by Eugene Ionesco and Etienne Delessert.

2. Perhaps the best-told history of CRISPR-Cas9 discovery, along with lucid explanations of its biological mechanisms, comes from the source: see Jennifer Doudna's insightful 2017 book written with her colleague Samuel Sternberg, *A Crack in Creation*.

3. Cilia are tiny hair-like structures on the outside of the bacterial cell wall that help it move; flagella are tail-like structures that rotate and whip around to help the cell swim, similar to how mammalian sperm cells do. Ribosomes are the "little organs" (i.e., organelles) within a cell that convert mRNA into protein.

4. The guide RNA in natural type II CRISPR-Cas9 systems is made up of a crRNA and a tracrRNA, where "cr" is yet a further abbreviation of CRISPR and "tracr" stands for "trans-activating crRNA." In engineered versions of CRISPR-Cas9 systems, biologists synthesize a single chimeric guide RNA molecule that fuses the crRNA and tracrRNA with a linker

loop. Type I and type III CRISPR-Cas systems that are found in some microbes work slightly differently in the natural bacterial immune responses to phage in that they use a single crRNA rather than the dual-RNA guide of type II systems.

5. Cas9 stands for CRISPR-associated protein 9. Genes encoding Cas proteins are found in the genomes of about 50% of bacteria and in about 87% of archaea. Consequently, CRISPR-Cas microbial immunity is more stereotypical of archaea than of bacteria. Archaea are single-celled microbial organisms superficially similar to bacteria, but that define a third domain of life distinct from both bacteria and eukaryotes (animals, plants, fungi, etcetera). Makarova, K., et al. 2015. An updated evolutionary classification of CRISPR–Cas systems. *Nature Reviews Microbiology*. 13: 722–736. https://doi.org/10.1038/nrmicro3569; Hille, F., et al. 2017. The biology of CRISPR-Cas: backward and forward. *Cell*. 172: 1239–1259. https://doi.org/10.1016/j.cell.2017.11.032.

6. Jinek, M., et al. 2012. A programmable dual-RNA–guided DNA endonuclease in adaptive bacterial immunity. *Science*. 337: 816–821. https://dx.doi.org/10.1126/science.1225829.

7. "dsDNA" refers specifically to "double-stranded" DNA, the kind that makes up our genomes. Quotation from p. 820 in Jinek, M., et al. 2012. A programmable dual-RNA–guided DNA endonuclease in adaptive bacterial immunity. *Science*. 337: 816–821. https://dx.doi.org/10.1126/science.1225829.

8. Komor, A.C., et al. 2017. CRISPR-based technologies for the manipulation of eukaryotic genomes. *Cell*. 168: 20–36. https://doi.org/10.1016/j.cell.2016.10.044.

9. In the years since CRISPR-Cas9 genome editing hit the scene, molecular biologists have devised myriad detailed layouts and protocols for constructing and using custom repair templates. Some of the schemes are tailored simply to ease later verification steps; others optimize laboratory efficiency for inducing genome edits of a particular size or to encourage cells to use HDR instead of NHEJ.

10. Li, Q., et al. 2020. Perspectives of CRISPR/Cas-mediated cis-engineering in horticulture: unlocking the neglected potential for crop improvement. *Horticulture Research*. 7: 36. https://doi.org/10.1038/s41438-020-0258-8.

11. *UCP1* stands for the seemingly enigmatically named gene that encodes "uncoupling protein 1." This protein separates, or uncouples, the production of ATP from the transfer of protons across the inner membrane of mitochondria to help animals produce heat and, in so doing, also influences the production of body fat. A deletion mutation destroyed the function of the UCP1 gene in pig genomes about 20 million years ago during a period in their evolutionary history when they inhabited only warm, tropical parts of the world. Zheng, Q., et al. 2017. Reconstitution of UCP1 using CRISPR/Cas9 in the white adipose tissue of pigs decreases fat deposition and improves thermogenic capacity. *Proceedings of the National Academy of Sciences USA*. 114: E9474. https://doi.org/10.1073/pnas.1707853114; Berg, F., et al. 2006. The uncoupling protein 1 gene (UCP1) is disrupted in the pig lineage: a genetic explanation for poor thermoregulation in piglets. *PLoS Genetics*. 2: e129. https://doi.org/10.1371/journal.pgen.0020129.

12. Naeem, M., et al. 2020. Latest developed strategies to minimize the off-target effects in CRISPR-Cas-mediated genome editing. *Cells.* 9: 1608. https://doi.org/10.3390/cells 9071608.

13. Proteins with "zinc fingers" have a series of amino acids that get stabilized by zinc ions to form a three-dimensional structure that adheres to grooves in the spiral shape of the DNA double helix. The flashy term *TALEN*, spelled out, is *transcription activator-like effector nuclease.*

14. The FokI protein gets its name from the initials of the rod-shaped bacterial species it was isolated from: *Flavobacterium okeanokoites.*

15. Note also that some Cas proteins like "type I" Cas1 and Cas2 are involved specifically in the bacterial immunity step of recording virus sequences into memory, the dictionary or wall of mug shots, writing the record in the so-called protospacers that make up the cluster of regularly interspaced short palindromic repeats of a CRISPR-Cas region in a bacterial or archaeal genome. It is the "type II" Cas9 proteins that group with Cas9 that are key to genome editing.

16. I say "presto change-o" as if it were as simple as waving a wand, but, in reality, the protocols in the lab require a great deal of technical know-how. In practice, one could expect a successful genetic edit about 50% of the time, depending on the details of the genetic manipulation and the experience of the researcher. Dickinson, D.J., and B. Goldstein. 2016. CRISPR-based methods for *Caenorhabditis elegans* genome engineering. *Genetics.* 202: 885–901. https://doi. org/10.1534/genetics.115.182162.

17. Gurumurthy, C.B., et al. 2019. Creation of CRISPR-based germline-genome-engineered mice without ex vivo handling of zygotes by i-GONAD. *Nature Protocols.* 14: 2452–2482. https://doi.org/10.1038/s41596-019-0187-x.

18. Ruan, J., et al. 2015. Highly efficient CRISPR/Cas9-mediated transgene knockin at the H11 locus in pigs. *Scientific Reports.* 5: 14253. https://doi.org/10.1038/srep14253; Moro, L.N., et al. 2020. Generation of myostatin edited horse embryos using CRISPR/Cas9 technology and somatic cell nuclear transfer. *Scientific Reports.* 10: 15587. https://doi.org/10.1038/ s41598-020-72040-4.

19. Lee, J., et al. 2019. Direct delivery of adenoviral CRISPR/Cas9 vector into the blastoderm for generation of targeted gene knockout in quail. *Proceedings of the National Academy of* Sciences USA. 116: 13288–13292. https://dx.doi.org/10.1073/pnas.1903230116; Ballantyne, M., et al. 2021. Direct allele introgression into pure chicken breeds using Sire Dam Surrogate (SDS) mating. *Nature Communications.* 12: 659. https://doi.org/10.1038/s41467-020-20812-x.

20. If you've seen news reports of how a duck sperm can lead to an egg that develops into a chicken, it uses the same principle—injecting chicken PGCs into a duck egg to make a chicken-duck chimera with some of its sperm cells having chicken DNA, despite all other cells having duck DNA. Liu, C., et al. 2012. Production of chicken progeny (*Gallus gallus domesticus*) from interspecies germline chimeric duck (*Anas domesticus*) by primordial germ cell transfer. *Biology of Reproduction.* 86: 1–8. https://doi.org/10.1095/biolreprod.111.094409.

7

Genetic welding

CRISPR-Cas9 genome editing truly is amazing in how it has unleashed the full passions of human creativity in biological research and in clinical approaches to treating disease. But, on its own, it is simply an efficient manipulative tool. On its own, it does not enter the realm of fundamental forces of evolution. There is, however, a twist in how the set of tools in the CRISPR-Cas9 toolkit can be arranged that allows it to take on an evolutionary life of its own. The twist to evolution that I am talking about is the CRISPR-Cas9 gene drive.[1]

A gene drive is a genetic element that disproportionately transmits itself to the next generation by subverting the usual rules of inheritance. The subversion has to do with which set of DNA gametes get, because each sperm cell or egg cell only gets one copy of each chromosome. Which copy goes in a given gamete? This is the usual rule: when someone, say, me, has two alternate versions of a gene in their genome, one genetic variant from my Mom and a slightly different one from my Dad, then each gamete I make will have a 50-50 chance of including inside its nucleus the Mom-like variant. Equally, there is a 50-50 chance that it was the Dad-like variant that ended up in a given gamete. That flip-of-the-coin chance is known as Mendel's law of segregation, and it is genetics' way of being fair. Gene drives break that law. They bias the coin toss in their own favor.

To see how gene drives accomplish this sleight of hand, let's take a closer look at what you can do with one of those "optional custom templates" in our core protocol of CRISPR-Cas9 genome editing. The design of that custom template contains the details of what will get edited, that is, the DNA sequence that will get inserted. Now, imagine what will happen if that template encodes a DNA sequence that can express the Cas9 protein itself. Next, imagine that the template also encodes a guide RNA. When the original cocktail of CRISPR-Cas9 ingredients gets injected into a cell, the template will insert the Cas9 and guide RNA information into the DNA of that single targeted spot in the cell's genome. The cell's genome will have been edited to encode everything it needs to edit genomes all by itself. The Cas9 and guide RNA aren't just floating in the cell's cytoplasm, they're now part of the genome itself (Figure 7.1). A genome copy with such a gene drive encoded will automatically edit other genome copies that find themselves in the same cell, as happens with fertilization of egg with sperm. The genomic copy of the gene drive element will insert itself into other genomes if they don't already have one. No injection necessary.

Evolving Tomorrow. Asher D. Cutter, Oxford University Press. © Asher D. Cutter (2023). DOI: 10.1093/oso/9780198874522.003.0007

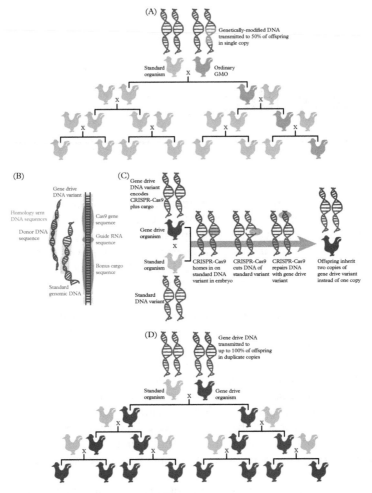

Figure 7.1 *Genetic transmission of ordinary genetic modifications differ from gene drive genetic modifications. (A) An ordinary genetic modification to a genome will be transmitted according to the standard rules of Mendelian inheritance. An ordinary genetic modification will originally be present on just one of the two genomic copies that an individual organism has. When that genetically modified organism (GMO) mates with a non-GMO standard organism, the engineered genetic variant will get transmitted to half of the offspring, on average. (B) Gene drive variants encode DNA sequence to make Cas9 protein, guide RNA, and an optional "bonus cargo" sequence as part of the "donor" portion of the engineered genetic sequence (see also Figure 6.2). (C) Gene drive genetic modifications that employ CRISPR-Cas9 genome editing act to automatically make themselves present in both of the genomic copies of an organism, rather than just one of the copies. (D) As a result, the gene drive gets transmitted to more than half of offspring (up to 100% of offspring), each of which automatically gets its genome edited so that both genome copies encode the gene drive genetic variant, even when the gene drive organism (GDO) reproduces with a non-GDO standard organism partner. Refer to Figure 6.2 for illustration of the initial genome editing that takes place; once encoded in the genome, a gene drive no longer requires the "injection" step to initiate the genome editing.*

This property of a gene drive is so important that I want to say it again in a different way. By hijacking the cell's own tools for how it repairs breaks to DNA, a gene drive converts the genome from having the gene drive element in just one copy to having the element present on both copies of its genome. This ability is sometimes called *homing*.[2] In genetic terms, gene drives convert heterozygotes into homozygotes, but only in the spot on the chromosome where the gene drive DNA is encoded.

CRISPR-Cas9 gene drives are composed of genetic elements that have the following basic structure: DNA encoding the Cas9 protein and DNA encoding a guide RNA that targets the sequence of DNA directly adjacent to where the gene drive element has been inserted (Figure 7.1). We'll soon explore how additional cargo as a third component can enhance the gene drive's capabilities, designed as part of that optional custom template of the genome editing protocol. The nascent gene drive can edit any naive chromosome that it encounters, and edit it in a way that simply inserts a copy of itself. In fact, the gene drive will encounter a naive chromosome in the very next generation, when the transgenic organism mates with another individual that has a standard genome. This capability of gene drives to reproduce by themselves gives them the genetic analog to self-awareness. It allows a gene drive to be genetically autonomous, free of the constraints of Mendel's law of segregation.

7.1   A new force of evolution

Gene drives encode a way of giving themselves greater than a 50-50 chance of ending up in any given sperm cell or egg cell. In the case of CRISPR-Cas9 gene drives, they can transmit themselves to essentially 100% of the gametes instead of just 50%. That kind of intense transmission advantage from one generation to the next is faster even than natural selection can usually work to increase the abundance of one genetic variant over another.

At this point, I need to zoom out for a moment. Rather than focusing on the nitty-gritty molecular mechanisms within the cell, let's look at what happens to a population or species. Let's think about what would happen should you set loose a gene drive on a naive population of organisms, a population with genomes composed entirely of standard chromosomes. In short, the gene drive genetic element can invade, spreading in what is sometimes termed a mutagenic chain reaction to embed itself into the genomes of every single individual in the species.[3]

To show how fast such genetic invasion can happen, let's walk through a numerical example. There are two types of genetic variants at the spot along a chromosome that we are interested in: the variant with the gene drive element and the standard variant that lacks the gene drive element. Imagine that a new individual with the gene drive genetic variant arrives in a population of 1000 standard individuals, arriving due to migration or experimenter introduction. The gene drive variant thus is present in 0.1% of the genomes, and the standard variant is present in the remaining 99.9%. Presuming the population has a stable size, then each individual will co-parent two offspring on average by transmitting one set of chromosomes through their sperm or eggs to each offspring. The embryos produced from mating with that newcomer, however, will inherit one drive variant and one standard variant. These are the offspring where something special happens: the gene drive will automatically convert the standard variant that cohabits with it in the embryo into another gene drive variant. Now, instead of just one of the two copies of the chromosome in that embryo having the drive variant, both do (Figure 7.1). Like Star Trek's Borg, the standard variant gets assimilated.

This conversion of "standard" into "gene drive" versions of that segment of DNA happens whenever the two distinct types get put together, which happens inside of the single individual cell of the most primitive stage of an embryo, the zygote. Consequently, in that second generation after the arrival of the newcomer, the incidence of the gene drive variant will have doubled to 0.2%. It will keep on doubling in subsequent generations: 0.4%, 0.8%, 1.6%, 3.2%, 6.4%, 12.8%, 25.6%. After just eight generations, that single individual can lead to more than a quarter of the species having the gene drive in their genomes. If more than a single individual had been introduced at the beginning, then this genetic invasion would happen even quicker. With an initial abundance of 10% instead of 0.1%, the gene drive could be found in just about *everyone* after only four or five generations. That kind of evolutionary change is lightning fast.

★

The concept of genetic elements breaking Mendel's law is not entirely new. And CRISPR-Cas9 isn't actually required to break Mendel's law. So-called selfish genetic elements arise in nature on a regular basis: about 45% of our own genomes encode a cornucopia of them in the form of transposable elements, also known as jumping genes.[4] Jumping genes, however, copy themselves into semirandom locations across the genome, not in a single targeted spot.

There are other types of selfish genetic elements, too, such as toxin-antidote systems. That roundworm species *C. elegans* that is the darling of developmental biologists, and my own research team, encodes one such genetic narcissist: the PEEL-ZEEL toxin-antidote element.[5] This genetic package has two genes linked close together that work hand-in-hand to ensure their mutual propagation at the expense of genome copies without the element. First, the selfish genetic package leads the animals to deposit the toxic protein PEEL-1 into sperm cells. It will poison the developing embryo that forms upon fertilization by the sperm. The genetic package, however, also plays the role of savior. In those embryos that inherit a copy of the element in their genomes, the embryo can express the ZEEL-1 protein which acts as the antidote. Like a genetic incarnation of Munchausen syndrome by proxy, ZEEL-1 comes to the rescue of the poisoned baby worms.

In nature, however, some genomes of *C. elegans* are missing the PEEL-ZEEL genetic package. As a consequence, when a male with the element encoded in one copy of its genome goes on to mate with a partner that lacks it entirely, then half of their hundreds of babies will die of sperm-derived PEEL-1 poison because they lack the ZEEL-1 protein. All of the survivors will be those that inherited a copy of the PEEL-ZEEL genetic element, thus perpetuating the selfish propagation of the genetic factor. As it turns out, such two-faced toxin-antidote and other selfish DNA systems appear to occur rampantly in animal genomes.[6] Synthetic gene drives, however, and synthetic CRISPR-Cas9 gene drives engineered to carry specific genetic cargo, in particular, take it to a whole other level.

The other level is defined by the formidable potential to engineer different possible genetic cargoes within a gene drive. Adding a specific genetic cargo to a gene drive gets its start at the "optional custom template" phase of CRISPR-Cas9 genome editing. The idea is to include a bonus element along with the basic components of the gene drive structure. With this bonus cargo included, the gene drive genetic element will then include DNA that encodes expression of the Cas9 protein, the guide RNA, and the bonus cargo. The cargo could be a beneficial new gene function or an innocuous DNA sequence that allows, say, green fluorescent protein to be expressed in neurons.

Novel genetic variants introduced by such "modification gene drives" into a species seem, superficially, entirely analogous to novel genetic variants introduced by mutation. The impact on the genome, however, differs profoundly. To illustrate why this is so, let's first walk through the evolutionary trajectory of a novel genetic variant that arises by mutation and that imbues those organisms lucky enough to encode it in their genomes with some beneficial effect.

A beneficial effect of a new mutation means that there is positive selection, either natural or artificial. Such positive selection will favor an increase in the abundance of the variant from one generation to the next. If there were 1000 individuals in the species, each with two copies of a given gene inherited from its two parents, then we can calculate the initial frequency of the brand-new variant to be 0.05%.[7] Eventually, we'd expect selection to increase that frequency to the maximum abundance possible, 100%, such that the once-novel variant becomes the only variant around. It now represents a fixed feature of the genome and the population.

But remember that the mutation happens to a gene embedded within the long string of DNA letters that make up a chromosome, a chromosome with thousands of other genes. What of the genes that happen to be nearby the one gene targeted by selection? Those other genes also will have genetic variants because of their history of chance mutational events. Perhaps those other genes' genetic variants are selectively neutral or perhaps they have effects that are less pronounced than those of the beneficial mutation that is our focus. Whichever variants of those linked genes that had the luck to be encoded on the same string of DNA will hitchhike along, increasing in abundance along with the selected variant. If they are close neighbors, then they too may get fixed at 100% frequency in the population.

In this way, natural selection and artificial selection cause some gene variants to surf the wave of increasing abundance of their neighbor, spreading throughout the species. They surf to high abundance by riding the coattails of that selectively favored neighbor, not because of their own intrinsic beneficial properties. This is the genetic hitchhiking that we discussed in Chapter 5.

This genetic hitchhiking that occurs during such a so-called selective sweep, however, comes with a catch. Another way of looking at what it means to say that a genetic variant achieves 100% frequency is to say that it has destroyed genetic diversity. Selective sweeps thus eliminate genetic variability not only at the gene targeted directly by selection. Selective sweeps eliminate genetic variability at *all* genes that happen to be linked nearby on the chromosome.[8] How far the span of hitchhiking genes extends along a chromosome will depend on how much recombination occurs before the beneficial mutation sweeps through the species. The amount of recombination that can happen depends on the number of generations that elapse. Stronger selection makes faster work of the selective sweep, meaning fewer generations elapse. Fewer generations means fewer opportunities for recombination to genetically separate the gene that is the direct target of selection from the rest of the DNA encoded on the chromosome. Strong artificial selection in dogs, for example, has caused selective sweeps to deplete genetic variability for spans up to millions of DNA nucleotides along chromosomes within a given dog breed.[9]

Now, gene drives. Gene drives can spread through a species even faster than mutations that get subjected to traditional natural selection or artificial selection. But gene drives don't create genetic hitchhiking. Consequently, gene drives don't decimate the genetic diversity that already exists elsewhere in the genomes of the species. The reason is that the CRISPR-Cas9 machinery inserts the gene drive cassette into every chromosome copy that it encounters, not just the chromosome copy on which it was originally engineered into as a transgene. The homology directed repair (HDR) mechanism co-opted by the CRISPR-Cas9 gene drive simply copies DNA from one place to another without waiting around for crossovers between distinct chromosomes to rearrange their genetic configurations. As a result, when a genetic novelty gets introduced into a species by a gene drive, it does not come at the expense of lost genetic diversity elsewhere in the genome.

★

For gene drives to operate as a force of creative evolution, we must use so-called modification gene drives.[10] These modification gene drives hold the ability to create and propagate novel genetic variants and novel combinations of genes that confer extraordinary features and, perhaps, to create novel species. A CRISPR-Cas9 gene drive's bonus cargo need not carry green fluorescent protein. It could encode DNA that alters animal development in any way we choose to design. Gene drives embody a means of reshaping the genetics of a population of creatures in an unsupervised way. Following the instantiating moment when a self-propagating designer genetic element gets dropped into a population, it will catalyze a cascade of genetic change.

Here's one practical example of a modification gene drive. Remember those pyrethrin-resistant lice? It turns out that many insects have evolved similar resistance to insecticides because of so-called kdr mutations of voltage-gated sodium channel (VGSC) genes. It might be convenient to restore sensitivity to pesticides into pest species, so that judicious use of pesticides would once again work as a treatment option. One way to compel a species into being genetically sensitive to insecticide would be to use a gene drive to modify a VGSC gene from a resistant genetic version back into a version that is susceptible. As proof of principle, that is exactly what Bhagyashree Kaduskar, Ethan Bier, and their colleagues did with experimental populations of the laboratory fruit fly *Drosophila melanogaster*.[11] As expected for gene drive evolutionary dynamics, the genetically engineered version of the VGSC gene that conferred sensitivity to insecticidal chemicals invaded the cage populations of flies within just a few generations. Similar logic has been proposed to drive improved immunity to pathogens into populations of endangered species as a kind of genetic vaccination.[12]

I've mostly been talking about modification gene drives that have the capacity for unrestricted spread within a species, so-called self-propagating global gene drives. Through clever genetic engineering, however, you can design the gene drive element to be geographically restricted or restricted in its persistence through time. Such "self-limiting" gene drives are designed to eventually get purged from the species, and so the evolutionary change would be transient.[13] They are a genetic form of planned obsolescence. Even for self-propagating global modification gene drives that invade a species to become present in the genomes of 100% of individuals, however, persistence is not permanent. At that point of fixation within the species, the drive pressure is released. New mutations are free to arise that may disable the cargo of the gene drive. The combined influence of mutation and genetic drift would ensure that, eventually, any cargo that does not confer a benefit would get eliminated, eventually. Natural selection may actually favor those disabling mutations without the gene drive's pressure to retain them, leading to their more rapid elimination. This process of elimination, however, would be much slower than the original speed of invasion, and the gene drive's modification would represent a feature of the species for hundreds of generations at the minimum.

Deployed in the laboratory, CRISPR-Cas9 gene drives are the manifestation of conceptual curiosity. Deployed in nature, gene drives take on new power. Gene drives in the wild hold the power to alter the fundamental core of life on Earth, to alter the DNA of entire species in ways that can modify their shape, their senses, their behavior, and their physiology. Gene drives in the wild will constitute a new force of evolution, an invention of the Anthropocene, and I call this new evolutionary force *genetic welding*.

★

Declaration of new forces of evolution doesn't happen every day. Let's put this rookie force of genetic welding into perspective. After all, the venerable natural selection dates to 1859 and discovery of recombination hails from 1865, albeit only entering biological canon in 1900.[14] The new kid on the block is genetic drift, now hitting its one hundredth birthday since getting delivered in 1921.[15]

It is true that a couple of other forces of evolution have joined the team over the years with roles that I have not yet emphasized. Most notably is the force of GC-biased gene conversion and, perhaps, genetic "draft."[16] These two newcomers describe ways that recombination, or lack thereof, can create patterns in DNA change that mimic some aspects of natural selection and genetic drift. GC-biased gene conversion results from HDR of DNA. It describes how DNA repair machinery tends to use genetic variants that have a G or C nucleotide to make the repair in preference to those variants with a T or A nucleotide. This bias usually gets introduced during the normal recombination process within cells, as recombination intentionally creates breaks in DNA in a carefully controlled manner during cellular meiosis. This subtle use of HDR and templates may sound familiar, as they are the stuff of genome editing as well. Arguably, GC-biased gene conversion describes a delicate kind of natural genetic welding. Genetic welding with CRISPR-Cas9 gene drives is not so subtle.

Just as gene drive genetic welding shares affinities to recombination and GC-biased gene conversion, it also harkens to mutation and gene flow via horizontal gene transfer, that bacteria-inspired means of moving genetic material from one evolutionary lineage to another. Like mutation, genetic welding can introduce genetic novelty never seen before in nature, albeit not as a simple side-effect of the imperfect error correction of cellular repair mechanisms. Like gene flow, genetic welding can introduce genetic factors that owe their origins to another branch of the tree of life, giving genetic history an element of crisscross. But genetic welding does these things and acts to drive up the genetic element's abundance in a population. Genetic welding lets us give a new gene the capacity to invade the molecular heart of a species whether it's good for the species or not.

Among the ancient forces of evolution, natural selection is special—together with sexual selection and artificial selection—in that it is the evolutionary driver of novel complex features of organisms. Genetic drift and mutation can't accomplish this feat on their own,[17] and neither can gene flow and GC-biased gene conversion. The evolution of complex traits has been the domain of selection alone since life originated on Earth nearly four billion years ago. Genetic welding, however, shares this capacity. The new evolutionary force of genetic welding sits alongside selection as distinctive in being able to push complex features throughout a species faster than could occur from the chance fluctuations in genetic composition by genetic drift and mutation.

Natural selection is amazing in its demonstrated power to produce the world's great multitude of species with novelty of form. And yet, its ability to shape the trajectory of a population's genetic composition through time, from one generation to the next, is subject to a wide array of speed bumps. The ability of natural selection and artificial selection to shape species is constrained by which mutations happen to arise, by populations composed of few individuals, by the linkage of genes to one another along chromosomes, by genetic correlations and pleiotropy that conflate multiple features together. Genetic welding can let evolution jump over all of those hurdles.

7.2   Genetic time bombs

When a gene drive spreads some genetic cargo throughout a species, we could engineer it to encode a change to how the organisms look or behave. It could influence a growth factor to increase or decrease size, or it could influence a gene in the melanin pathway to alter their coloration. Or the cargo could encode a genetic time-bomb.

Biologist Austin Burt first outlined the insight for how gene drives could be commandeered to deploy time bombs.[18] He envisioned this as a genetic form of biocontrol over humanity's greatest biotic enemies, such as malaria-spreading mosquitoes. Inspired by Burt's logic, a flurry of biologists have begun to explore how to engineer CRISPR-Cas9 gene drives to suppress mosquito and other pest populations or even to drive them extinct.[19] The idea is that a gene drive could force a cargo to invade the population even if it is costly, where costly means that it reduces the fitness of individuals who have the gene drive element in their genomes. This counterintuitive capability can arise because of that transmission advantage: by breaking Mendel's law, the gene drive can offset the cost. In some cases, the cost might only appear after the gene drive variant has reached some threshold abundance in the population.

Take, for example, the idea of a toxin-antidote genetic element packaged within a CRISPR-Cas9 gene drive. Instead of a protein poison like *C. elegans'* PEEL-1, the "toxin" effects would be caused by a guide RNA leading Cas9 to disrupt an essential gene within the animal's genome, which would induce mortality in the absence of an antidote. The "antidote" would be provided by a bonus cargo that recodes the stretch of DNA targeted by the guide RNA into a version that won't get disrupted by Cas9, perhaps along with some additional bonus cargo. Such a gene drive would, generation after generation, convert standard versions of the gene target into recoded versions. This is the logic of a so-called cleave-and-rescue (ClvR, pronounced "cleaver") gene drive and an alternate version called the toxin-antidote recessive embryo (TARE) gene drive.[20]

One convenient property of such a strategy for genetically modifying a population is that such ClvR and TARE gene drives are robust to mutations that can cause genetic resistance to gene drive spread. Resistance mutations in a conventional CRISPR-Cas9 gene drive will arise inevitably as a byproduct of the Cas9 cut to DNA, a fact that limits the potential of the gene drive to invade a species. Occasionally, those DNA breaks will get repaired by the mutation-prone quick n' dirty nonhomologous end joining (NHEJ) mechanism rather than the high-fidelity HDR mechanism that makes use of an engineered custom template within the gene drive's genetic cassette. Ordinarily, those NHEJ repairs to DNA could create genetic variants that are resistant to subsequent insertion of the gene drive element. They would no longer be "standard" versions of a chromosome, and so could deflect the homing ability of the guide RNA and Cas9 of the gene drive. This outcome becomes especially prevalent when the bonus cargo exacts a fitness cost on those individuals with genomes that encode the gene drive element. ClvR or TARE gene drives outwit this potential outcome: any NHEJ repairs would not survive because they would disrupt an essential gene. By having disrupted that biologically essential target gene, such a NHEJ repair to DNA is doomed. As a result, NHEJ won't create genetic resistance to the gene drive's charms.

A second convenient property of a TARE gene drive is that it could be restricted in space. You can engineer it to be contained within a given region so that it won't spread throughout the entire range of a wild species. This regional targeting can hold out even if there is some gene flow through migration between populations. The reason for such restricted geography is that a TARE gene drive needs to be present at a relatively high abundance to begin with for the genetic spreading to be effective; rare migrants wouldn't do the trick. A TARE gene drive needs to have a critical mass in order to spread. Being able to confine the genetic spread of a gene drive is a convenient property, especially if one hopes to avoid spreading the genetic modifications to an entire species or to other species through occasional hybridization.

A good reason to prevent such genetic wildfire—to prevent gene drive spread beyond the intended bounds, to other populations or other species—is when the gene drive is engineered for the purposes of population suppression. Population suppression sounds innocuous enough, but in its purest form it is truly sinister: to smite a species to extinction with a genetic time-bomb.

★

Mosquitoes are the poster child for human interest in population suppression. Despite the annoyance of itchy bites on water-adjacent vacations, and as much as humans hate them because of it, mosquitoes are actually not the problem. Mosquitoes are just the messenger or, more formally, the vector. Mosquitoes transmit nasty diseases, and those diseases are the problem for hundreds of millions of people around the world,[21] diseases like Zika and Dengue fever (caused by viruses) as well as malaria (caused by *Plasmodium*, single-celled eukaryotic creatures with a baroque life cycle that involves growing inside multiple distinct species of animal host). Because the disease-causing agents, viruses and *Plasmodium*, live inside the messenger, a shoot-the-messenger strategy has a long history. Biocontrol from insecticide chemicals sprayed in wafting clouds out of fogging trucks provides perhaps the most familiar approach. Gene drives engineered for population suppression offer an alternative.[22]

One way to engineer a CRISPR-Cas9 gene drive for the goal of suppressing mosquito populations is by messing with sex. In mosquitoes, there is a gene called *dsx* that gets transcribed and translated to form one protein in males and a different, longer version of the protein in females. In 2018, biologists at Imperial College London devised a gene drive that targets *dsx* and disrupts only the part of the gene that gets expressed in females.[23] Lest this strike you as unabashedly sexist, consider the pivotal role that female mosquitoes play in population growth and disease transmission. Thinking about female mosquitoes is essential, not because they are the only ones that bite you, but because of why they bite you. They bite you to feed on protein-rich blood to provide to their eggs, and it is their ability to make lots of eggs following a blood meal that expands the population of vectors from a pool to an ocean of disease-transmitting blood-suckers.

When *dsx* gets disrupted by the gene drive, male mosquitoes develop normally but female mosquitoes develop anomalies in the formation of their reproductive organs that make them infertile. The developmental anomalies are such that the animals are termed intersex rather than female, per se. The intersex mosquitoes also are unable to take a blood meal. Despite this profound influence on the reproductive potential of animals that ordinarily would develop as females, the gene drive cassette can spread through males as they go on to mate with the remaining female mosquitoes, all of whom lack the gene drive element. Eventually, the gene drive DNA will become so prevalent in having genetically invaded the population that all the bugs end up being either male or sterile intersex mosquitoes, with no females at all. At this point, as Porky Pig would say, "That's all folks."

A biological system composed entirely of males is a biological system doomed. The mosquito population would be rendered fully suppressed and incapable of further reproduction, which is to say, extinct. In laboratory populations used to illustrate the potential for such a CRISPR-Cas9 gene drive, researchers observed total population collapse within experimental mosquito cages in less than 12 generations. With a life cycle of just one to two weeks, in principle, one could witness the annihilation of a mosquito population in the span of a single summer.

Is species extinction by gene drive more insidious than extinction by hunting or habitat destruction? This and other ethical dilemmas will get our full scrutiny later in this book. Right now, our concern lies more squarely with the ability of gene drives to create rather than destroy. Gene drives as a creative force requires the persistence of populations and so genetic welding involves modification gene drives, not suppression gene drives.

★

Genetic welding as a distinct force of evolution is a product of the Anthropocene, this here-and-now epoch that we, humanity, have defined by our profound intervention in shaping the natural world. As Elizabeth Kolbert wrote recently, "In a world of synthetic gene drives, the border between the human and the natural, between the laboratory and the wild, already deeply blurred, all but dissolves."[24] Whether genetic welding truly manifests as a fully fledged force of evolution will depend on whether we unleash gene drives to remodel the genetic fabric of wild creatures. This is what we must confess should we pursue the creation of self-sustaining populations of new and extraordinary species, genetic exotics forged with imagination and inventiveness. This is what we must grapple with ethically. We must acknowledge that the legend on the map now reads, everywhere, "here be dragons," when we release into the wild the fruits of genome editing to manifest in the world as genetic welding.

In an upcoming chapter, we'll delve into some details of what new features we might want to choose to evolve in a species through genetic welding, and why. The options range as wide as your imagination, from the outward appearance or behavior of an organism, to how it senses and withstands its environment, to how it conducts its internal physiology and metabolism. Genetic welding could confer fluorescent coloration or extraordinary visual acuity; the capacity to digest noxious compounds or biochemically synthesize vitamins or tolerate environmental stress. Later chapters also will deliberate the ethical considerations of applying this force to wild animals, and to us.

★

One exceptional form of genome editing is the design of CRISPR-Cas9 gene drives. Gene drives are capable of replicating themselves to genetically invade a species. Gene drives can be engineered to contain cargo that can confer novel characteristics to a species. When introduced into wild populations, such evolution with modification gene drives constitutes a new force of evolution, genetic welding. Gene drives also can be designed as a destructive force, as genetic time bombs. Such suppression gene drives can reduce the abundance of a population or push it to extinction. Genetic interventions can clearly alter the measurable characteristics of a species, but can they help define a new species?

Notes

1. Esvelt, K.M., et al. 2014. Emerging technology: concerning RNA-guided gene drives for the alteration of wild populations. *eLife*. 3: e03401. https://dx.doi.org/10.7554/eLife.03401.

2. Homing endonuclease genes are encoded naturally in the genomes of some microbes, operating as selfish genetic elements. A gene drive represents a customizable variation on the homing endonuclease theme that can be engineered to target a specific genomic location. Stoddard, B.L. 2011. Homing endonucleases: from microbial genetic invaders to reagents for targeted DNA modification. *Structure*. 19: 7–15. https://doi.org/10.1016/j.str.2010.12.003.

3. Gantz, V. M., and E. Bier. 2015. The mutagenic chain reaction: a method for converting heterozygous to homozygous mutations. *Science*. 348: 442–444. https://dx.doi.org/10.1126/science.aaa5945.

4. Lander, E.S., et al. 2001. Initial sequencing and analysis of the human genome. *Nature*. 409: 860–921. https://dx.doi.org/10.1038/35057062.

5. The name of the PEEL-1 protein stands for paternal-effect epistatic embryonic lethal, which aptly describes what it does: induce death of embryos upon the protein being transmitted via sperm cells. For embryos to survive, they must have the DNA to be able to make ZEEL-1 protein, which stands for zygotic epistatic embryonic lethal. This kind of selfish genetic element is a variation on the theme first shown in animals for *Tribolium* flour beetles with Medea (maternal-effect dominant embryonic arrest) factors. Beeman, R.W., et al. 1992. Maternal-effect selfish genes in flour beetles. *Science*. 256: 89–92. http://www.jstor.org/stable/2876732; Seidel, H.S., et al. 2011. A novel sperm-delivered toxin causes late-stage embryo lethality and transmission ratio distortion in *C. elegans*. *PLoS Biology*. 9: e1001115. https://dx.doi.org/10.1371/journal.pbio.1001115.

6. Hurst, G., and J. Werren. 2001. The role of selfish genetic elements in eukaryotic evolution. *Nature Reviews Genetics*. 2: 597–606. https://doi.org/10.1038/35084545; Ben-David, E., et al. 2021. Ubiquitous selfish toxin-antidote elements in *Caenorhabditis* species. *Current Biology*. 31: 990–1001.e1005. https://doi.org/10.1016/j.cub.2020.12.013.

7. $0.05\% = 100 \times (1 / (2 \times 1000))$ for a new "heterozygous" mutation, in contrast to 0.1% for a new "homozygous" gene drive variant in a single individual.

8. The concept of genetic hitchhiking of gene variants in connection with the force of selection was laid out in 1974 by mathematical evolutionary biologists John Maynard Smith and John Haigh. Maynard Smith, J., and J. Haigh. 1974. Hitch-hiking effect of a favorable gene. *Genetical Research*. 23: 23–35. https://doi.org/10.1017/S0016672300014634.

9. Population bottlenecks and inbreeding also contribute to the long, linked stretches of DNA in dogs, in addition to the influence of artificial selection. Sutter, N.B., et al. 2004. Extensive and breed-specific linkage disequilibrium in *Canis familiaris*. *Genome Research*. 14: 2388–2396. https://dx.doi.org/10.1101/gr.3147604.

10. As will become clear later in this chapter, here we are not talking about "suppression gene drives" like those intended for mosquito biocontrol.

11. Kaduskar, B., et al. 2022. Reversing insecticide resistance with allelic-drive in *Drosophila melanogaster*. *Nature Communications*. 13: 291. https://doi.org/10.1038/s41467-021-27654-1.

12. Rode, N.O., et al. 2019. Population management using gene drive: molecular design, models of spread dynamics and assessment of ecological risks. *Conservation Genetics*. 20: 671–690. https://doi.org/10.1007/s10592-019-01165-5; Kosch, T.A., et al. 2022. Genetic approaches for increasing fitness in endangered species. *Trends in Ecology & Evolution*. 37: 332–345. https://doi.org/10.1016/j.tree.2021.12.003.

13. Long, K. C., et al. 2020. Core commitments for field trials of gene drive organisms. *Science*. 370: 1417–1419. https://doi.org/10.1126/science.abd1908.

14. Charles Darwin introduced the world to selection, both natural and artificial, in his magnum opus *On the Origin of Species by Means of Natural Selection* in 1859. Alfred Wallace gets some credit, too, having landed on the idea of natural selection while in the haze of a malarial fever during a biological collection foray in southeast Asia around the same time. Summaries of the logic from both men were presented simultaneously at an otherwise famously drab scientific conference prior to publication of Darwin's book. Gregor Mendel brought us the idea of recombination, but nobody noticed until 1900 when Hugo Devries, Carl Correns, and Erich von Tschermak-Seysenegg happened upon his old research when independently arriving at the same notion when interpreting their own data. von Tschermak-Seysenegg, E. 1951. The rediscovery of Gregor Mendel's work. *Journal of Heredity*. 42: 163–171. https://doi.org/10.1093/oxfordjournals.jhered.a106195.

15. Referred to originally as the Hagedoorn effect, it earned its own name "drift" a decade later thanks to the mathematical theory championed by biologist Sewall Wright. Plutynski, A. 2007. Drift: A historical and conceptual overview. *Biological Theory*. 2: 156–167. https://doi.org/10.1162/biot.2007.2.2.156.

16. GC-biased gene conversion occurs as an incidental byproduct of DNA repair mechanisms that tend to "fix" single-nucleotide DNA mismatches in favor of the genetic variant that happens to be a guanine (G) or cytosine (C). Lamb, B. 1984. The properties of meiotic gene conversion important in its effects on evolution. *Heredity*. 53: 113–138. https://doi.org/10.1038/hdy.1984.68; Pessia, E., et al. 2012. Evidence for widespread GC-biased gene conversion in eukaryotes. *Genome Biology and Evolution*. 4: 675–682. https://doi.org/10.1093/gbe/evs052. Genetic draft was introduced by John Gillespie in 2000 to describe how mutation, recombination, and selection on a gene can interact to induce random fluctuations in the relative abundance of genetic variants that happen to be nearby to the gene.

17. For some kinds of cellular traits or genetic features, however, one can argue that "developmental system drift" or "constructive neutral evolution" can lead to complexity in a runaway molecular bureaucracy as a byproduct of genetic drift. Gray, M.W., et al. 2010. Irremediable complexity? *Science*. 330: 920–921. https://doi.org/10.1126/science.1198594; Haag, E.S., and J.R. True. 2021. Developmental system drift. In L. Nuño de la Rosa and G. B. Müller (eds). *Evolutionary Developmental Biology: A Reference Guide*. Cham: Springer International Publishing: 99–110. https://doi.org/10.1007/978-3-319-32979-6_83.

18. Burt, A. 2003. Site-specific selfish genes as tools for the control and genetic engineering of natural populations. *Proceedings of the Royal Society of London* B. 270: 921–928. http://dx.doi.org/10.1098/rspb.2002.2319.

19. Dhole, S., et al. 2020. Gene drive dynamics in natural populations: the importance of density dependence, space, and sex. *Annual Review of Ecology, Evolution, and Systematics*. 2020. 51:505–31. https://doi.org/10.1146/annurev-ecolsys-031120-101013.

20. Oberhofer, G., et al. 2019. Cleave and rescue, a novel selfish genetic element and general strategy for gene drive. *Proceedings of the National Academy of Sciences*. 116: 6250. https://doi.org/10.1073/pnas.1816928116; Champer, J., et al. 2020. A toxin-antidote CRISPR gene drive system for regional population modification. *Nature Communications*. 11: 1082. https://doi.org/10.1038/s41467-020-14960-3.

21. Franklinos, L.H.V., et al. 2019. The effect of global change on mosquito-borne disease. *The Lancet Infectious Diseases*. 19: e302–e312. https://doi.org/10.1016/S1473-3099(19)30161-6.

22. Burt, A. 2014. Heritable strategies for controlling insect vectors of disease. *Philosophical Transactions of the Royal Society B*. 369: 20130432. https://doi.org/10.1098/rstb.2013.0432; Champer, J., et al. 2016. Cheating evolution: engineering gene drives to manipulate the fate of wild populations. *Nature Reviews Genetics*. 17: 146–159. https://doi.org/10.1038/nrg.2015.34.

23. The gene name *dsx* is an abbreviation for "doublesex." This gene name alludes to the consequences of laboratory mutations of this gene in *Drosophila* fruit flies, which also depend on *dsx* for sexual development. Mutations to *dsx* cause fruit flies and mosquitoes to develop features of both males and females. There are numerous other ways for a gene drive to manipulate sex ratio, such as by inducing developing sperm cells to destroy any copies of the X chromosome; this approach is termed the "X-shredder." All the fertile sperm cells made by a male with an X-shredder gene drive will carry the Y chromosome and, consequently, sire male offspring only. Kyrou, K., et al. 2018. A CRISPR-Cas9 gene drive targeting doublesex causes complete population suppression in caged *Anopheles gambiae* mosquitoes. *Nature Biotechnology*. 36: 1062–1066. https://doi.org/10.1038/nbt.4245; Hammond, A., et al. 2021. Gene-drive suppression of mosquito populations in large cages as a bridge between lab and field. *Nature Communications*. 12: 4589. https://doi.org/10.1038/s41467-021-24790-6.

24. Elizabeth Kolbert, CRISPR and the splice to survive, *The New Yorker*, January 18, 2021. https://www.newyorker.com/magazine/2021/01/18/crispr-and-the-splice-to-survive. Accessed May 7, 2021.

8

On the origin of species

It wasn't quite dawn when I woke to the sound of an eerie howl. It reverberated through the morning rainforest mist that blanketed the canopy of the valley below the inselberg, the "island mountain" peering over the rainforest.[1] The sound ebbed and flowed, ghostly, guttural, tinged with an almost mechanical vocal fry, as if a collective of wilderness monks were chanting through a scratchy megaphone. A troupe of howler monkeys was staking their territory just a few kilometers away along La Crique Nouragues, the shallow French Guianan stream that ran below the isolated field research station that was, for the time being, my South American home base (Figure 8.1). I opened my eyes and peered through the permethrin-impregnated mosquito netting of my hammock—my bed—strung between joists of the open air *carbet*, a structure that was little more than a wall-less wooden platform with a plank floor and posts to hold up the corrugated iron roof panels. Inside is outside, without the direct rainfall. When you breathe deep, when you breathe the moist fog of the rainforest, you share it with capybara and tapir, with poisonous blue dart frogs and scarlet macaws, with morpho butterflies and leaf-cutter ants, with giant tarantula and amblypygid whip scorpions, with every single species.

It was August 17, 2014. I didn't know it yet, but I was to discover a new species that day.

A new species is the epitome of biological novelty: a creature so genetically distinct that one must define it as something separate from everything else that exists. Nature creates new species on a regular basis, an incidental byproduct of populations getting separated from one another and evolving in parallel to their distinct life circumstances. In birds, over the last hundred million years or so, for every 9 species that go extinct there would be 10 new species that arise so that, on average, they've shown net positive diversification that's yielded about one additional species every one hundred thousand years.[2]

Unfortunately, the last 500 years of the Anthropocene has been less kind. It has witnessed birds driven extinct at a rate of more than one every four years,[3] resulting in a substantial loss in biodiversity. Mammals are, relative to their total species count, even worse off. A recent analysis concluded that it will require three to seven million years to evolve new species diversity sufficient to offset the current degree of Anthropocene extinction.[4] For a goal of maximizing species biodiversity, we already know what is, by far, the most effective near-term strategy: stop destroying wild habitat in order to reduce

Evolving Tomorrow. Asher D. Cutter, Oxford University Press. © Asher D. Cutter (2023). DOI: 10.1093/oso/9780198874522.003.0008

(A)   (B)

Figure 8.1 *Tropical field research in the Nouragues Nature Reserve of French Guiana to collect and identify new species of nematode roundworms.*
Photo by the author.

extinction rates. If we could work out nature's tricks to making new species, however, perhaps we might also nudge it along in partial recompense for our extreme acceleration of extinction over recent centuries.

★

We named the new species from French Guiana *Caenorhabditis astrocarya*. You can't find *C. astrocarya* just anywhere. You have to look inside the bounty of brilliant yellow pollen of the flowers from the meter-tall floral inflorescence of *Astrocaryum paramaca* palms and in the piles of pollen that they dump unceremoniously beneath the floral stalk. Or that is what I discerned with my friend and colleague Christian Braendle after peering under the microscope at what creatures lived amidst the nutrient-rich plant pollen.[5] The leafy fronds of these trunkless palms are coated in long spines, in the midst of which the plants raise a long hairy frond that opens to form a tough sheath from which emerges an elaborate candelabra of dozens of droopy yellow pollen-bearing fingers. We traversed many kilometers of rough trails through the rainforest, exploring as many novel microhabitats as we could, but this alone was where *C. astrocarya* chooses to make its home.

At just 1 mm long (1/25th of an inch) with a transparent body, the nematode round-worm *C. astrocarya* is, literally, a small discovery. And yet, *C. astrocarya* is nonetheless a distinct biological entity never before known to science. Given its diminutive form and inconspicuous lifestyle, I'd also hazard a guess that it was also never before known to Indigenous knowledge in the period prior to the arrival of European colonists.

Christian and I gave a brief research show-and-tell for the French Justice Minister and her bodyguards after they arrived one afternoon to see what tax dollars were being put to. They came by helicopter, the same way we did, as no roads lead to the Nouragues Nature

Reserve. Our attempt at flattery—perhaps to name a new species after the Minister—was rebuffed with throaty laughter, however, upon our response to her question: what kind of animal is it? Apparently, politicians are especially sensitive to allying their name to a worm.

8.1 What makes a species

Humans and chimpanzees are different species. I trust that this fact is not news to you. We know this is true, in part, because the range of measurable characteristics among all humans differs so substantially from the range of measurable characteristics of all chimpanzees. In other words, we look different. We also behave differently in such a way that humans and chimpanzees are mutually disinterested in attempting to reproduce with one another. Such separation of reproductive gene pools imprints itself in the DNA of our genomes.

As new mutations pop up with each new birth of a bouncing baby, some of those mutational variants will go on to increase in abundance across the generations, as we have learned, increasing because of natural selection. They increase by natural selection because of some benefit they confer to survival or reproduction, or increase because of genetic drift's chance influences on gene frequencies. The same thing happens in chimpanzees, independently, on their own set of new mutations. As a result, the genomes of humans and chimpanzees continually become more and more distinct. We can measure this divergence between genomes by counting DNA differences and recording the observable trait differences that those DNA differences cause.

Nature measures this divergence, however, by whether it prevents those different genomes from coming into contact within the cellular confines of the same individuals. One readout for nature's yardstick is behavioral disinterest in cross-breeding. Another readout, if cross-breeding does occur, is the suite of consequences (e.g., infertility, developmental defects, or failure to thrive) for the resulting hybrid organisms that contain the different genomes within each of their cells.

Not all species reject out of hand the sexual overtures of another, as we would reject a frisky chimpanzee. The cues and courtship that many species use to identify prospective mates may only incompletely weed out members of the opposite sex from other species. This situation can be especially common when species shared a very recent common ancestor, perhaps isolated for a few hundred millennia on opposite sides of a river's changed trajectory or of a now-receding glacier or on previously separated islands when ocean levels were higher. When the geographic ranges come back into contact, whether in nature, in zoos, or in laboratories, we can observe the outcome of mixing diverged genomes within the bodies of hybrid individuals.

Sometimes hybrid individuals develop dramatically dysfunctional features. The liger and tigon make for a pair of outlandish and frequently cited examples from zoo cross-breeding, the hybrid offspring between lions and tigers that involved lion fathers or tiger fathers, respectively. Ligers, in particular, grow to exceptionally large sizes because of misregulation of growth factors. Male hybrids of both types, however, are almost always sterile. Such sterility of males among hybrids, in particular, is so widespread across the animal kingdom that it gets its own name: Haldane's rule, after the eminent early twentieth-century evolutionary biologist J. B. S. Haldane who first noted the pattern in 1922.[6]

Mice can form hybrids, too, and in nature rather than zoos. Should you encounter a house mouse while in France, almost assuredly it will be *Mus domesticus*. A house mouse in Poland, on the other hand, will be *Mus musculus*. These species evolved from a common ancestor about 500,000 years ago, with their ranges having come to abut one another in Central Europe following the end of the last glaciation.[7]

What happens in the swathe of Europe in between France and Poland? Eastern Germany south down through Austria and Slovenia marks the zone where the two species meet. This zone where they meet forms a hybrid zone, where the genomes of the mice are a mix.[8] The hybrid mice display Haldane's rule, with the males tending to be sterile: they have small testes and defects in sperm production. If that weren't bad enough, hybrid mice also are sexually unattractive to other mice. The 20-km-wide (12-mile-wide) hybrid zone persists in Europe because these substantial costs to fitness, which eliminate hybrids and the mixed genomes inside of them, are counterbalanced by the continual creation of new hybrid individuals through cross-breeding between *Mus domesticus* and *Mus musculus* at the boundary of the species ranges.

The genomic differences between species can become so vast, however, that fertilization like that seen between *Mus* species can't even be achieved. The molecular pillow-talk between sperm cell and egg cell gets whispered in such mutually unintelligible dialects that fusion to form an embryo becomes an impossibility. Or, if gamete fusion does take place, embryonic cell division arrests prematurely so that no hybrid baby is ever born. These two possibilities are the usual outcomes between species of *Caenorhabditis* roundworms, and how Christian and I were able to determine that *C. astrocarya* was, in fact, a new and distinct species from all others.

★

Why do problems of viability and fertility arise in hybrids? After all, each parental species could survive and reproduce perfectly well on its own, as could the shared common ancestral species from which they both arose. The answer lies with the fact that genomes do not encode a singular item.

Genomes encode thousands of different proteins that provide a collection of building blocks that must work together to allow an organism to emerge from their mutual interaction through development. For simplicity, let's consider just two genes, the *A* gene and the *B* gene. In this thought experiment, the proteins made by *A* and *B* work together to help a creature to develop properly. Every individual in the ancestral species gets copies of *A* and *B* from both Mom and Dad, so that we can summarize its genetic makeup as *AABB*.

Now, think of how evolution proceeds after this ancestral species gets split into two separate populations (Figure 8.2). Each will evolve independently, say, because a subgroup of intrepid souls migrated to a distant island and established a separate colony for

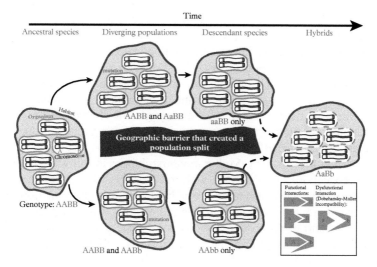

Figure 8.2 *Separation of an ancestral population into distinct descendant populations can lead to the evolution of genetic incompatibilities that can act to prevent gene flow, known as Dobzhansky-Muller incompatibilities. A population of individual organisms, each with two copies of each chromosome in its genome, may encode genes that must interact for the organisms to develop, survive, and reproduce properly (genes "A" and "B"). The evolutionary forces of mutation, selection, and genetic drift can lead to different gene variants evolving in each of the descendant populations. Should the divergent descendant populations come into subsequent reproductive contact, the hybrid genotypes may experience reduced fitness because of dysfunctional interactions between the novel genetic variants.*

a few hundred thousand generations. Mutation and natural selection may lead to a new version of gene *A* to predominate in the island population, called *a*. This is a common and inevitable reality in nature, as we see when we look at DNA changes over time. We can now summarize the genetic makeup of the island creatures as *aaBB*. Meanwhile, mutation and natural selection may lead to a new version of gene *B* to evolve in the mainland population, called *b*, so their genetic makeup becomes *AAbb*. All is well. The *A* and *b* gene versions work well together, as do the *a* and *B* gene versions. Should these new derivative populations both then colonize a third island and cross-breed with one another, the hybrids would have *AaBb* genomes.

The trouble for the *AaBb* hybrids is, however, that the *a* gene version never before got the chance to interact with the *b* gene version. They may not function properly together. This possibility is called nontransitivity: the facts that the *A-B* and *A-b* and *a-B* interactions create a properly developing organism is no guarantee for what happens with the *a-b* interaction. You get along with your new boyfriend and you get along with your father, but there is no guarantee that your boyfriend will get along with your father. When interacting genes create dysfunctional interactions, they will disrupt development in a way that it can cause hybrid animals to be sterile, have reduced survival, or exhibit mismatches to the environment.

Now consider the fact that animal genomes have tens of thousands of genes that must interact, not just two. The potential for such nontransitive evolution to occur becomes enormous. Nontransitivity of interactions such as this are called, in genetic terms, *Dobzhansky-Muller incompatibilities*.[9] In house mice, the *Prdm9* gene forms part of a Dobzhansky-Muller incompatibility to cause hybrid male mice to be sterile, representing, so to speak, the gene with *A* and *a* variants in *Mus domesticus* and *Mus musculus*.[10] The more time that elapses, the more genetic changes distinguish the separated populations, and the more Dobzhansky-Muller incompatibilities there will be that will reveal their detrimental effects, should hybrid animals be formed.

The evolution of Dobzhansky-Muller incompatibilities in this way allows for species to evolve into collections of individuals with irreversibly distinct genetic compositions. In other words, new species form as an inevitable byproduct of populations getting separated for long enough that mutations arise and become fixed distinguishing features of their DNA. This kind of reproductive isolation of gene pools is a genetically intrinsic feature of their genomes. Anything that accelerates that DNA sequence divergence will accelerate the speciation process.

★

This is a good place to point out that Dobzhansky-Muller incompatibilities are an impossibility in some kinds of organisms. For such genetic interactions to arise through evolution and to have the potential to exert an effect in hybrids, the organisms must have two properties. First, they must reproduce through the genetic union of two distinct individuals. Second, they must receive a copy of their genetic material from each of two parents.[11]

Bacteria and viruses don't fit these criteria. Most fundamentally, they don't fit because of their single genomic copy.[12] In addition, most of the time, bacterial cells reproduce by cloning themselves without sex, and this on its own means that bacterial species arise by a different set of rules than most animals and plants. Defining what is a bacterial species is made all the harder by the fact that vastly different forms of bacteria are still capable of exchanging genetic material. This exchange results from the horizontal gene transfer mode of introgression in bacteria through microbial mechanisms akin to recombination that are entirely foreign to animals and plants (transformation, transduction, and conjugation).[13] We can, nonetheless, quantify these peculiar forms of bacterial recombination. By comparing genome sequences for tens of thousands of bacterial or viral isolates, biologists can discriminate distinct groups that exchange genetic material on a regular basis from those bacterial types that don't.[14] Despite bacteria seeming like inscrutable microscopic creatures, this procedure defines about 85% of them as corresponding to evolutionarily distinct biological species in a directly analogous way to how we think of any given animal species as the collection of individuals that reliably share the same gene pool.

★

The negative effects that Dobzhanky-Muller incompatibilities impose on hybrid organisms is a genetic means of enforcing species boundaries. They lock species identity into place. But species can operate as independent gene pools even without the genetically draconian influence of Dobzhansky-Muller incompatibilities. Sometimes hybrids develop successfully, but their characteristics differ from both parents in such a way that they make for a poor match to the environment that they find themselves in. In essence, the island species may be adapted to perform well on islands, and the mainland species well-adapted to mainland conditions, but their hybrids perform worse than their parents in their respective habitats. The ecological conditions lead natural selection to exterminate such hybrid individuals disproportionately, even if their intrinsic genetically programmed development proceeded perfectly fine.

On the rocky shores of the North Atlantic, in that no-man's-land between land and sea that defines the intertidal zone according to the saltwater whims of high and low tide, there lives a little snail, the rough periwinkle *Littorina saxatilis*. On the coast of Galicia, the thumb of Spain atop Portugal that pushes west into the Atlantic Ocean as if to vie for the westernmost promontory of continental Europe, Spanish snails feel the tide's 3-m (10-foot) vertical flux twice per day. They feel its effects acutely because they crawl at only a snail's pace: a rate of just 1 to 3 m (3 to 10 feet) per month.[15] *Littorina* at the bottom of the intertidal zone suffer the costs of the most intense wave action. *Littorina* at the top of the intertidal zone avoid the worst of the wave stress but are more exposed to the longer dry spells of low tide and the deadly green claws of *Pachygrapsus marmoratus* crabs.

These natural selection pressures have shaped evolution of the characteristics of the snails at one end of the intertidal zone from those at the other end, just 60 m away (200 feet). The *Littorina* snails on the upper shore are bigger, with a shell that is ridged and with a smaller opening for the snail's foot, the better to seal up tight away from crabs and desiccating dryness. The smaller snails on the lower shore are smooth and with a big opening for their foot, the better to adhere to wet rocks in the rough-and-tumble of waves. The ridged versus smooth snails each prefer to mate with their own type, too. This assortative mating gets further encouraged by their slow pace of movement and the fact that they carry their eggs with them to hatch as crawl-away babies rather than releasing them into the water as float-away babies.

The snail's mating preferences are imperfect, however, and the middle portion of the intertidal zone is also a hybrid zone. The hybrid males have lower sperm viability, and the hybrid females make fewer eggs. These defects, however, are relatively weak. As a result, the potential for distinct species of *Littorina* to emerge at the high and low ends of the intertidal zone depends on the continued ecological and environmental pressures imposed by those distinct subhabitats.

The catch to species remaining distinct by virtue of distinct ecological pressures of natural selection is that those ecological pressures must hold steady. If the ecological circumstances of the environment change, then the divergent groups of individuals may simply hybridize willy-nilly and reconstitute a single gene pool where previously there had been two.

This extinction by fusion, or reverse speciation, happened in the late 1990s to what seemed to be an emerging pair of stickleback fish species within Enos Lake of Canada's

Vancouver Island in British Columbia.[16] Prior to that, two morphs of *Gasterosteus* stickleback had emerged in the lake. One was a population of heavy-set fish that specialized in eating invertebrates like snails and insect larvae along the lake bottom. The other morph was a population of slender fish that specialized in eating zooplankton in the open water, microscopic shrimp-like creatures.

Sexually, each fish morph preferred its own kind, and so the two types of fish were able to co-exist as biologically distinct gene pools despite sharing the same pool of water. They could, that is, until the signal crayfish *Pacifastacus leniusculus* arrived on the scene. These invasive crayfish disrupted the fish's normal mating rituals through their territoriality and, in making the water murkier, interfered with the fish's visual mate recognition. Without any support from Dobzhansky-Muller incompatibilities to reinforce how natural selection had favored extreme forms and disfavored hybrids, speciation reversed course. The emerging two distinct fish stocks mated with one another and collapsed down into a single interbreeding population that lacked the pair of distinctive specialized forms.

Wholesale merger of two species like in Enos Lake is an extreme outcome of hybridization. More commonly, occasional hybrid formation, even in the presence of Dobzhansky-Muller incompatibilities, can result in those hybrid individuals subsequently reproducing with members of the parental species stock.[17] These reproductive links allow gene variants to bridge the divide from one species to another, gene variants migrating from the gene pool of one species into another. A portage of a DNA canoe from one gene pool to another.

This genetic transmission of segments of DNA from one species to another through occasional interbreeding and back-crossing is called *introgression*, that evolutionarily long-distance form of gene flow. Thanks to a little bit of hybridization once upon a time, our own genomes today are a little bit Neanderthal, with introgressed fragments of Neanderthal DNA being especially common on our ninth chromosome.[18] People with African ancestry tend to have much less of this introgressed Neanderthal DNA than other humans, however. Folks with African ancestry typically have 0.3% or less compared to the roughly 2% of the genomes of people with non-African ancestry—approximately 60 million base pairs of DNA. This fact points to the conclusion that interbreeding between *Homo neanderthalensis* and *Homo sapiens* took place primarily in Europe, the Middle East, and Asia after migrating out of Africa and before modern humans populated the rest of the world. The genomes of people with Melanesian and native Australian ancestry also show substantial introgression of Denisovan *Homo denisova* DNA, which can make up 2% or more of their genomes.

One result of hybridization is that it can act as a source of novel genetic variants that one species' gene pool had never before encountered. Such extreme gene flow, in effect, replaces the role of mutation as a source of genetic novelty. As a consequence, genetic introgression can sometimes provide a source of beneficial variants. While sometimes beneficial, when natural selection differentiates between a species' native DNA and DNA encountered through hybridization, more typically it will act to eliminate the non-native DNA variants. This negative selection occurs because of Dobzhansky-Muller incompatibilities or genetic variants that are relatively maladaptive.[19] The fact of rare imports of DNA from other gene pools does not in itself mean that there is simply one bigger gene pool of fewer species, however. So long as the introgression pales in comparison to the gene exchange within the smaller gene pool of each species proper, we still consider the groups to represent distinct biological species.

8.2   Radiations

The ecological and environmental pressures of distinct habitats can be so strong and so abundant with opportunity that they drive an explosion of new species, an adaptive radiation. The multitude of Caribbean islands have given just those circumstances to the anoles. These lizards, formally known as *Anolis*, have diversified in size and shape to an astonishing degree. Competition between ancestrally generic lizards created selection to carve up different parts of the habitat, to exploit its riches with adaptive specializations, the islands largely unconstrained by the community of predators that are found on the mainland of Central and South America. The anole's unusual toepads, in particular, may have predisposed them to radiating. Anole toes have especially dense microscopic hairs called setae. These toepad setae number in the tens of thousands and give anoles the special ability to adhere to and climb smooth surfaces.[20]

What is especially striking about the anoles on each of the largest islands of the Greater Antilles is that each island independently evolved sets of species that fill equivalent habitat specializations. That is, the islands of Cuba, Jamaica, Puerto Rico, and Hispaniola (the Dominican Republic plus Haiti) each have a similar set of *Anolis* ecomorphs that arose from separate genetic stocks. Take, for example, the Crown Giant anole ecomorph. These lizards all are big and green, with big heads and big toes, and are at home at the tops of trees with a variety of other similarities in behavior and morphology that match them to perform better than other types of anoles in this particular habitat. Cuba has *A. baracoae*, Jamaica has *A. garmani*, Hispaniola has *A. ricordii*, and Puerto Rico has *A. cuvieri*. When you look at their DNA, however, you find that they are not any more related to one another than to any other type of anole: each Crown Giant species of *Anolis* evolved its similar set of characteristics separately on each island, starting from a totally distinct progenitor.

The Greater Antilles islands also have distinctive species for each of five other specialized habitats, giving rise to anoles with the Trunk, Trunk-ground, and Trunk-crown ecomorphs; the Grass-bush ecomorph; and the Twig anole ecomorph. Superficially, the different species in a given ecomorph look extremely similar, as with the excessively long tails of the Grass-bush anoles, or the short-and-stubby legs of the Twig anoles that improve their grip on the skinny tips of tree branches as they slowly and methodically hunt for cryptic insects to eat. This so-called convergent evolution in lizard form is possible because of the similar range of habitats on each island to which a colonizing species might adapt. Anole convergent evolution illustrates how evolution can be predictable, in some respects, how the evolutionary tape of life sometimes *can* repeat itself.[21]

There are many ways one might think about what it is that makes the characteristics of a species evolve more rapidly in one way than in another, despite a common ecological pressure for increased survival and reproduction. Let's think about it genetically. For an evolutionary response to a source of natural selection, we know that there must be genetic variants that create a link between inheritance and fitness. But there will be more genetic variability for some traits than for others, and some traits will correlate more strongly than others with fitness. The path that evolution will take, it turns out, is the path of least resistance, as pointed out by Canadian biologist Dolph Schluter.[22] What this means is that a characteristic with lots of genetic variability that is only partially correlated with survival and reproduction might evolve more rapidly than another characteristic with a strong association with fitness but that has little genetic variability. In the adaptive radiation of Caribbean anoles, divergence in lizard skeletal characteristics appears to have followed just such a set of least-resistance genetic paths.[23]

Caribbean islands that are smaller than the big four of the Greater Antilles have fewer ecomorphs and fewer species, some with just a single representative. Or, following the decimation of especially minute islands on the tail end of hurricane season, zero species of anole. All told, the adaptive radiation of Caribbean anoles peaked over 20 million years ago to give the world today's 150 species that are sprinkled across a constellation of tropical islands,[24] each and every one the envy of us Canadian residents in winter.

★

A species is not a singular biological quantum. A species is made up of a population, a group of individual organisms that each act with reproductive agency. Any one body may give us the gist of what the species is all about. But it is the collection of bodies with their similar yet distinct genetic forms, their similar yet distinct outward forms that grow, interact, and reproduce to support a viable population. A viable, sustainable population is essential. What is it that makes a population and, consequently, a species, self-sustaining?

In simplest terms, to be viable, the population must maintain a birth rate that is greater than or equal to the death rate. It must be able to occupy habitat with sufficient resources to support at least a minimally sized population. There are four key natural risks that define how small a population can be in order to remain viable, that is, to avoid extinction.[25]

The first two risks differ primarily in severity: variability over time of a normal magnitude versus of an extreme magnitude in basic environmental characteristics that influence food availability, competition, predation, and disease—the extremes creating unpredictable catastrophes like floods, fires, or droughts. Next is the variability over time in the number of individuals, with such demographic stochasticity arising inevitably even under benign conditions from chance fluctuations in survival and reproduction. Finally, there are genetic risks that are due to the detrimental consequences of inbreeding and the fact that small populations have little genetic diversity with which to adapt to ecological changes. Low genetic diversity also makes them more vulnerable to infectious diseases.

Whether there is one group of individuals occupying a single large habitat patch versus multiple smaller groups in multiple smaller patches ties into these risk factors in complex ways to make overall persistence of the species as a whole either more or less likely. When the total population is already small, each of these risk factors can overwhelm a species and bring it to extinction.

How small is too small? Conservation biologists talk about the minimum viable population size that a species should have to avoid the aforementioned challenges to persistence. As a yardstick, there is the notion of a 99% chance of persistence for 40 generations (summing to 100 to 1000 years, depending on assumptions about what generation time to use). Analysis of vertebrate animals suggests a rule of thumb: a minimum of about 7000 adult individuals are needed for such durations of persistence.[26]

For any given species, however, the minimum size required to avoid extinction depends on the potential growth rate. Slower reproductive rates need a larger number of individuals to avoid a worrisome risk of dying off completely as a species. Slow reproductive rates are most often found in species that take a long time to reach adulthood, produce few offspring per reproductive episode, and reproduce infrequently over their lifetime. These kinds of animals often devote a lot of energy to each individual offspring, perhaps even providing parental care. Consequently, the likelihood of survival of each offspring is quite high, which can offset the low reproductive output under ordinary circumstances. This lifestyle is especially common in large animals. Ecologists sometimes call this a K-selected reproductive strategy.[27] You are familiar with many such species; after all, you are one of them.

This special K lifestyle contrasts with the so-called r-selected reproductive strategy, in which huge numbers of eggs may be produced during reproduction, potentially leading to a huge number of offspring. Usually, though, most of the babies quickly meet a mortal fate. In such creatures, there is little parental investment in any given offspring. They are playing the evolutionary lottery, gambling that some lucky few in their shotgun blast of reproduction will land in suitable conditions to grow to adulthood or that it will be a jackpot year when most survive. It's sometimes called a *boom-and-bust lifestyle*. Both the K and r strategies are good ones. How good they are in a given environment depends on how likely the different life stages are to experience mortality or restricted resources.

Species, in reality, don't fall neatly into one extreme or the other, usually falling somewhere in between these simplified r and K archetypes. Those species tending toward the r-selected approach, however, will have a higher potential reproductive rate that often allows them to tolerate a smaller minimum population size to be viable in the long term.

★

Speciation in nature from start to finish—population splitting, mutation and natural selection, and genetically enforced separation of gene pools—often can be fast on a geological time scale, but still take many thousands or millions of years. In some cases, however, it can be nearly instantaneous. Sunflowers are a shining example. The common sunflower, *Helianthus annuus*, grows over vast swathes of central and western parts of the United States, partially overlapping with the more disjointed range of *H. petiolaris*, the prairie sunflower. Despite the superficial geographic overlap, they like different soils, clay-laden for *H. annuus* versus sandy for *H. petiolaris*, a fact that tends to keep them separated into neighboring habitats. Given the broad geographic ranges of these species, the landscape contains other soil types where one can also find sunflowers: dunes, desert floors, and salt marshes. It turns out that sunflowers in each of these three more extreme environments arose from hybridization of *H. annuus* with *H. petiolaris*.[28]

My favorite of the three hybrid species is *H. anomalus*. This hybrid species is found amid the inland dunes of Arizona and Utah, their flowers brilliant beacons of yellow in a sea of sand. Experiments that recreated hybrids in the greenhouse to mimic the natural hybrid species recovered characteristics even more extreme than seen in the progenitor species. What is especially interesting about such extreme characteristics is that natural selection could then act on those extreme features to allow adaptation to novel ecological circumstances. This hints at how novel gene combinations of drastically different gene variants from the species merger of *H. annuus* and *H. petiolaris* allowed the hybrids to colonize extreme habitats, to exploit a niche like sand dunes that was inaccessible to either of the progenitor species.[29] The genomes of the hybrid species show radical rearrangement of genes along chromosomes compared to *H. annuus* and *H. petiolaris*, which partly explains why they can't easily interbreed with either of the two progenitor species to collapse down into a single gene pool. Further analysis of their genomes showed that it took somewhere between 10 and 60 generations for the wild hybrid populations to establish themselves as distinct species. In this way, hybrid speciation allows hybrids to spawn a brand new, third species distinct from each of the two progenitors in just tens of generations.

Nature isn't the only place where hybridization creates something biologically new. New breeds of house cats derive from the intentional hybridization of domestic cats (*Felis catus*) with wild feline species. Since 1986, hybrids formed with African servals (*Profelis serval*) to make Savannahs are officially recognized by The International Cat Association (TICA) in competitions, and TICA recognizes Bengals, as of 2001, which derive from hybridization with Asian leopard cats (*Prionailurus bengalensis*).[30] Not surprisingly for species combinations that shared their last common ancestor over seven million years ago, the initial male hybrids are sterile.[31] The domestic Bengal and Savannah breeds result from back-breeding the hybrid females to male domestic cats over the course of multiple generations. All the while, the generations of offspring were getting subjected to human-mediated artificial selection for desirable domestic characteristics.

★

Pigeons, as we have seen, have diversified radically in size and plumage and behavior through selective breeding. But there is more evolution than meets the eye, at least without the aid of a hand lens or magnifying glass. Different *Columba livia* pigeon breeds, it turns out, represent distinct ecological niches for minuscule inhabitants: feather lice. Named by Carl Linnaeus in 1758 in what can only be described as a nomenclatural tongue twister, the rock dove feather louse *Columbicola columbae*, Carl claimed, chronically crawls on *Columba livia*.

The feather lice live out their lives on a single bird, growing and mating, generation after generation, usually only emigrating to new pigeons that snuggle close as mating partners or new chicks. The pigeons, of course, do not appreciate *C. columbae* chewing on the downy nether regions of their feathers, and so preen with their beaks to remove them. Preening favors those lice that can avoid getting preened, by running quickly between feather barbs and being small enough to squeeze between feather barbs.[32] Small lice run slower on bigger breeds of birds, like on the triple-sized Giant Runt, as a result of the more difficult terrain of 20% longer distances between the barbs of their larger feathers. Small lice also lay fewer eggs. A team of biologists in Utah experimentally transferred lice onto a collection of louse-free Giant Runts and louse-free wild-caught feral pigeons. After about 60 louse generations, four years in the aviary, the *C. columbae* lice from the Giant Runts had evolved to be about 75% bigger than the *C. columbae* from the other pigeons, as a result of genetic changes.

Such rapid evolutionary change would be captivating enough as-is, especially when assuaged that pigeons are the only host to be infested by giant *C. columbae*; this species does not colonize mammals. But there is more. The giant lice on the Giant Runts were ineffectual in mating with those lice from the smaller feral pigeons: they had become reproductively isolated as a result of size differences. As Scott Villa and his colleagues wrote matter-of-factly, "when males are either too large or too small, relative to the female, they have difficulty copulating."[33] All of the male lice were eager to mate with females of all sizes, indicating that they still recognized them as potential partners. But mismatched pairs spent 70% less time going at it, and such couplings were much less likely to yield eggs. The evolution of body size had, as a byproduct, initiated the evolution of reproductive isolation because of mating incompatibility. In a matter of just tens of generations, they had taken an early step on the path to speciation into distinct biological entities. And doesn't the world need more species of louse?

★

Species represent distinctive, genetically cohesive biological entities. To be sustainable, a species needs a large enough number of individuals, typically including many thousands of adults at the minimum. Species can nonetheless hybridize under certain circumstances, potentially shuttling genetic variants between otherwise separate gene pools. Intrinsic genetic barriers to species merger and genetic fusion arise inevitably in the form of genetic incompatibilities as a result of DNA sequence divergence between separated populations of organisms. Natural selection is an important contributor to such genetic divergence and, as a consequence, to speciation.

Notes

1. An *inselberg* is a lone mountain or hill protruding up from an otherwise more-or-less flat landscape, often made of erosion-resistant granite; the word comes from the German for "*island mountain*." In this case, the inselberg is nestled deep within the tropical rainforest of the Nouragues Nature Reserve of French Guiana. The French Centre National de la Recherche Scientifique (CNRS) operates a pair of remote field stations inside the research

zone of the Reserve, at which I have had the good fortune of conducting biodiversity studies with my colleague Christian Braendle.

2. This average calculation, however, obscures the fact that bird species accumulated in an approximately exponential fashion through pre-Anthropocene times to reach nearly 10,000 species today from a diversification rate of roughly 0.1 per lineage per million years. Yu, Y., et al. 2021. Deep time diversity and the early radiations of birds. *Proceedings of the National Academy of Sciences USA.* 118: e2019865118. https://doi.org/10.1073/pnas.2019865118; Jetz, W., et al. 2012. The global diversity of birds in space and time. *Nature.* 491: 444–448. https://doi.org/10.1038/nature11631.

3. Johnson, C. N., et al. 2017. Biodiversity losses and conservation responses in the Anthropocene. *Science.* 356: 270–275. https://dx.doi.org/10.1126/science.aam9317.

4. Davis, M., et al. 2018. Mammal diversity will take millions of years to recover from the current biodiversity crisis. *Proceedings of the National Academy of Sciences USA.* 115: 11262–11267. https://doi.org/10.1073/pnas.1804906115.

5. Ferrari, C., et al. 2017. Ephemeral-habitat colonization and neotropical species richness of *Caenorhabditis* nematodes. *BMC Ecology.* 17: 43. https://doi.org/10.1186/s12898-017-0150-z.

6. Haldane's rule is, in fact, slightly more nuanced than simply hybrid male sterility. The pattern applies to hybrid individuals that have distinct sex chromosomes, such as the X and Y sex chromosomes of mammals or the Z and W sex chromosomes of birds. Consequently, it is sterility in the females of bird hybrids that contribute to the Haldane's rule pattern. As it turns out, plants that have sex chromosomes tend to follow Haldane's rule, as well. Haldane, J.B.S. 1922. Sex ratio and unisexual sterility in hybrid animals. *Journal of Genetics.* 12: 101–109. https://doi.org/10.1007/BF02983075; Delph, L.F., and J.P. Demuth. 2016. Haldane's rule: genetic bases and their empirical support. *Journal of Heredity.* 107: 383–391. https://dx.doi.org/10.1093/jhered/esw026.

7. Latour, Y., et al. 2014. Sexual selection against natural hybrids may contribute to reinforcement in a house mouse hybrid zone. *Proceedings of the Royal Society B.* 281: 20132733. http://doi.org/10.1098/rspb.2013.2733.

8. Note that some researchers refer to these house mice as subspecies, using the names *Mus musculus domesticus* and *Mus musculus musculus.* Baird, S., and M. Macholán. 2012. What can the *Mus musculus musculus*/*M. m. domesticus* hybrid zone tell us about speciation? In M. Macholán, et al. (eds), *Evolution of the House Mouse (Cambridge Studies in Morphology and Molecules: New Paradigms in Evolutionary Bio).* Cambridge: Cambridge University Press. pp. 334–372. https://dx.doi.org/10.1017/CBO9781139044547.016.

9. Theodosius Dobzhansky and Herman Muller formulated the logic of genetic incompatibilities in this way in the 1930s and 1940s. Johnson, N. 2008. Hybrid incompatibility and speciation. *Nature Education.* 1: 20. https://www.nature.com/scitable/topicpage/hybrid-incompatibility-and-speciation-820/.

10. Mukaj, A., et al. 2020. *Prdm9* intersubspecific interactions in hybrid male sterility of house mouse. *Molecular Biology and Evolution.* 37: 3423–3438. https://doi.org/10.1093/molbev/

msaa167; Turner, L.M., and B. Harr. 2014. Genome-wide mapping in a house mouse hybrid zone reveals hybrid sterility loci and Dobzhansky-Muller interactions. *eLife*. 3: e02504. https://doi.org/10.7554/eLife.02504.001.

11. Organisms that receive genetic material from each of two parents are usually termed *diploid*. Some species, however, have experienced genome duplications in their evolutionary history that make their genomes have more than two copies. Such polyploid species include tetraploid wheat, with four copies of each chromosome (a given wheat plant inherits two genome copies from each of their two parents).

12. *Ploidy* is the genetic term for how many genomic copies a cell has. Most cells of an animal or plant have two copies, one each inherited from each parent, and so are referred to as *diploid*. Sperm cells have one genomic copy, and so are termed *haploid*. Bacteria also have just one copy of their genome, and so are referred to as *haploid*.

13. Recall our discussion of horizontal gene transfer (HGT) and introgression from Chapter 3.

14. Bobay, L.-M., and H. Ochman. 2017. Biological species are universal across life's domains. *Genome Biology and Evolution*. 9: 491–501. https://doi.org/10.1093/gbe/evx026; Bobay, L.-M., and H. Ochman. 2018. Biological species in the viral world. *Proceedings of the National Academy of Sciences USA*. 115: 6040–6045. https://doi.org/10.1073/pnas.1717593115.

15. Galindo, J., and J.W. Grahame. 2014. Ecological speciation and the intertidal snail *Littorina saxatilis*. *Advances in Ecology*. 2014: 239251. https://doi.org/10.1155/2014/239251.

16. Velema, G.J., et al. Effects of invasive American signal crayfish (*Pacifastacus leniusculus*) on the reproductive behaviour of threespine stickleback (*Gasterosteus aculeatus*) sympatric species pairs. *Canadian Journal of Zoology*. 90: 1328–1338. https://doi.org/10.1139/z2012-102.

17. Especially at the early stages of speciation, the genetic variants that would create Dobzhansky-Muller incompatibilities may not yet have been fixed within each incipient species' population. This heterogeneity in the intrinsic genetic barriers could facilitate hybridization and back-crossing by limiting the severity of fitness problems in hybrid individuals.

18. Notably, however, natural selection has acted to eliminate stretches of archaic DNA of hybrid origin from our genomes over the eons. Slatkin, M., and F. Racimo. 2016. Ancient DNA and human history. *Proceedings of the National Academy of Sciences USA*. 113: 6380–6387. https://doi.org/10.1073/pnas.1524306113; Chen, L., et al. 2020. Identifying and interpreting apparent Neanderthal ancestry in African individuals. *Cell*. 180: 677–687.e616. https://doi.org/10.1016/j.cell.2020.01.012.

19. Abbott, R., et al. 2013. Hybridization and speciation. *Journal of Evolutionary Biology*. 26: 229–246. https://doi.org/10.1111/j.1420-9101.2012.02599.x.

20. Stroud, J.T., and J.B. Losos 2020. Bridging the process-pattern divide to understand the origins and early stages of adaptive radiation: a review of approaches with insights from studies of *Anolis* lizards. *Journal of Heredity*. 111: 33–42. https://doi.org/10.1093/jhered/esz055.

21. Stephen J. Gould popularized the thought experiment of "replaying life's tape" to explore the repeatability of evolution in his 1989 book *Wonderful Life* (Norton Press). Blount, Z. D., et al. 2018. Contingency and determinism in evolution: Replaying life's tape. *Science*. 362: eaam5979. https://dx.doi.org/10.1126/science.aam5979.

22. In his words, "adaptive differentiation occurs principally along 'genetic lines of least resistance' such that 'patterns of quantitative genetic covariance bias the direction of evolution'": p.1766 and 1772 in Schluter, D. 1996. Adaptive radiation along genetic lines of least resistance. *Evolution.* 50: 1766–1774. https://doi.org/10.1111/j.1558-5646.1996.tb03563.x.

23. McGlothlin, J.W., et al. 2018. Adaptive radiation along a deeply conserved genetic line of least resistance in *Anolis* lizards. *Evolution Letters.* 2: 310–322. https://doi.org/10.1002/evl3.72.

24. Losos J.B., and C.J. Schneider. 2009. *Anolis* lizards. *Current Biology.* 19: R316–R318. https://dx.doi.org/10.1016/j.cub.2009.02.017; Sherratt, E., et al. 2015. Amber fossils demonstrate deep-time stability of Caribbean lizard communities. *Proceedings of the National Academy of Sciences USA.* 112: 9961–9966. https://doi.org/10.1073/pnas.1506516112.

25. Shaffer, M.L. 1981. Minimum population sizes for species conservation. *BioScience.* 31: 131–134. https://doi.org/10.2307/1308256; Nunney, L., and K.A. Campbell. 1993. Assessing minimum viable population size: demography meets population genetics. *Trends in Ecology & Evolution.* 8: 234–239. https://doi.org/10.1016/0169-5347(93)90197-W.

26. Reed, D.H., et al. 2003. Estimates of minimum viable population sizes for vertebrates and factors influencing those estimates. *Biological Conservation.* 113: 23–34. https://doi.org/10.1016/S0006-3207(02)00346-4.

27. Modern ecological theory about population dynamics takes a more sophisticated and nuanced approach than the simple dichotomy of K-selected and r-selected strategies. Nevertheless, many of the same conceptual themes still apply. Reznick, D., et al. 2002. r- and K-selection revisited: the role of population regulation in life-history evolution. *Ecology.* 83: 1509–1520. https://doi.org/10.1890/0012-9658(2002)083[1509:RAKSRT 2.0.CO;2.

28. Rieseberg, L.H. 2006. Hybrid speciation in wild sunflowers. *Annals of the Missouri Botanical Garden.* 93: 34–48. https://doi.org/10.3417/0026-6493(2006)93[34:HSIWS]2.0.CO;2.

29. The occurrence of characteristics of hybrids being more extreme than either parent because of the merger and recombination of their divergent genomes is termed *transgressive segregation*. Dobzhansky-Muller incompatibilities represent special cases of transgressive segregation such that the extreme feature is detrimental.

30. https://www.tica.org/breeds/browse-all-breeds. Accessed March 22, 2021.

31. Davis, B.W., et al. 2015. Mechanisms underlying mammalian hybrid sterility in two feline interspecies models. *Molecular Biology and Evolution.* 32: 2534–2546. https://doi.org/10.1093/molbev/msv124.

32. Villa, S.M., et al. 2019. Rapid experimental evolution of reproductive isolation from a single natural population. *Proceedings of the National Academy of Sciences USA.* 116: 13440–1344. https://www.pnas.org/content/116/27/13440.

33. Quotation from p. 13441 of Villa, S.M., et al. 2019. Rapid experimental evolution of reproductive isolation from a single natural population. *Proceedings of the National Academy of Sciences USA.* 116: 13440–1344. httsp://www.pnas.org/content/116/27/13440.

9

Do-it-yourself speciation

Natural selection plays a profound role in the origin of species. Might artificial selection prove similarly profound? That is, could we ourselves make a new species?

We have seen how artificial selection is potent in evolving organisms with extreme characteristics, as for the extraordinary milk production of Holstein heifers of about 30 liters (8 gallons) per animal per day or the adorable fluffy-looking feathers of Frizzle and Silkie chickens.[1] The bigger question is, could such selective breeding also lead to animals' reproductive isolation from their closest relatives in the wild? The radical evolution in dog domestication, however, hasn't eliminated gene flow with other canines to make their genomes perfectly separated gene pools. Hybridization has remained an ongoing feature of dog interactions with coyotes and wolves.[2] As we have seen, however, some degree of hybridization does not preclude them from being defined as distinct biological species.

If we intend to evolve a new species, however, it would be prudent to prevent unintended impacts on the gene pools of wild populations. Ideally, this would mean that genetic factors would help enforce species boundaries: Dobzhansky-Muller incompatibilities encoded in the genome of sufficient strength to make hybrids inviable or sterile (Figure 8.2). The difficulty in realizing this ideal lies in the disconnect between the genetic material that is most easily used by selection to produce divergent creatures and the genetic material that is most likely to generate Dobzhansky-Muller incompatibilities. Let me explain why.

To begin a program of artificial selection, we must begin with a founding stock, a base population. Artificial selection, just like natural selection, will then act on whatever genetic variants are present in that founding stock, discriminating between their effects and potentially driving rapid evolutionary change. Usually, the genetic variants of one gene will interact well with the genetic variants of other genes in the genome: natural selection would have weeded out those combinations of genetic variants that didn't play nice long before we had established the base population.

In other words, the pre-existing genetic variants aren't a reliable source of gene variants to create Dobzhansky-Muller incompatibilities. A new population evolved by artificial selection is unlikely to be genetically distinct from the progenitor species in ways that keep genes from playing nice inside the cells of hybrids that might arise between the new and the old populations. That is, the little *a* and little *b* gene variants interact just

Evolving Tomorrow. Asher D. Cutter, Oxford University Press. © Asher D. Cutter (2023). DOI: 10.1093/oso/9780198874522.003.0009

fine because they were already interacting just fine in the population of the progenitor. We'll need to look elsewhere if we are to aim to evolve new species to have strong genetic barriers to interbreeding. Artificial selection alone just won't cut it in the near term.

★

It is new combinations of new mutations, combinations that nature has never put together before, that have the greatest potential to form negative genetic interactions and to thus provide the stuff of Dobzhansky-Muller incompatibilities. One approach to create new mutations is to boost traditional artificial selection with genetic engineering. We can use genetic engineering and genetic welding, in particular, to evolve new and distinct traits. We also can apply genetic engineering to evolve reproductive isolation to form genetically distinct species.

Some such strategies employ clever permutations of CRISPR-Cas9. Instead of letting the Cas9 protein cut DNA, this method uses a modified version of Cas9, called "dead" dCas9 that can't cut DNA. When you deploy dCas9, it sniffs around the genome like a molecular hound dog with a guide RNA as its nose, tracking down specific locations in the genome without actually biting into them. The dCas9 doesn't actually cut DNA, despite the fact that the guide RNA lets it home in on the right spot in the genome.

This modified dCas9 protein has one other special tweak: the protein sequence is fused to a motif known as a VPR expression activator. When this dCas9-VPR protein snuggles up to a gene in the genome, it causes that gene to be expressed, in this scenario, overexpressed. The dCas9-VPR leads to too much production of the standard protein encoded by the gene that got sniffed out by the guide RNA. In this scheme, in fact, too much of that protein will disrupt embryonic development to be 100% lethal.[3] But we'll first engineer a nonstandard version of the gene in the set of animals of our newly made population. This nonstandard version of the gene will let their genomes evade the dCas9-VPR and its guide RNA. These newly engineered animals are designed to be immune to the lethal effects of dCas9-VPR and its guide RNA. Hybrids that have one copy each of a standard and nonstandard gene version, however, will not be immune.

As a result, even if any mating were to occur between the engineered and wild animals, the hybrid offspring will not be viable: dCas9-VPR will induce lethal overexpression of the crucial embryonic gene copy inherited from the nonengineered individuals. Geneticists term this effect *underdominance*. The engineered organisms are now a reproductively isolated and evolutionarily distinct biological entity. CRISPR-Cas9 genome editing provides a path to engineered speciation and to preventing the transmission of genome edits to wild populations. A difficulty with this technique is that it depends on a severe bottleneck to create the nonstandard gene version plus the CRISPR-dCas9-VPR genetic cassette. This bottlenecking may create problems in evolving other novel genetic features. As proof of principle, however, researchers used this approach to create from scratch eight mutually distinct "synthetic species" of *Drosophila* fruit flies in the lab.

★

Evolutionary responses depend most fundamentally on there being genetic differences among individuals that cause the organisms to develop distinct traits. This dependence is true for artificial selection through selective breeding just as it is for natural selection. In the absence of genetic differences, selection is impotent to produce lasting genetic change. Mutation produces genetic variants where none existed before, and introgression through hybrid crosses can contribute genetic variants, as well. And, of course, genome engineering can create very specific new genetic variants.

Take, for example, teeth. Birds don't have them.[4] But could they? Part of the problem is that the key genes responsible for making dentin and enamel have accumulated inactivating mutations, rendering the gene copies functionless in bird genomes.[5] One solution is to create genetic variants that facilitate the resurrection of bird teeth by employing CRISPR-Cas9 genome editing. A start would be to recode the nonfunctional copies of genes like dentin sialophosphoprotein (DSPP) and enamelin protein (ENAM), among others, with versions that make functional proteins to confer on a bird the capacity to make dentin and enamel.

The next problem is to initiate tooth formation in the first place, such that the right gene pathways get turned on and off at the right times and places during development. Tissue grafts and ectopic expression of activated β-catenin protein show that bird jaws have the potential to instigate tooth development, and a genetic variant of the talpid2 gene does in fact initiate tooth development in chicken embryos—although one severe additional effect of the talpid2 variant includes failure to survive hatching, another example of pleiotropy.[6] At present, there is no clear sightline to engineer the developmental pathway to make fully functional bird teeth. Hen's teeth remain as rare as hen's teeth. This assortment of findings, however, suggests that further research, and genetic engineering ingenuity, may uncover mechanisms capable of reactivating avian tooth development.[7]

★

Biochemists already use this kind of approach in industrial settings, often referred to as directed evolution. By combining genetic engineering and selection at a microscopic scale, biochemists will evolve new forms of proteins that are optimized to efficiently carry out a particular biochemical reaction. Recently, I even received an unsolicited email advertisement from a company that offers directed evolution services to provide engineered proteins to researchers, for a fee. Using computers and genetic engineering in protein design "can initiate sequence space exploration from starting points that are inaccessible to evolutionary processes originating from naturally existing genes," as Michael Packer and David Liu wrote about it, "as a result, it has the potential to expedite the evolution of completely novel protein functions."[8] In other words, so-called rational design coupled to evolutionary principles lets us choreograph the trajectory of change to create entirely unprecedented biological forms.

In principle, this directed evolution approach is simple and now conceptually familiar: apply mutation to the DNA encoding a given protein to create a population of unique variants to then screen for desirable properties to propagate. This procedure can then be iterated to optimize the properties of a protein for a given application. But the conceptual and the practical can sometimes butt heads: for even a minuscule protein of just 10 amino acids in length, there are over 10,000,000,000,000 ($= 10^{13} = 10$ trillion) possible amino acid combinations and over 10^{18} (1 quintillion) possible DNA sequence encodings of those proteins. Most protein enzymes are over 300 amino acids long, which makes for an incomprehensibly large number of possible variations. Needless to say, that is too many to quantify with individual experiments. The trick lies in devising strategies, of which clever experimentalists have proved adept, to let natural selection in a test tube do the hard work of distinguishing among all the distinct genetic variants.

Take, for example, vanilla. Do you like the smell of vanilla as much as I do? If so, we can thank the phenolic aldehyde $C_8H_8O_3$, a chemical otherwise known as vanillin. Originally extracted from vanilla beans, the seed pods of *Vanilla planifolia* orchids, chemists figured out how to synthesize vanillin in the lab in the 1870s. As an intermediate compound for pharmaceuticals and as the biggest flavor additive in use in the world today, it is economically valuable to make it efficiently. Efficiency is helped by discriminating distinct components of the biosynthetic pathway that leads to vanillin. To do just that, Aditya Kunjapur and Kristala Prather melded genetic engineering with directed evolution to obtain a set of enzyme biosensors that permit sensitive tests for the presence of structurally similar, but distinct, byproducts in vanillin biosynthesis to assist in the efficiency of industrial production.[9]

Even more fundamentally novel, other researchers applied genetic engineering and directed evolution to evolve an enzyme that catalyzes a chemical reaction for which no protein catalyst exists in nature. Through the inclusion of directed evolution, the kinetics of this reaction that transfers protons from carbon atoms, known as Kemp elimination, experienced a 400-fold improvement over the originally engineered enzyme.[10] Still others use directed evolution in the creation of proteins so novel that they incorporate so-called unnatural amino acids into their structure, that is, proteins that successfully make use of amino acid molecules other than the 20 found in the standard genetic code of life.[11]

★

Biochemists use directed evolution with phage viruses and bacteria and single-celled yeast. But the principles apply to larger organisms, too. Instead of microbial tricks of molecular biology, there is CRISPR-Cas9 genome editing of animals or plants. When engineered into a gene drive, you can inject a specific genetic novelty into the genetically variable population of genomes of a large collection of animals. That animal population contains a natural library of potential genetic modifiers to then tune the engineered novelty through artificial selection. The novel genetic variants to inject into the directed evolution milieu could be entirely new genes, synthesized from scratch with a computer algorithm. Or they could be genes composed of DNA sequence extracted from another species that branches off elsewhere in the tree of life. The genes could even be sourced from sequences of ancient DNA exhumed from permafrost and subfossil bone marrow. We'll hear more about this idea in upcoming chapters.

We can imagine trying this approach to evolving hen's teeth. Chicken genomes encode a large collection of genes that are integral to signaling cell growth in the jaw. Chicken populations contain huge quantities of genetic variants encoded in other parts of the genome, some of which are likely to influence those signaling pathways. In particular, those natural genetic variants could act as modifiers of the effect of genetically engineered changes, either amplifying or dampening the consequences of the genetic edit. The engineered edit to the genome would give the animals a new starting point from which subsequent evolutionary change can take place. This chain of interaction would set the stage for artificial selection to work to enhance the outcome of genetic engineering.

★

Directed evolution integrates these two seemingly distinct things: artificial selection on a large population of individuals and genetically engineered variants that confer novel effects on an organism's characteristics. The evolutionary result would truly be pulled from the roots of meaning of that word "creature": something created.

It's a marriage of convenience, of sorts. It sings the old rhyme about bringing to the wedding day something old, something new, something borrowed, something blue. The old thing is the species as founding stock that gets used as the base population for directed evolution. The new is the introduction of mutations and engineered genetic elements. Borrowed are the sequences read from other species that hint at what they might create in a creature without them. And then there is blue. The blue hazard of unexpected outcomes and unintended consequences.

Some of these blue hazards are technical in nature. The many practical details of getting a novel genetic cargo in a gene drive to push its way through a population may prove trickier than a paper plan might otherwise suggest.

Other blue hazards present as code blue alarms. A given genetic edit might predispose the creature to establishing itself as an invasive species in undesired habitats. After all, invasive species are responsible for astounding biodiversity loss, meting out extinctions of local species through predation, competition, disease, or disruption of habitat. Hybridization is also on the radar of concerned citizens and policymakers as a mechanism that provides a potential genetic bridge between bystander populations of organisms and engineered organisms. Some genetic modifications, in particular, may be predisposed to introgression into nontarget populations or species—with undesirable consequences. In upcoming chapters, we will expose these ecosystem, health, and ethical hazards to more fulsome investigation.

★

Despite the importance of natural selection in speciation, artificial selection alone is unlikely to permit new species formation on short timescales. It is possible to use genetic engineering to create reproductive barriers between populations. Genetic engineering can create a new genetic starting point for a population that allows novel trajectories of evolutionary change and the evolution of unique properties. Genetic engineering and directed evolution, in combination, offer a path to exploring the birth of reproductively isolated populations of creatures with entirely novel biological properties.

Notes

1. The genetic variants responsible for Frizzle and Silkie chicken feathers were traced to genes involved in keratin production and affecting feather development in skin cells. Ng, C.S., et al. 2012. The chicken frizzle feather is due to an α-keratin (KRT75) mutation that causes a defective rachis. *PLoS Genetics.* 8: e1002748. https://doi.org/10.1371/journal.pgen.1002748; Feng, C., et al. 2014. A *cis*-regulatory mutation of PDSS2 causes silky-feather in chickens. *PLoS Genetics.* 10: e1004576. https://doi.org/10.1371/journal.pgen.1004576; Patton, J., et al. 2007. Relationships among milk production, energy balance, plasma analytes, and reproduction in Holstein-Friesian cows. *Journal of Dairy Science.* 90: 649–658. https://doi.org/10.3168/jds.S0022-0302(07)71547-3.

2. Pacheco, C., et al. 2017. Spatial assessment of wolf-dog hybridization in a single breeding period. *Scientific Reports.* 7: 42475. https://doi.org/10.1038/srep42475; Monzón, J., et al. 2014. Assessment of coyote-wolf-dog admixture using ancestry-informative diagnostic SNPs. *Molecular Ecology.* 23: 182–197. https://doi.org/10.1111/mec.12570.

3. In practice, one would engineer multiple guide RNAs for dCas9-VPR to target multiple genes. This multigene reproductive isolation would ensure that simple mutations in other populations would not lead to genetic leakage through hybridization. Buchman, A., et al. 2021. Engineered reproductively isolated species drive reversible population replacement. *Nature Communications.* 12: 3281. https://doi.org/10.1038/s41467-021-23531-z.

4. It's true that males of the tooth-billed hummingbird *Androdon aequatorialis* look toothy, but the serrations on their 4-cm (1.5- inch) beaks are not true bony teeth, instead being barbs made of keratin on the edge of their beaks. Rico-Guevara, A., et al. 2019. Shifting paradigms in the mechanics of nectar extraction and hummingbird bill morphology. *Integrative Organismal Biology.* 1: oby006. https://doi.org/10.1093/iob/oby006; Schuchmann, K.-L. 1995. Taxonomy and biology of the tooth-billed hummingbird (*Androdon aequatorialis*). *Mitteilungen aus dem Zoologischen Museum in Berlin.* 71: 109–113.

5. Meredith, R.W., et al. 2014. Evidence for a single loss of mineralized teeth in the common avian ancestor. *Science.* 346: 1254390. https://dx.doi.org/10.1126/science.1254390.

6. A deletion mutation to the C2CD3 gene is responsible for the talpid2 genetic variant, which imposes a diversity of effects on cranio-facial development, another example of pleiotropic influence of single genetic changes. The way that the bone morphogenic protein 4 (BMP4) gets expressed in jaw development is also especially influential in the start and stop in the development of bird dental structures. Harris, M.P., et al. 2006. The development of archosaurian first-generation teeth in a chicken mutant. *Current Biology.* 16: 371–377. https://doi.org/10.1016/j.cub.2005.12.047; Chang, C.-F, et al. 2014. The cellular and molecular etiology of the craniofacial defects in the avian ciliopathic mutant talpid2. *Development.* 141: 3003. https://dx.doi.org/10.1242/dev.105924; Chen, Y., et al. 2000. Conservation of early odontogenic signaling pathways in Aves. *Proceedings of the National Academy of Sciences USA.* 97: 10044–10049. https://doi.org/10.1073/pnas.160245097.

7. Louchart, A., and L. Viriot. 2011. From snout to beak: the loss of teeth in birds. *Trends in Ecology & Evolution.* 26: 663–673. https://doi.org/10.1016/j.tree.2011.09.004.

8. p. 391 in Packer, M., and D. Liu. 2015. Methods for the directed evolution of proteins. *Nature Reviews Genetics*. 16: 379–394. https://doi.org/10.1038/nrg3927.

9. Kunjapur, A.M., and K.L.J. Prather. 2019. Development of a vanillate biosensor for the vanillin biosynthesis pathway in *E. coli*. *ACS Synthetic Biology*. 8: 1958–1967. https://doi.org/10.1021/acssynbio.9b00071.

10. Khersonsky, O., et al. 2011. Optimization of the in-silico-designed Kemp eliminase KE70 by computational design and directed evolution. *Journal of Molecular Biology*. 407: 391–412. https://doi.org/10.1016/j.jmb.2011.01.041.

11. Brustad, E.M., and F.H. Arnold. 2011. Optimizing non-natural protein function with directed evolution. *Current Opinion in Chemical Biology*. 15: 201–210. https://doi.org/10.1016/j.cbpa.2010.11.020.

10

Ongoing evolutionary outcomes

On one particular bright afternoon in March, I learned about teeth. From 20 m down, I gazed up through dive goggles at the toothy maw of a gray reef shark. Its jaws hung agape beneath a hard-set pair of stony eyes, pushed through the water by the effortless to-and-fro swish of a long tailfin. The shark's hydrodynamic and cartilaginous body skulked along in what now seemed an impenetrable column of water between me and the surface of the Coral Sea. Saltwater snorkelers and SCUBA divers will tell you about swimming with sharks. The teeth of sharks. Teeth that are sharp-tipped and multitudinous.

Let me use all those teeth as an opportunity to tell you some short stories of evolution. Let me tell of how teeth in all their unsung variety speak to how the forces of evolution can leave their mark, and how you can hear what they say, if you know how to listen.

10.1   Show me your teeth

"Show me your teeth," as the founding father of paleontology George Cuvier is credited to have put it, "and I will tell you who you are."[1] Despite the old adage that you are what you eat, the evolution of teeth, it turns out, involves many more activities than just chowing down. Teeth aid in digging and burrowing, in grooming, in making sounds and displays for communication, in moving oneself around and in moving other objects, including the most precious of cargo: offspring.[2] Males of the tooth-walking sea horse, or so it translates from the binomial *Odobenus rosmarus*—more commonly known as the walrus—will use their nearly 1-m-long (3-foot-long) tusks to help pull up onto ice-shelves and to smash holes into ice for ready access to the sea. They also use their tusks in battle with other males in sexual contests, a toothsome role shared with many other tusked mammals akin to how horned mammals use their headgear in mating displays and combat. In the narwhal, what looks like a horn is the single sinistral tusk, an elaborate 3-m-long (10-foot-long) twisted tooth, its pointy spiral impregnated with nerve endings and capable of detecting diverse changes to the outside environment. Teeth, as Cuvier claimed, tell us who you are—what evolutionary pressures you and your ancestors had to face.

Teeth, then, are not just the darlings of dentists. Being 97% mineral, these prefabricated fossils are darlings of evolutionists, as well. In paleontological and archaeological

Evolving Tomorrow. Asher D. Cutter, Oxford University Press. © Asher D. Cutter (2023). DOI: 10.1093/oso/9780198874522.003.0010

remains, their size, their orientation, their crenulations all provide clues to function and adaptation. Each of the hundreds of teeth in a great white shark's mouth features serrations along each tooth edge, an adaptation to effective slicing of flesh. The mouths of sharks continually shed and regrow a fresh phalanx of sharp snags throughout their life,[3] the next set and the 5-to-50 next after that tucked one behind the other in a Rolodex of dentition (Figure 10.1).

★

What other ways can teeth tell us about who we are? The dentition of humans with our succession of two sets of teeth, infant and permanent, makes us the evolutionary recipients of a so-called diphyodont mouth ("double toothed" from the Greek).[4] In rodents, the temporary first mouthful come and go in utero. The permanent set of teeth then arrive by birth. The so-called gliriform incisors are the famed buck teeth of beavers and rats that continue to grow throughout life, worn down to a stable length through perpetual gnawing. The long canines of carnivorous mammals like lions and wolves serve to hold and tear prey, whereas the ridge-like arrangement of sharp cusps among their rear carnassial teeth are specialized for slicing. By contrast, an elephant's tooth, one of just four inside its mouth, plus the tusks, is the size of a cobblestone. Its broad surface is patterned with a topography of coarse cusps like the sole of a hiker's boot and ready-made for grinding rough vegetation. Its tusks, on the other hand, up to 2 m

Figure 10.1 *Jaws of a shark showing its multiple rows of teeth, set on a bed of fossilized shark teeth that I collected as a child from beaches on the East Coast of North America.*
Photo by the author.

(7 feet) long, are maxillary multifunctional tools: prybar, battering ram, sexual display, and weapon.

Evolution's trajectory in other animals led to them growing just one set of teeth. The technical term is monophyodonty. Monophyodonts, like sperm whales and orcas, sprout a single mouthful of nearly conical peg-like teeth to last their entire long lives. Their brethren the baleen whales begin to grow teeth as embryos, teeth that then resorb, as if their developmental program evolved the ability to press a biological "unsend" button on an ill-conceived "reply all" message. An embryonic humpback's fetal jaws, after that brief developmental dabble in mineralized dentition, subsequently give way to development of a different hard secretion from the upper mandible. A baby humpback whale cries its way into the world with jaws that sprout baleen plates composed of keratin, that dense yet pliable protein in fingernails and hair and rhino horn and beaks.

Like the baleen whales, the evolution of bird mouths has eschewed mineralized calcium. They've put aside the perks of hydroxyapatite that make a tooth so very hard after its deposition in forming the thick inner layer of dentin and outer layer of enamel. Birds have no teeth at all, despite their dinosaur ancestry. And yet, they still successfully chew and slice and crush their food. Birds' horny beaks are made of keratin, forming a structure known by the flock of the in-crowd as the rhamphotheca. In raptorial birds like osprey and eagles, of course, the beak replaces the role of sharp teeth, and the gizzard in their guts substitutes for the grinding of molars. Even toothless plants have their own evolutionary paths to getting in on tooth-like action: the cagey jaws of the Venus flytrap enable the capture of insect prey. From the mouths of a red hawk and a red fox on the outskirts of town, they might tell you, after snagging a feral *Felis catus*, there is more than one way to skin a cat.

Within just a single species even, we can see the drastic evolutionary change in the morphology of the mouth. Case in point: the snouts of dogs. Perhaps you are on the best of friendly terms with a Bull Terrier and its characteristically downward-sloping muzzle? It was shaped by breeders selecting on what turned out to be a gene known as Runx-2, altering the skull's shape over a period of some 45 years. Dog breeders favored, unbeknownst to them, particular gene variants of this Runx-2 protein, a protein known as a transcription factor that alters the expression of a cascade of other genes to shape cranial bone development to confer that special snout shape of today's Bull Terrier.[5] The short-faces of Pugs and Pomeranians is due, in large part, to human-mediated artificial selection, too. In this case, the flat-faced brachycephaly involves a genetic difference associated with the SMOC2 gene.[6] How might you instead, should you so choose, produce such abbreviated snouts in, say, a Saluki greyhound—an exceptionally long-nosed breed? You could go through the tedious task of cross-breeding and artificial selection, as dog aficionados have done for centuries, imposing the human-mediated counterpart to natural selection. Or, these days, one could induce heritable changes even faster, simply by performing genome editing on genes like Runx-2 and SMOC2 in Saluki embryos.

Nature may be red in tooth and claw, but teeth, at least, are painted with an entire rainbow of evolutionary colors.

★

Animals and plants can inherit changes, whether through genetic engineering or through the traditional evolutionary forces. The change may be uncontrolled and uninfluenced by human activity, or it may be intentional and directed. Change may even accrue in an unintentionally directional way, as a side-effect of human-mediated perturbations to the living circumstances of a creature—as we'll see in Chapter 14 with the evolution of tusklessness among the elephants of Mozambique subjected to poaching. Such evolution can rapidly diversify and differentiate any characteristic, not just jaws and teeth. Form and physiology, body size and behavior—all of it is fair game for evolutionary change.

10.2   Inputs and outputs

In Book II of his 350 BCE treatise on the soul, *De Anima,* Aristotle touted the existence of five fundamental senses that scope the adjectives of perception and lived experience. Sight, sound, smell, taste, and touch. This convenient pentad, however, is both too few and too many. Too few because modern philosophers count over twenty. What Barry Smith calls the "symphony of senses" includes such things as equilibrioception (sense of balance), nociception (sense of pain), thermoception (sense of temperature), proprioception (sense of your body parts in space), and agency (sense of control over one's actions).[7]

We could also, however, distill the senses down to just three fundamentals. These three types correspond to the three basic kinds of sensory receptors connected to animal nervous systems, the cells and biomolecules that detect and discriminate among the diverse physical attributes of the world. This triad is made up of electromagnetic receptors (sensing light or magnetic fields), mechanoreceptors (sensing physical perturbations), and chemoreceptors (sensing gaseous or liquid chemicals). The key feature of all these receptors is that they convert diverse kinds of inputs into a common language: the electrochemical language of neurons. The neurons speak with different dialects to other neurons and to muscle cells and to specialized cells in organs to arrive at an internal understanding of the world around us and to induce responses to it. Animal development, shaped by evolutionary change, acts as a translator between the language of DNA and this language of neurons.

★

The electromagnetic receptors are responsible for perceiving the various forms of light energy. This includes the photoreceptors in eyes, which contain the best characterized sensory receptor proteins of any yet examined: the opsin proteins that respond to light and that let us see.[8] Opsin proteins nest within a large family of proteins found throughout animals known as G-protein coupled receptors, a.k.a. GPCRs. GPCRs are distinctive in having a physical structure that passes sinuously from one side of a cell membrane to the other, snaking in and out of the cell surface with a stereotypical set of seven so-called transmembrane domains (Figure 10.2). By sitting on the boundary of the inside and outside world, they are perfectly positioned to convert external signals from the environment into biochemical activity within the cell.

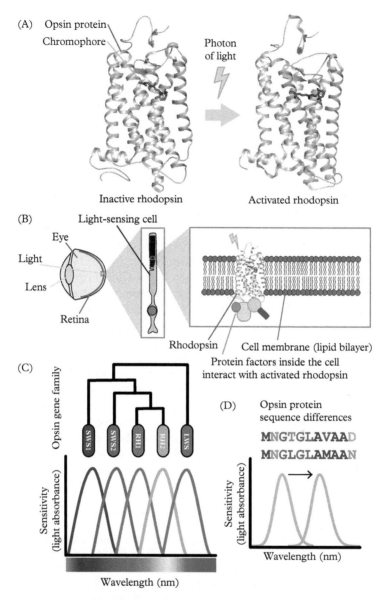

Figure 10.2 *(A) Opsin proteins are a special kind of G-protein coupled receptor that can trigger biochemical activity by absorbing photons of light, by virtue of the protein's interaction with a retinal chromophore molecule. The model of the opsin molecule shown here illustrates the seven transmembrane domains within it that each form a spiral shape, known as an alpha-helix. (B) The combined molecule of the opsin protein bound to the chromophore, termed rhodopsin, changes shape to create a biomolecular cascade that leads to visual perception upon rhodopsin's absorption of a photon of light, after the light*

continued

Opsin proteins gain their light sensitivity by bonding with pigment molecules called retinal chromophores. This tag-team duo, together termed rhodopsin, absorbs photons of specific wavelengths of light. When rhodopsin gets hit by a photon of the right wavelength, it jerks its posture into a new conformation like an at-ease soldier getting hit by a drill sergeant's holler of "attention." Rhodopsin is so sensitive that two out of every three photons that hit it will elicit that conformation change. Absorbing a photon makes the protein alter its shape in a way that triggers a cascade of amplifying biochemical interactions. This cascade climaxes in the perception of light by our brains.[9]

Different opsins and chromophore molecules can link up together to produce combinations that are tuned to respond to different wavelengths of light. These distinct combinations allow animals to interpret those distinct wavelengths as distinct colors.[10] One thing that my colleague Belinda Chang has taught me is that opsin experts like her measure rhodopsin sensitivity with a metric called λ_{max} ("lambda max"), the peak wavelength of light that can get absorbed. Many vertebrates, including some birds and fish, have extreme versions of the violet short-wavelength-sensitive opsin protein SWS1. For example, canaries and budgies have vision beyond violet, in the UV-A spectrum, because of λ_{max} values as low as 360 nm. In insects, the thrips *Caliothrips phaseoli* is even more sensitive, sensing UV-B wavelengths as short as 290 nm.[11] At the upper, red end of the spectrum, the λ_{max} of the long-wavelength-sensitive opsin protein typically hovers near 560 nm, permitting sensitivity to scarlet hues approaching 700 nm.

Why is vision focused primarily on wavelengths between 300 nm and 700 nm? There may be a good reason, it turns out. These wavelengths are the most abundant and biologically accessible wavelengths provided by the Sun. Of all the electromagnetic radiation from the Sun that hits the Earth's surface, roughly 42% correspond to this visible spectrum, which also matches the spectrum used by chlorophyll in plants to make chemical energy from sunlight.[12] Most of the remainder is made of lower-energy longer wavelengths, primarily in the form of infrared heat energy (wavelengths up to about 1 mm). Even more extreme wavelengths make up just a small fraction of the

---

Figure 10.2 *continued*

passes through the eye to the light-sensing photoreceptor cells of the retina in which many such rhodopsin molecules are embedded into cell membranes that form disc-like structures. (C) Opsin genes are part of a family of genes in animal genomes that are sensitive to different wavelengths of light. Humans have a copy of RH1 ("rod opsin" used for dim-light vision), SWS1 (short-wavelength sensitive), and two slightly different duplicate versions of LWS (long-wavelength sensitive, one is often referred to as mid-wavelength sensitive MWS; both are encoded on the X chromosome). Humans do not have versions of RH2 or SWS2 that some other animals have. (D) Mutations that create different opsin protein sequences influence what wavelength of light can be detected by that opsin protein, and different species can encode different numbers of opsin proteins in their genomes that permit them greater or lesser sensitivity across the visible spectrum.

Images modified with permission from Hauser, F.E., and Chang, B.S.W. 2017. Insights into visual pigment adaptation and diversity from model ecological and evolutionary systems. *Current Opinion in Genetics & Development.* 47: 110–120. https://doi.org/10.1016/j.gde.2017.09.005.

wavelengths emitted by the Sun, which mostly get absorbed, reflected, and scattered by Earth's atmosphere.

Animals like sperm whales have genomes that encode just a single opsin gene. They can see just fine, but in a way analogous to the shades of gray of an old black-and-white television. We humans have three opsins used in bright-light "trichromatic" color vision, plus one used for vision under very low-light conditions. Most day-active vertebrates, however, have a total of five, like chickens, with their tetrachromatic color vision giving finer sensitivity to a broader range of wavelengths of light.[13] A rainbow under our rainbow.

Some animals have dozens of opsin genes. The silver spinyfin fish *Diretmus argenteus*, a 10-cm (4-inch) disc fish that looks rather like a dented chrome hubcap, has built up dozens of opsin genes over the course of its evolutionary history. Each additional copy arose from a mutation that duplicated the DNA sequence of an existing opsin gene, a mutation that then survived the chance fluctuations of genetic drift during its initial rarity to grow in abundance across generations by conferring an adaptive benefit on those fish that expressed that extra opsin, until, eventually, it became a fixed feature of the genome of the species. All those extra opsins give *Diretmus* sensitive color vision of bioluminescence in the dim light of the deep sea, thought to have enhanced their ability to detect minute bioluminescent crustacean prey.[14]

★

At the bottom of the sea, there is very little light. Red wavelengths of light, especially, get absorbed by the thick water column between the surface and the seafloor. Consequently, deep-sea creatures often have lost all sensitivity to red light. Evolutionary losses like this arise when natural selection can no longer notice the effects that new mutations have on some aspect of an organism's biology. Mutations arise by chance in opsin genes—just like in any other kind of gene. Some of those mutations disrupt their function to, say, impede the detection of red light. If organisms suffer no ill effects with that disrupted function, say, because they do not normally encounter red light, then the genetic variant won't be purged from the population by countervailing natural selection. Such so-called relaxed selection ensues to give genetic drift a greater say over the evolutionary fate of mutations. Genetic drift along with additional disabling mutations can then lead to permanent elimination of that gene from the genome of all members of the species. This evolution by gene loss is a common phenomenon for genes involved in red-light sensitivity among species that live at the bottom of the ocean.

The stoplight flashlight loosejaw dragonfish—they have an impressive repertoire of nicknames—have exploited the fact that most deep-sea creatures can't see red. These predators have evolved astonishing vision enhancement. Like nearly 5% of fish species, they have bioluminescent glands that can emit light. But the bioluminescent glands of *Malacosteus* dragonfish are positioned below their eyes, and these glands emit red light: not as a beacon or lure, but as a spotlight. Their retinas, unusual for deep-sea creatures, can detect the red light that reflects back off of prey. Their retinas contain a photoreceptor with a photosensitizing pigment that gives an exceptionally high λ_{max} value of 673 nm that may push light detection to nearly 800 nm, in the realm of the infrared.[15] This far-red spotlight vision puts prey within striking distance of their loosejaw mouth, but the prey themselves are blind to the gleam of red bioluminescence.

The jaws of these sit-and-wait predatory dragonfish have evolved into a kind of living harpoon. *Malacosteus* jaws are so very modified that they have a complete "lack of a floor to the oral cavity," meaning that there is no tissue connecting the two sides of their lower jawbones beneath their chins.[16] If you tried to scratch their chin, you'd tickle the roof of their mouth. This skeletal jaw structure minimizes hydrodynamic drag as it opens and closes its mouth, an open mouth meaning that the lower jaw has jutted forward an alarming distance from the rest of the body. Dragonfish like *Malacosteus niger* can eject their lower jaw forward incredibly rapidly at a gape angle opening up to 120° wide, their upper mandible and skull tipping upward to widen the gape. This retractable harpoon of a jaw is built for speed, its form shaped as if drawn from sci-fi imaginings, a reminder of the extremes that evolution can take, more arresting even than the jutting chin of the Habsburg jaw among seventeenth century aristocrats. The fang-tipped jaws of *Malacosteus* make up to 30% of the length of the fish and allow them to eat prey in one gulp that are half as big as they are.

★

An intrepid movie-goer around 1960 might have had the opportunity to infuse their film thrills with not just sight and sound, but scent as well. Despite the power of olfaction in defining immersive experiences and to trigger deep memory, however, Smell-O-Vision and Odorama never took off as a complement to enhancing the movie-watching experience in the theater.[17] Maybe a paper card with 10 synthetic scratch-n-sniff odors just isn't sophisticated enough to make it worthwhile. After all, the human nose can distinguish more than a trillion odorant stimuli. At least, some of us can. People differ nearly one sextillion-fold in their olfactory discrimination ability.[18] This sensitivity, and variation in it, is due in large part to the expansive set of olfactory receptors that we express deep inside of our noses. Chemosensory olfactory receptor proteins are similar physically to opsin photoreceptor proteins, in that they are also G-protein coupled receptor (GPCR) proteins. Unlike the fistful of opsin genes encoded in our genomes, however, the human genome contains nearly 400 functional olfactory receptor proteins. Before you pat yourself on the back for having such a robust odorant repertoire, however, consider the African elephant with more than 2500 functional olfactory receptor genes in its genome.[19]

These odorant receptor proteins work by a kind of chemical lock-and-key mechanism. A chemical from the environment slips into the lock of the olfactory receptor to trigger a cellular latch that instructs the nervous system that it has detected the presence of that chemical. Most of these olfactory locks will take many differently shaped chemical keys, however. That is, an olfactory receptor can respond to many different chemicals, making them biochemically promiscuous and broadly tuned to multiple odorants. And our noses have millions of such cellular locks, so the distinctive combination of which shapes of keys unlatching which set of cell's locks gives a given combination and concentration of chemical compounds its distinctive scent. This is the stoichiometric chemical key to stink.[20]

Many animals can navigate by smell alone. A mouse may twitch its nose to and fro and scurry back and forth to learn the direction of an odorant source. Sitting still, snakes smell in stereo. The aggregation of chemoreceptors in the tips of each side of a snake's forked tongue informs the animal about subtle differences in left versus right concentrations of scents in the air.[21] Tongues and noses aren't the only organs that one might use to smell and taste, however. You might use your teeth, at least if you were a narwhal. That long unicorn horn of a tusk extending 2.5 m (8 feet) in a long left-handed spiral is actually a highly modified canine tooth, usually the left upper canine. The externally erupted *Monodon monoceros* narwhal tusk is infused with sensory nerve cells capable of detecting fluctuations in salinity, and perhaps other environmental features like temperature, pressure, and water chemistry. But it is males that are tusked, for the most part, and bigger male narwhals have disproportionately long tusks—an example of hyper-allometry that implicates sexual contests or mate choice as a driving force in its evolutionary elaboration.[22]

The ultimate organs for olfactory sensation, however, are the baroque antennal forms of insects.[23] As elaborately enormous as antlers—relative to their body size—these feelers of airborne chemistry often are most attuned to the perfumes of sex. Olfactory receptors of the fluffy white and petite *Bombyx mori* male silkmoths, for example, sniff the air, so to speak, with an odor acuity 100 times more sensitive than a Bloodhound. Resting with the contemplative look of Yoda, their pair of antennal combs orient in long drooping arcs alongside their head to pull out the spice of a female's bombykol sex pheromone in concentrations as low as three molecules per microliter, roughly one in a hundred trillion molecules of air.[24] A chemical key rarer than a needle in a haystack, but infinitely more precious, if you are a silkmoth.

★

All of the sensory modalities of animals are geared toward communing with the world, whether to eavesdrop on what the environment can tell us or to listen to one another. We communicate with our own species and with others, emitting intentional and unintentional transmissions of information to teach, so to speak, and to learn. Animals listen in a multimodal way for the unintentional transmissions: for the sound of footsteps and flapping, the flitting of flippers or flash of light, the stinky funk of musk or fresh stems, the warmth of flesh. And we animals *intentionally* send signals. Despite the gloriously diverse ways that animals use sight, sound, touch, and smell to talk to one another, the gist of an intentional message often boils down to just a few simple statements and questions. Go away. Come here. Where are you? Be careful. Feed me. Follow me. Fornicate with me.

You can say those things in song, if you are a songbird or a whale, or with a roar, with clicks and croaks, with varying pitch and loudness. You can say those things with your physical form or behavior. The mane of a male lion says one thing, the elaborate dance choreography of a male manakin bird with his bright-bodied boy band buddies says another. Sometimes touch is more eloquent, or so think the prospective mates of spiders and flies both as they feel the vibratory elocutions of lovelorn courting males of their own respective species. The smell of sulfur thiols on a twig marked by a fox tells his neighbors, from a distance, to keep their distance; the musk of urine from an elephant in estrus speaks the poetry of her sexual allure;[25] and the marks of an army ant scout make her a veritable Pied Piper to recruit millions of her sisters to trace her footsteps.

★

Sometimes we use sensory transmission to do the opposite of communicating, instead producing masking signals to evade sensory detection. Crypsis instead of conspicuousness. You may want to hide from predators to avoid being eaten. You may want to hide from prey, to creep up close, or to let them unintentionally creep up close to you. Crypsis can exploit any sensory modality, but we often find the most astonishing cases in the visual realm.

Perhaps the most austere way to avoid being seen is to not have any color at all. To be transparent, from a human vantage, is a superpower. But transparency is not so uncommon: it is found in baby eels and in the glassy wings of clearwing butterflies, in the comb jelly ctenophores and shrimp-like amphipods of the sea, in the glass octopus, and in the belly skin of glass frogs. Invisibility is a behaviorally passive trick to evade predators by forging a body without opacity. Optically transparent animals operate on the anti–Field of Dreams principle that, if you don't build it, they won't come.

There are, however, other ways to hide. In the trinket shops of seaside towns along the Gulf Coast of the Florida panhandle, you will find tourist mementos emblazoned with declarations that "life's a beach." For the beach mouse *Peromyscus polionotus*, it's just a statement of fact. The picturesque and brilliant white sand of the dunes and barrier islands along Florida's coast are inhabited by several subspecies of these mice, mice with pale fur, a blanched white over much of their body. The mice of Santa Rosa Island—for those in the know, called *Peromyscus polionotus leucocephalus*—are especially pale. Inland, away from the pure white quartz sand deposits, *P. p. subgriseus* mice, instead, have a darker tan pelage which blends in with the brown loamy soil of meadows and oldfields. In experiments with clay model mimics of mice, dark mice on white sand and light mice on brown dirt get attacked by predators at a rate triple that of mice that matched the color of the ground.[26] Predators like an easy meal. Mouse prey that evolved cryptic background matching make for an effective way to avoid such a fate.

Much of the difference in this local adaptation of mouse color, Hopi Hoekstra and her research team discovered, develops as a consequence of which set of genetic variants a given mouse has in the melanocortin-1 receptor gene, abbreviated as MC1R. In this case, it's due to a single C to T mutational difference in their DNA that controls the identity of the amino acid at position 65 in the MC1R protein, encoding either arginine or cysteine. Two copies of the cysteine variant makes a mouse pale by depressing the ability of MC1R to induce melanin pigment production in hair follicles. Among mice on the beach, a whopping 95% of the genome copies encode the light-conferring genetic variant with a cysteine at position 65 in MC1R. The crystalline quartz sand dunes only became part of the Florida landscape in the last 4000 to 6000 years, it turns out. This geological timeline tells us that the biological divergence, the local adaptation in cryptic coloration that matches beach mouse to beach environment, arose in just a short span of recent millennia.

Regardless of the way that such incredible fits of form to an organism's circumstances end up, they trace the simple genetic path of all of the natural world's adaptations. They start with mutation. Some of those mutations influence the shape of their body or the expression of color in a particular part of the body. Those genetic variants that altered shape or color in a way that led the animal to survive and reproduce better were preserved, and the trait got elaborated further by subsequent mutation and natural selection. The mutations that proved themselves to be beneficial increased in abundance in the population to become a defining genetic feature of the genome of the entire species. When multiple genetic variants in different genetic locations occurred in the species at the same time, recombination shuffled their combinations, sometimes creating combinations with higher fitness than previous combinations. Repeated again and again—mutation, selection, recombination, fixation—the genomes within each individual came to encode the ability for the body to match the features of the environment with incredible exactitude.

★

What lessons might we take from the sensory systems that living creatures have evolved as we think about how one might shape the course of evolution in creating a new species? The molecular basis for detecting light and chemicals depends crucially on genes that encode GPCR proteins, whether opsins or odor receptors. Perhaps genetic engineering of the molecular properties of individual opsins or an increased repertoire of GPCRs in a genome, sourced as transgenes from other species, could modulate the sensory capabilities of an organism of interest. Those properties could modify the range of light wavelengths detectable to values beyond what the organism currently is capable of or provide finer discrimination ability to distinguish all the shades of red or to detect the faintest and most sophisticated bouquet of scents.

10.3   Big and small

The size of an animal depends, in large part, on the size and number of cells that accumulate over the course of its development.[27] The proliferation in number and the expansion in size of individual cells, in turn, depend on growth-factor molecules that usually get produced elsewhere in the body. Mammalian pituitary glands, for example, make and secrete growth hormone that then induces other organs to make another protein hormone, insulin-like growth factor 1 (IGF1). IGF1 then initiates a cascade of molecular commotion inside still other cells by triggering the phosphatidylinositol-3 kinase protein (PI3K) to make those cells grow. Mitogen molecules activate proteins known as cyclin-dependent kinases to cajole cells into dividing by progressing through their cell cycle more quickly. Higher concentrations of these activator proteins, in combination with lower concentrations of inhibitory proteins, lead to faster proliferation of tissues. As tissues get bigger, however, they produce more and more inhibitors, creating negative feedback that slows growth by counterbalancing the activators.

The size of a tissue that results, and of an animal as a whole, is the balance point in a biochemical tension. Size is the balance between the simultaneous activity of molecules that promote growth and those that suppress growth. Disruption of just a single gene can sometimes shift this balance, as for the example of the myostatin protein in mammals.[28] Cows and dogs and sheep and pigs, and even humans, can have one or both copies of the myostatin gene rendered nonfunctional by mutations. Myostatin's normal role in muscle cells is to inhibit growth, and so its absence tips the balance toward excessive muscle development. Termed *double-muscled*, cattle with these variants will genetically turn a humble cow's "dad-bod" physique into what looks like a steroid-enhanced bovine bodybuilder's. Unchecked proliferation and growth usually, however, is not so much a recipe for gigantism as it is a recipe for tumors and cancer.

Given the extraordinary cell proliferation that went into making a whale or sauropod dinosaur, and that each cell division comes with the risk of mutations leading to cancers, do such behemoth and long-lived creatures experience a greater incidence of cancer? The answer, apparently, is "no."[29] This lack of a link between body size and cancer risk across species, known as Peto's paradox, is thought to result, at least in part, from the parallel evolution of more potent mechanisms of tumor suppression along with the evolution of large size. This evolutionary relief doesn't apply, however, to organisms of different size *within* a species. In humans, for example, every 10 cm (4 inches) of extra height comes with a 10% extra risk of developing cancer.[30]

★

Growth that produces big animals can't help but capture our attention.[31] The biggest bird, the common ostrich, seems much less common when up close and personal as you face the beak at the front of a specimen that is 2.8 m (9 feet) tall atop 150 kg (330 pounds) of bone, muscle, and claws. An ostrich is impressively big even if its wingspan can't rival the 3.5 m (11.5 feet) of a *Diomedea exulans* wandering albatross or the largest bird ever, which was nearly twice as big: the now 25-million-years extinct albatross-like *Pelagornis sandersi* that achieved a 6.4-m (21-foot) wingspan.[32] The biggest fliers in history, though, were behemoth species of pterosaur like *Quetzalcoatlus* that could boast wing spans up to 11 m (36 feet).[33] On land, they walked on all fours, using their three normal fingers at the midpoint of their wing as front feet, stilt-legged almost like a giraffe.[34] The landlubber Komodo dragon hits 3 m (10 feet) in length as these longest lizards alive hunt amid the hot grasslands of a smattering of Indonesian islands. And who could forget those big South American rodents of unusual size, the capybara *Hydrochoerus hydrochaeris*, as heavy as a man in a rat suit?[35] The largest living land mammal is, of course, the African bush elephant, which weighs in at 8000 kg (17,600 pounds) and 4 m (13 feet) at the shoulder.

The oceans house even more massive creatures: 30-m-long (100-foot-long) blue whales,[36] 19-m-long (60-foot-long) whale sharks, 13-m-long (43-foot-long) giant squid, and the 30-m-long (100-foot-long) or more tentacles trailing the 2-m (6.5-foot) diameter bell of a lion's mane jellyfish. The plant kingdom has its giants, too. The height of a coast redwood *Sequoia sempervirens* of North America can reach 115 m (380 feet), though Asia and Australia also have tree species that grow to 100 m (330 feet) tall.

Their true size, of course, would also include the underground root mass. A single root system of quaking aspen is thought to connect about 47,000 trunks into a single organism of *Populus tremuloides*, a "tree" nicknamed "Pando." Pando covers an area over 436,000 square meters (108 acres), making it the largest known organism on Earth.[37]

What is it that drives the evolution of such gigantism? Two things continuously favor big size in most animal species: natural selection on females to make more offspring and sexual selection on males to outcompete mating rivals.[38] The idea is that female animals that are larger have more resources to invest in producing offspring, whether through larger litters or clutches, more frequent reproduction, or larger individual offspring that would have higher likelihood of survival. On the male side, larger males experience greater mating and reproductive success in many species, either as a result of competition between males or from female preference. For both sexes, then, characteristics will be adaptive that allow the more effective acquisition of food to attain larger size.

Usually, however, these benefits of being big are counterbalanced, which is why most species don't evolve to grow into giants. For one thing, growing to be large usually takes a long time. This long juvenile development means greater risk of getting parasitized, eaten by predators while young, or suffering starvation because of a food shortage. Slower development also means that it takes longer to reach reproductive maturity, and so it delays when offspring can start being delivered to the next generation. This effect is akin to compound interest: it is financially more advantageous to earn a smaller 50 cents on the dollar every 6 months than to double your money every 12 months. Larger size may also come with diminished agility and greater conspicuousness, which can compromise an individual's ability to evade predators, to sneak up on prey, or to counter the mating advances of smaller and more furtive rivals.

Over the long term, species with gigantic individuals tend to have lower population abundance. Low abundance imposes a higher risk of extinction when faced with population bottlenecks because of disease, environmental perturbations, or new sources of predation. It is a sad truth that large-bodied species show disproportionately high extinction rates, especially since humans colonized the globe.[39]

Despite such downsides to being big, some circumstances clearly push the balance in favor of the evolution of gigantism. These conditions are most likely when a food resource is extremely abundant, widespread, and reliable. It also requires a large geographic range.[40] Krill would be a good example of such a food resource. One of the 80-odd species of these finger-long shrimp-like crustaceans *Euphausia superba* has an estimated census number of 800 trillion individuals (800,000,000,000,000) to give it, perhaps, the greatest combined biomass of any animal in the world.[41] After the evolutionary innovation of baleen as a filtering mechanism to trap a literal ton of the midocean crustaceans in each gulp,[42] even the smallest baleen whale, the pygmy right whale, still reaches 6 m (20 feet) at maturity. Technically, such baleen whales, or rorquals, are predatory carnivores since krill are fellow animals. Curiously, however, rorquals have evolved multichambered guts similar to their terrestrial ruminant relatives, like cows, that are vegetarians.

It's not just gut shape, but what lives inside whale guts that matters. The microbes from whale intestines differ sharply depending on whether you examine filter-feeding rorquals or their purely predatory toothed-whale brethren like orcas and dolphins. The kinds of microbes you find in a rorqual gut, it turns out, are capable of fermentation, similar to the microbial communities in the guts of herbivores on land.[43] The genomes in the community of bacteria contain distinct gene combinations that can drive sophisticated digestive biochemistry. The microbes of rorqual and ruminant guts (such as hippos, cattle, deer, and giraffes) allow them to extract extra nutrients through fermentation from abundant but otherwise nutritionally inaccessible resources. For example, the plant cell wall is rich in polysaccharide complex sugars in the form of cellulose, and crustacean shells are made of the polysaccharide chitin. To you and me, these potential energy sources are insoluble roughage. Wildebeest and whales, by contrast, have intestinal pacts with microbes to mine their wealth.

We have mostly neglected the contributions that such microscopic assemblages of internal bacteria have on our livelihoods. The right microbial soup in a gut, however, might just be nature's secret sauce in fostering gigantic body sizes. As a result, among animal giants, herbivorous giants are bigger than carnivorous giants, owing to the eightfold higher food consumption demands of top herbivores and their inner microbes.[44]

The evolution of gigantism can sometimes be favored when organisms encounter a changed environment that reduces some of those costs of large body size. For example, colonization of an island can be an evolutionary paradise: it can release a population from predation pressure and from competitors. This kind of rebalancing appears to have led to the avian gigantism of the South Island giant moa of New Zealand, with its head up to 3.5 m (11.5 feet) high. The tragedy of these great birds, now extinct, is that they subsequently fell prey to a countervailing change in predation pressure when the island later got colonized by a predator with a taste for large quarry—humans.

Moas are a recent extinction, in the grand scheme of life on Earth. It is deeper time extinction that the fossil record captures with the most hard-rock proof of animal gigantism. Dinosaurs. More specifically: the sauropod dinosaurs grew to truly titanic proportions. Specimens of *Argentinosaurus* and other titanosaur sauropods are thought to have been longer even than modern day blue whales, perhaps stretching over 40 m (130 feet) long and 80,000 kg (175,000 pounds).[45] Despite being globally widespread long ago, none of these gargantuan creatures survived past the end-Cretaceous mass extinction 66 million years ago that decimated all dinosaurs, save birds.

★

While the giants catch the eye most quickly, equally impressive are the extremes of size at the other end of the scale. So-called dwarf forms. Any walk in the park will expose you to a menagerie of tiny toy dog breeds—Chihuahuas, Pomeranians, Bichon Frise—the culmination of human-imposed artificial selection. More extraordinary still, the Maltese pygmy elephant evolved down to roughly the size of a very large dog: skeletons of *Palaeoloxodon falconeri* point to adults of a creature less than 1 m (3 feet) tall having lived 500 thousand years ago.[46] It is hard not to lament the extinction from Mediterranean islands of all the many species of dwarf elephants that resided there in prehistory.

The Etruscan shrew, *Suncus etruscus*, and the bumblebee bat, *Craseonycteris thong-longyai*, however, vie for the smallest mammals. At only one to two inches (3–5 cm) long and one to two grams in weight, they fall into the same weight category as a 2.5-gram dime coin. The genes inside of you and me share nearly everything in common with such miniature beings, creatures the size of a cicada, as well as with the giants of the Earth that are up to 100 million times more massive. The details of the genetic differences, however, make all the difference for the world's astonishing diversity.

10.4   Air and ground

Wings are an inspiration. Wings for powered flight, however, are an evolutionary rarity. How *do* they do it? Pull out your binoculars; let's inspect more closely some of the world's airborne animals.

Active flying takes muscle and energy and, from what we know of vertebrates, demands the use of modified arms. The anatomy and evolution of a wing, however, varies substantially in how bone[47] and muscle can promote aerodynamic lift (Figure 10.3). Pterosaurs, sadly now extinct, used a long skinny finger to stretch out the leading edge of their wingspan, the longest pinky fingers in history, extending 2 m (6.5 feet) or more. Bats do something similar, but with multiple finger phalanges that shape the membrane of skin of the outer wingtips from front to back. Another membrane, each one officially known as a patagium, connects a bat's hindlegs to its tail for even more surface area.

Birds, on the other hand, have stubby finger bones. You may have noticed the inelegant nubs of bird forelimbs when tearing into a bucket of hot wings at a local sports bar. They make up for it, though, with haute couture: birds wrap their stubby arms in feathers, and they exploit the stiff shaft, the rachis, of long flight feathers to create a wide wing. It is feathers, too, that create a lightweight airfoil when layered one upon another, linked together delicately with microscopic Velcro-like barbs.

The effectiveness of feathers in enabling bird flight, however, appears to be a co-opted function. After all, feathers are important for lots of things: insulation and thermoregulation, repelling water and shading skin from the Sun, and in communicating signals, whether to attract mates or repel rivals or deter would-be predators. The evolutionary origin of feathers even predates the origin of birds, with fossilized feathers now known from some other dinosaur relatives.[48] The leading hypothesis for an adaptive benefit to primitive feather-like structures is that they profited small animals by helping to maintain body temperature. Feathers as insulation then set the stage for additional selection pressures to elaborate their form and function.

Fully fledged powered flight, however, demands more than skin flaps or feathers. All vertebrates that are capable of powered flight have evolved, over time, numerous specialized structures to their skeletons. We've seen how exaggerated finger bones are important for bats and pterosaurs, but active fliers share other distinctive aspects of their skeletons. Pterosaurs and birds both evolved a keeled sternum for chest muscle attachment that helps power wing movements, whereas bats have disproportionately

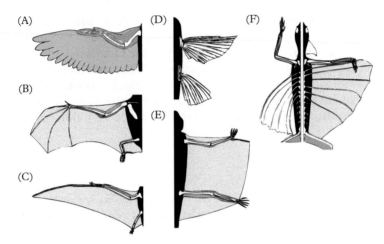

Figure 10.3 *Diverse anatomical forms of wing shapes among vertebrate animals. (A) Bird wings have most of their surface area covered in feathers (orange) with a substructure of bone (white) and muscle (gray). (B) The surface area of bat wings is made of membranes of skin, termed* patagia *(yellow), that extend between the foreleg, elongated finger bones, hindleg, and tail. The calcar bone (red) is a novel bone that evolved in bats that helps shape the tail patagium. (C) Pterosaur wings also have large patagia between the forearm and leg, with bones from a single elongated finger (red) extending to the outer front edge of the wing. (D) Flying fish "wing" musculature lies within the body, with the bony rays visually but not anatomically reminiscent of the bony supports in* Draco *flying lizards. (E) Flying squirrels also have a single long finger (red) that controls the shape of the leading edge of the patagia. (F)* Draco *flying lizard patagia occur between the elongated rib bones that animals extend and grab with their forelimbs during passive flight (right half of diagram).*

Images redrawn from Dehling, J.M. 2017. How lizards fly: a novel type of wing in animals. *PLoS One.* 12: e0189573. https://doi.org/10.1371/journal.pone.0189573.

large clavicle collarbones and scapula shoulder blade bones for similar functions. All three groups also have homeothermic endothermy in common, meaning that they are warm-blooded, capable of generating and maintaining a constant body temperature. This extremely energy-expensive kind of physiology may nonetheless be a precondition for powered flight in large animals, due to how it enhances aerobic capacity and stamina.[49] Some other features associated with flight are unique to each group. The pteroid bone is unique to pterosaurs, helping to shape the leading edge of the wing. The calcar bone in bats helps shape the skin flap that extends from the hindlegs to their short tail. The evolution of entirely new bones like these, rather than co-opting pre-existing bones for new functions, is quite rare.

★

Not all animals with wings will flap to fly. There is another way to soar: gliding. Gliding animals have evolved even more varied mechanisms than powered fliers, as gliding demands less in the way of specialized structures. Approximately 50 species of flying squirrels glide thanks to skin flaps much like you find on bats. They may not soar transcontinental distances, but they can float between trees that are nearly 100 m apart. If that isn't exciting enough, then simply shine a UV light on them: their fur will glow fluorescent pink.[50]

Flying *Draco* lizards also have skin flaps on the sides of their bodies, but they take a distinct approach to harnessing the air with this skin cape. Their skeleton has evolved long, slender rib bones that stick out on either side to make the membrane rigid, which they can retract to the sides of their bodies[51] (Figure 10.3). This structure is reminiscent of the elegant pectoral fins of flying fish with the slender bony lines of their rays radiating outward to shape the wing-like structure. When jumping between trees, the lizard's front hands grasp the leading edge of the extended rib-reinforced skin cape on either side to control the direction of flight. The elaborate use of ribs in *Draco* is a distinctive evolutionary invention, circumventing the usual limitations of the tetrapod body plan for the presence of wings or other airfoils. *Draco* can conveniently keep its limbs unencumbered by clumsy flaps of loose skin as they climb up and down tree trunks in the shade.

It could be said that flying snakes—yes, there are such creatures—also use their ribs: when leaping from a tree, the *Chrysopelea* paradise tree snake flattens out its long ribcage to help with gliding, undulating its body to keep itself upright and to avoid simply tumbling tooth-over-vent.[52] But *Chrysopelea* has no skin flaps whatsoever. It simply flattens its form in a way reminiscent of how your uncle sucks in his gut for a photo at the family pool party.

Gliding as a rudimentary way of getting airborne is the leading candidate for how the evolution of fully powered flight got its start. This arboreal "trees-down" hypothesis is easy to imagine but difficult to prove for certain. Consequently, a competing idea has survived for over a hundred years now, the so-called cursorial ground-up hypothesis.[53] The "ground-up" logic suggests that flapping forelimbs helped to speed up two-legged running, perhaps to escape predators or to reach prey, perhaps to assist in rapid scaling of inclined surfaces like tree trunks. Analysis of robotic mimics, however, undermines the idea that such ground-up benefits would be sufficient to boost animal performance enough to drive the evolution of powered flight in the absence of gliding precursors.[54]

★

Then there are the animals with wings that are not used as wings at all. Flaring up one's wings or other thin membranes, it turns out, is a good way to make an animal look bigger and more intimidating at a moment's notice. This tactic can work on would-be predators, as well as would-be competitors, especially when coupled to the surprise of bright coloration. Frill-necked lizards don't have wings, but the expansive skin flaps that ring their heads are effective as a signal to ward off trespassers and potential predators on their territory.[55] The spur-winged goose is one of a variety of birds, along with jacanas and the alarmingly named screamers, that escalate aggression beyond mere display. They have sharp, horny spurs an inch or more long in the wrist-region in the middle of their wings.[56] These are fighting weapons, and the aggressive spur-winged goose is prone to injuring other waterfowl. The steamer ducks of Argentina take aggression with wings as weapons to the extreme. Rather than sharp claws, they use hard bright-orange knobs in the middle of their wings as clubs when delivering flapping beatings to other creatures, sometimes to the death, both their own and other species.[57]

Exaggerated wing displays can also play a role in attraction, in addition to repulsion. The ebullient tail of a male peacock is a familiar zoo example. Even flight-incapable ostriches still use their perfunctory wings in ritual courtship during mating. These birds show how the appeal of wings extends beyond a desire for flight, to the foundation of desire itself.

★

Flying organisms remind us of the many simultaneous biomechanical challenges that natural selection has juggled and the difficulty of evolving powered flight from scratch. Wings also do more than enable flight. It's taken over 50 million years for bats and over 150 million years for birds to achieve their current states since the origins of flight.[58] And powered flight has just three known origins in all vertebrate animals. As we dream of what capabilities one might endow upon a newly evolved species, it's hard not to consider flight. To engineer the evolution of an animal that is fully flight capable, however, you'd be hard-pressed to do so from square one. The logical alternative, as we will contemplate in the next couple of chapters, would be to use an existing flight-capable animal as a genetic starting point.

★

When we witness nature's astonishing matches between form and function, the adaptations that present themselves to the world, it seems to show a job well done. The job, however, is not done. It is never done. The perpetual wakefulness of the forces of evolution ensures that the DNA of all species, and the features that DNA encodes, is a perpetual work in progress. The evolutionary outcomes that we observe are simply a snapshot of ongoing biological dynamism. The features stay only so long as the balance of evolutionary forces and ecological circumstances retain some degree of stability from one generation to the next. Sometimes the ecological circumstances are exceptionally stable or the developmental constraints exceptionally strong or mutation rates exceptionally low: these give us the world's so-called living fossils. We know, of course, that their DNA has not been fossilized. Today's evolutionary outcomes are just the latest incarnation of ongoing heritable change, and we fully expect to see something different in the future.

Notes

1. The eighteenth-century biologist would, of course, have said it in French: «*Montrez-moi vos dents et je vous dirai qui vous êtes.*» Ungar, P.S., and H.-D. Sues. 2019. Tetrapod teeth: diversity, evolution, and function. In V. Bels and I.Q. Whishaw (eds), *Feeding in Vertebrates: Evolution, Morphology, Behavior, Biomechanics.* New York: Springer International Publishing. pp. 385–429. https://doi.org/10.1007/978-3-030-13739-7_11.

2. Gorman, C.E., and C.D. Hulsey. 2020. Non-trophic functional ecology of vertebrate teeth: a review. *Integrative and Comparative Biology*. 60: 665–675. https://doi.org/10.1093/icb/icaa086.

3. This form of dentition of continuous growth of new teeth is a condition termed *polyphyodonty*.

4. The first set of mammalian teeth go by many names: milk teeth, baby teeth, and deciduous teeth. This last name seems to imply that by age six years, we humans enter the autumn of our lives, shedding discolored teeth like so many deciduous maple leaves.

5. Genetic variants of the human version of the Runx-2 gene lead to cleidocranial dysplasia (CCD), a condition that affects development of teeth and bones, especially of the face and skull. Fondon, J.W., and H.R. Garner. 2004. Molecular origins of rapid and continuous morphological evolution. *Proceedings of the National Academy of Sciences USA*. 101: 18058–18063. https://dx.doi.org/10.1073/pnas.0408118101; Ryoo, H.-M., et al. 2010. RUNX2 mutations in cleidocranial dysplasia patients. *Oral Diseases*. 16: 55–60. https://doi.org/10.1111/j.1601-0825.2009.01623.x.

6. Marchant, T.W. et al. 2017. Canine brachycephaly is associated with a retrotransposon-mediated missplicing of SMOC2. *Current Biology*. 27: 1573–1584.e6. https://doi.org/10.1016/j.cub.2017.04.057.

7. https://www.irishtimes.com/culture/aristotle-got-it-wrong-we-have-a-lot-more-than-five-senses-1.3079639. Accessed June 16, 2021.

8. Shichida, Y., and T. Matsuyama. 2009. Evolution of opsins and phototransduction. *Philosophical Transactions of the Royal Society of London B*. 364: 2881–2895. https://doi.org/10.1098/rstb.2009.0051.

9. This property is referred to as *bistability*, meaning that the rhodopsin protein takes on one shape when absorbing a proton and a second distinct shape otherwise; its shape flips back and forth between these two possibilities. When light is treated as a wave, distinct energy levels and properties are measured by the distance between repeating sine waves of a given wavelength. Visible wavelengths are measured on a scale of a few hundred nanometers (nm), with there being a million nm in one millimeter (mm) and 10 million nm in one centimeter (cm).

10. Hauser, F.E., and Chang, B.S.W. 2017. Insights into visual pigment adaptation and diversity from model ecological and evolutionary systems. *Current Opinion in Genetics & Development*. 47: 110–120. https://doi.org/10.1016/j.gde.2017.09.005; Baldwin, M.W., and M.-C. Ko. 2020. Functional evolution of vertebrate sensory receptors. *Hormones and Behavior*. 124: 104771. https://doi.org/10.1016/j.yhbeh.2020.104771.

11. Mazza, C.A., et al. 2010. A look into the invisible: ultraviolet-B sensitivity in an insect (*Caliothrips phaseoli*) revealed through a behavioural action spectrum. *Proceedings of the Royal Society B*. 277: 367–373. https://doi.org/10.1098/rspb.2009.1565.

12. https://www.fondriest.com/environmental-measurements/parameters/weather/photosynthetically-active-radiation. Accessed June 16, 2021.

13. Some human females can experience a subtle form of tetrachromatic vision when they inherit two functionally different gene variants of either the red-sensitive or green-sensitive opsin gene that are encoded on the two copies of their X chromosome.

14. Musilova, Z., et al. 2019. Vision using multiple distinct rod opsins in deep-sea fishes. *Science.* 364: 588–592. https://dx.doi.org/10.1126/science.aav4632.

15. Where does this unusual photosensitizing compound come from? Known as a bacteriochlorophyll, it derives from photosynthetic bacteria that have made their way into the crustaceans that make up part of the *Malacosteus* diet. Rather than you are what you eat, the stoplight loosejaw dragonfish eat what they are. Douglas, R., et al. 2016. Localisation and origin of the bacteriochlorophyll-derived photosensitizer in the retina of the deep-sea dragon fish *Malacosteus niger. Scientific Reports.* 6: 39395. https://doi.org/10.1038/srep39395; Widder, E.A., et al. 1984. Far red bioluminescence from two deepsea fishes. *Science.* 225: 512–514. https://doi.org/10.1126/science.225.4661.512.

16. Kenaley, C.P. 2012. Exploring feeding behaviour in deep-sea dragonfishes (Teleostei: Stomiidae): jaw biomechanics and functional significance of a loosejaw. *Biological Journal of the Linnean Society.* 106: 224–240. https://doi.org/10.1111/j.1095-8312.2012.01854.x.

17. Given my tender age at the time, I managed to miss out on the special 1982 release of John Waters's film *Polyester* that featured scratch-and-sniff scent cards to be scratched and sniffed at momentous points in the story arc.

18. One sextillion is an actual number, a 1 with twenty-one zeros after it, which you get to by multiplying a billion times a billion and then multiplying *that* by another thousand. Bushdid, C., et al. 2014. Humans can discriminate more than 1 trillion olfactory stimuli. *Science.* 343: 1370–1372. https://dx.doi.org/10.1126/science.1249168.

19. I keep specifying "functional" olfactory receptors because up to half of the GPCR genes encoded in the genome of a given species may not actually encode functional proteins, as a byproduct the rapid pace of imperfect gene duplication mutations in this family of genes. Dogs have more than double the human repertoire of functional olfactory receptors, summing to 896 genes. Hughes, G.M., et al. 2018. The birth and death of olfactory receptor gene families in mammalian niche adaptation. *Molecular Biology and Evolution.* 35: 1390–1406. https://doi.org/10.1093/molbev/msy028.

20. In olfaction, smell, the chemical compounds arrive in the gas phase or as aerosols. Taste handles the subset of chemical concoctions that arrive at gustatory taste receptors in the solid and liquid phases. Taste receptors, made up of related kinds of GPCR proteins, work in a similar way to olfactory receptors, but with distinct subsets of chemical compounds that they detect, imbuing our brains with bitter and sweet, umami (savory), and salty and sour.

21. Schwenk, K. 1994. Why snakes have forked tongues. *Science.* 263: 1573–1577. https://dx.doi.org/10.1126/science.263.5153.1573.

22. Graham, Z.A., et al. 2020. The longer the better: evidence that narwhal tusks are sexually selected. *Biology Letters.*16: 20190950. https://doi.org/10.1098/rsbl.2019.0950; Nweeia, M.T., et al. 2014. Sensory ability in the narwhal tooth organ system. *Anatomical Record.* 297: 599–617. https://doi.org/10.1002/ar.22886.

23. Hansson, B.S., and M.C. Stensmyr. 2011. Evolution of insect olfaction. *Neuron.* 72: 698–711. https://doi.org/10.1016/j.neuron.2011.11.003.

24. This calculation presumes the density and molarity of air at sea level to be 0.001225 g/ml and 28.97 g/mol, respectively, the quotient of which we multiply by Avogadro's number to arrive at the number of molecules in 1 ml of air; there are 1000 microliters in 1 ml of air. Kaissling, K.-E. 2009. The sensitivity of the insect nose: the example of *Bombyx mori*. In A.Gutiérrez and S. Marco (eds), *Biologically Inspired Signal Processing for Chemical Sensing*. Berlin: Springer. pp. 45–52. https://doi.org/10.1007/978-3-642-00176-5_3.

25. The elephant sex pheromone has the sexy name of (*Z*)-7-dodecen-1-yl acetate, an unlikely candidate for marketing in a Parisian perfume boutique. Rasmussen, L.E.L. 1998. Chemical communication: an integral part of functional Asian elephant (*Elephas maximus*) society. *Écoscience*. 5: 410–426. https://dx.doi.org/10.1080/11956860.1998.11682469.

26. Vignieri, S.N., et al. 2010. The selective advantage of crypsis in mice. *Evolution*. 64: 2153–2158. https://doi.org/10.1111/j.1558-5646.2010.00976.x; Hoekstra, H.E., et al. 2006. A single amino acid mutation contributes to adaptive beach mouse color pattern. *Science*. 313: 101–104. https://doi.org/10.1126/science.1126121.

27. In some creatures, the amount of extracellular matrix and fluid is also important in defining body size. And, programmed cell death provides another key element governing body size, in addition to cell division and cell stasis. Conlon, I., and M. Raff. 1999. Size control in animal development. *Cell*. 96:235–244. https://doi.org/10.1016/S0092-8674(00)80563-2; Lui, J.C., and J. Baron. 2011. Mechanisms limiting body growth in mammals. *Endocrine Reviews*. 32: 422–440. https://doi.org/10.1210/er.2011-0001.

28. The myostatin protein (MSTN), which hints at the tissue most affected (muscle), is a member of the transforming growth factor beta (TGFβ) family of proteins, which hints at the activity most affected (growth). Another synonym for myostatin is growth differentiation factor 8 (GDF8). Aiello, D., et al. 2018. The myostatin gene: an overview of mechanisms of action and its relevance to livestock animals. *Animal Genetics*. 49: 505–519. https://doi.org/10.1111/age.12696.

29. Caulin, A.F., and C.C. Maley. 2011. Peto's paradox: evolution's prescription for cancer prevention. *Trends in Ecology & Evolution*. 26: 175–182. https://doi.org/10.1016/j.tree.2011.01.002; Tollis, M., et al. 2017. Peto's paradox: how has evolution solved the problem of cancer prevention? *BMC Biology*. 15: 60. https://doi.org/10.1186/s12915-017-0401-7.

30. Nunney, L. 2018. Size matters: height, cell number and a person's risk of cancer. *Proceedings of the Royal Society B*. 285: 20181743. https://doi.org/10.1098/rspb.2018.1743.

31. As stated by eminent paleontologist Geerat Vermeij: "very large plants and animals are functionally unlike their smaller counterparts: They are more likely to be top consumers or producers, to tolerate a greater range of environmental conditions (at least in the case of animals), to maintain internal homeostasis more effectively, to be less vulnerable as adults to lethal predation, to compete more successfully for mates (again mainly in animals) and to be more prone to extinction during times of crisis." Vermeij, G.J. 2016. Gigantism and its implications for the history of life. *PLoS One*. 11: e0146092. https://doi.org/10.1371/journal.pone.0146092.

32. Ksepka, D.T. 2014. Flight performance of the largest volant bird. *Proceedings of the National Academy of Sciences USA*. 111: 10624–10629. https://dx.doi.org/10.1073/pnas.1320297111.

33. Witton, M.P., et al. 2010. Clipping the wings of giant Pterosaurs: comments on wingspan estimations and diversity. *Acta Geoscientica Sinica.* 31: 79–81.

34. Witton, M.P. 2013. *Pterosaurs.* Princeton, NJ: Princeton University Press. pp. 250–257.

35. You may know the ROUS from the Fire Swamp of the 1987 film *The Princess Bride*, but I know them from the banks of the Aratai River in French Guiana. After dark, families of capybara would explore the grounds of the Parare field station just outside my hammock in the Nouragues Nature Reserve.

36. A blue whale's heart is as tall as a fully grown person, as I saw in the plastinated display at the Royal Ontario Museum some years ago, though I have yet to visit the Icelandic Phallological Museum to witness in person a portion of a kayak-length phallus. https://phallus.is/en/gallery. html. Accessed October 27, 2021.

37. DeWoody, J., et al. 2008. "Pando" lives: molecular genetic evidence of a giant aspen clone in central Utah. *Western North American Naturalist.* 68: 493–497. https://doi.org/10.3398/1527-0904-68.4.493.

38. Blanckenhorn, W.U. 2000. The evolution of body size: what keeps organisms small? *Quarterly Review of Biology.* 75: 385–407. https://doi.org/10.1086/393620.

39. Cardillo, M., et al. 2005. Multiple causes of high extinction risk in large mammal species. *Science.* 309: 1239–1241. https://dx.doi.org/10.1126/science.1116030.

40. Marquet, P.A., and M.L. Taper. 1998. On size and area: patterns of mammalian body size extremes across landmasses. *Evolutionary Ecology.* 12: 127–139. https://doi.org/10.1023/A:1006567227154.

41. Atkinson, A., et al. 2009. A re-appraisal of the total biomass and annual production of Antarctic krill. *Deep Sea Research I.* 56: 727–740. https://doi.org/10.1016/j.dsr.2008.12.007.

42. Goldbogen, J.A., et al. 2011. Mechanics, hydrodynamics and energetics of blue whale lunge feeding: efficiency dependence on krill density. *Journal of Experimental Biology.* 214: 131–146. https://dx.doi.org/10.1242/jeb.048157; https://www.discovermagazine.com/planet-earth/blue-whales-can-eat-half-a-million-calories-in-a-single-mouthful. Accessed October 13, 2021.

43. Sanders, J., et al. 2015. Baleen whales host a unique gut microbiome with similarities to both carnivores and herbivores. *Nature Communications.* 6: 8285. https://doi.org/10.1038/ncomms9285.

44. Burness, G.P., et al. 2001. Dinosaurs, dragons, and dwarfs: the evolution of maximal body size. *Proceedings of the National Academy of Sciences USA.* 98: 14518–14523. https://dx.doi.org/10.1073/pnas.251548698.

45. Sander, P.M., et al. 2011. Biology of the sauropod dinosaurs: the evolution of gigantism. *Biological Reviews.* 86: 117–155. https://doi.org/10.1111/j.1469-185X.2010.00137.x; Sander, P.M., and M. Clauss. 2008. Sauropod gigantism. *Science.* 322: 200–201. https://dx.doi.org/10.1126/science.1160904.

46. A newborn present-day African elephant is also about 1 m (3 feet) tall. A separate lineage of 2-m-tall pygmy elephants continue to persist in Borneo. Larramendi, A., and Palombo,

M.R. 2015. Body size, structure, biology and encephalization quotient of *Palaeoloxodon* ex gr. *P. falconeri* from Spinagallo Cave (Hyblean plateau, Sicily). *Hystrix, the Italian Journal of Mammalogy.* 26: 102–109. https://doi.org/10.4404/hystrix-26.2-11478.

47. The bone material of both birds and bats is unusually dense, and despite received wisdom about hollow and lightweight bird bones, the skeleton actually makes up a similar fraction of their total body weight as similarly sized flightless animals. That high density confers strength and rigidity to such aircraft-grade biomaterial. Dumont, E.R. 2010. Bone density and the lightweight skeletons of birds. *Proceedings: Biological Sciences.* 277: 2193–2198. https://doi.org/10.1098/rspb.2010.0117.

48. Benton, M.J., et al. 2019. The early origin of feathers. *Trends in Ecology & Evolution.* 34: 856–869. https://doi.org/10.1016/j.tree.2019.04.018.

49. Koteja, P. 2004. The evolution of concepts on the evolution of endothermy in birds and mammals. *Physiological and Biochemical Zoology.* 77: 1043–1050. https://dx.doi.org/10.1086/423741; Bennett, A.F., and J.A. Ruben. 1979. Endothermy and activity in vertebrates. *Science.* 206: 649–654. https://dx.doi.org/10.1126/science.493968.

50. Kohler, A.M., et al. 2019. Ultraviolet fluorescence discovered in New World flying squirrels (*Glaucomys*). *Journal of Mammalogy.* 100: 21–30. https://doi.org/10.1093/jmammal/gyy177.

51. Dehling, J.M. 2017. How lizards fly: a novel type of wing in animals. *PLoS One.* 12: e0189573. https://doi.org/10.1371/journal.pone.0189573.

52. Socha, J. 2002. Gliding flight in the paradise tree snake. *Nature.* 418: 603–604. https://doi.org/10.1038/418603a; Yeaton, I.J., et al. 2020. Undulation enables gliding in flying snakes. *Nature Physics.* 16: 974–982. https://doi.org/10.1038/s41567-020-0935-4.

53. Norberg, U.M. 1985. Evolution of vertebrate flight: an aerodynamic model for the transition from gliding to active flight. *American Naturalist.* 126: 303–327. https://dx.doi.org/10.1086/284419.

54. Peterson, K., et al. 2011. A wing-assisted running robot and implications for avian flight evolution. *Bioinspiration and Biomimetics.* 6: 046008. https://doi.org/10.1088/1748-3182/6/4/046008.

55. Shine, R. 1990. Function and evolution of the frill of the frillneck lizard, *Chlamydosaurus kingii* (Sauria: Agamidae). *Biological Journal of the Linnean Society.* 40:11–20. https://dx.doi.org/10.1111/j.1095-8312.1990.tb00531.x.

56. Spurs distinguish themselves from true claws by growing their horny keratin from someplace other than the tip of a digit. Rand, A.L. 1954. On the spurs on birds' wings. *Wilson Bulletin.* 66: 127–134. https://www.jstor.org/stable/4158290.

57. In recounting an attack by a steamer duck on a 675-g (1.5-pound) red shoveler duck, "the steamer-duck grabbed the shoveler by the neck and pounded its body with his wing knobs . . . At intervals, the male steamer-duck pulled the shoveler beneath the surface, then raised it up again and renewed the wing-beating. After approximately 2 min, the male steamer-duck was distracted by the female [steamer-duck] and displayed with her. Within 30 s, he returned to the shoveler, grabbed it by the neck, and again beat it another 15-20 times with its wings . . . The crippled shoveler eventually reached shore, where it died 15 min later. Examination

of the specimen disclosed several broken bones, hemorrhages in the lower neck region, and massive internal bleeding at the base of the right leg. The skin was not broken, but there was obvious subcutaneous evidence of bites on the head, back, tail and left hip." p. 89 from Nuechterlein, G., and Storer, R. 1985. Aggressive behavior and interspecific killing by flying steamer-ducks in Argentina. *Condor.* 87: 87–91. https://www.jstor.org/stable/1367137.

58. Also, pterosaur species persisted for a span of about 160 million years prior to their total extinction about 66 million years ago at the end of the Cretaceous period. Simmons, N., et al. 2008. Primitive Early Eocene bat from Wyoming and the evolution of flight and echolocation. *Nature.* 451: 818–821. https://doi.org/10.1038/nature06549; Xu, X., et al. 2014. An integrative approach to understanding bird origins. *Science.* 346: 1253293. https://dx.doi.org/10.1126/science.1253293; Ezcurra, M.D., et al. 2020. Enigmatic dinosaur precursors bridge the gap to the origin of Pterosauria. *Nature.* 588: 445–449. https://doi.org/10.1038/s41586-020-3011-4.

Part 2

Evolutionary futures

11

How to evolve a dragon

Dragons. Close your eyes and say it again: *dragons*.

The notion of dragons might be the most exotic idea of what a wild animal could be. How might they fit into today's world? Perhaps you'd foresee a world where building codes in San Francisco and Tokyo require skyscrapers to resist dragonfire in addition to earthquakes. A world where automobile safety standards demand dragon repellants to avoid being snatched off the highway, the occupants otherwise plucked and gobbled like juicy lychee fruits from inside the crusty husk of a minivan. A landscape with Statue of Liberty sized scarecrows popping out of Florida orange groves to fend off giant flying fruitarians. Would the rich lay down small fortunes to commission artist-scientists to design dwarf varieties as pets, the modern noble's corgi? Do the seas teem with serpentine devourers of watercraft, filling Loch Ness with a vibrant population of its rightful inhabitants and filling the Scottish village of Drumnadrochit even fuller with tourists? Or, instead, the tourist industry on the outskirts of Winnipeg reinvigorating the Manitoba economy with a new brand of paleo-park, a neo-park, fenced with high tech and squadrons of drones to ensure that all the extraordinary exotic beasts stay safe from poachers?

This world we live in, though, is a material world without living myth. Despite the evocative thought experiments of the Brothers Grimm and of Hollywood, and the preceding paragraph, we must abide physical laws. Through the forces of evolution, populations of all living organisms arose and change, arise and are changing, and will do so perpetually, until the Sun dies and snuffs us all out. This is the reality in which we must consider all the life around us. This is the reality to ground notions of extraordinary creatures. In *this* reality, can dragons evolve to gallop and fly among us? How would we go about making that happen? If some form of new animal could evolve, dosed with human intervention, would we want them to, and why, and what would be the consequences?

Evolving Tomorrow. Asher D. Cutter, Oxford University Press. © Asher D. Cutter (2023). DOI: 10.1093/oso/9780198874522.003.0011

11.1 No myth

"*Here be dragons*" is Western civilization's provocative label for the edge of knowledge. On the periphery of the maps of medieval times, the notion of unknown lands and uncharted waters was inscribed alongside evocative sketches of monsters and beasts (Figure 11.1). The Roman phrase inked onto maps was, in Latin, *HIC SVNT LEONES* ("*Here be lions*"). As mighty as *Panthera leo* demonstrated itself to be as a gladiatorial adversary inside the Colosseum, the subsequent substitution of LEONES with DRA-CONES in popular imagination surely serves to amplify the prickles of danger and the allure of discovery when faced with the unknown.

The unknown packs a promise. It packs the promise of teaching and enlightenment. It also promises fear and foreboding. And the unknown packs the promise of change. Change in the revolutionary sense that, in this case, can also change the course of biological evolution.

★

Taking what we've learned about the forces of evolution, genetic welding, and the diversity of animal traits, let's write down an algorithm, a recipe for how to evolve a dragon. We might say this recipe is figurative, that biotechnology could unleash some abstract existential crisis. Or is it literal? Might we apply biotechnology in a way to engage in

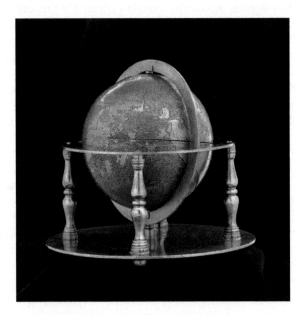

Figure 11.1 *The Hunt-Lenox Globe, an 11-cm-diameter globe fabricated out of copper in c. 1510 from a predecessor made of conjoined ostrich eggs, is one of the few historical maps inscribed with the phrase* here be dragons *in Latin.*

Image credit: Rare Book Division, New York Public Library, reproduced under the CC0 1.0 license.

human-assisted evolution of new life forms that we could touch and see and legitimately give a Latin binomial name? *Draco obscurus* has a nice ring to it. Well, we'd need to pick another because we can't use that epithet; it's already taken, a name conferred on a handsome orange-and-brown striped dusky gliding lizard from Borneo, in 1887, by the intrepid zoologist George Boulenger, one of over 2000 animal species that he named in his life.[1] Figurative or literal? I mean both.

By literal dragon, I mean a viable population of organisms, a distinct species with a novel and unique combination of characteristics not present in any of today's collection of organisms. I mean a hyperexotic species whose origins meld evolution through selective breeding, genome engineering, and genetic welding. This *wyrm* may very well arrive[2], however, with a posse of figurative dragons in tow.

We have explored, so far in this book, some of the key traits that make up an animal. We've seen some of the more impressive examples of those traits, where nature has elaborated them over the course of evolution. We also explored how evolution managed to do that with the team of forces at its disposal. We detailed some of the constraints to evolution and some of the factors that can bend those constraints. Now it is time to put all these parts together: pattern meets process meets human ingenuity.

*

One might aim to produce a dragon of myth. But which myth? Those of us with British roots know British myths—in my case, the roots most recently tracing back to the Old Country three generations ago in the shape of my great-grandfather George. They harken to the folklore surrounding the dragon-slaying exploits of a different George, Saint George, the patron saint of England.[3] In historical paintings and sketches, St. George's dragon is superficially reptilian: scaly, sharp-clawed, long-tailed and long-necked, occasionally web-toed, and with a spiky crest along its spine. But usually sporting wings, of dubious flight capability, and sometimes beaked instead of befanged; sometimes two-legged instead of four-legged, or even serpentine and legless. Perhaps it had protruding ears and horns and an elongated tongue, perhaps it had a fringe of feather or hair under its chin and along its legs. In mass and bulk, George's dragon lies somewhere between the size of a man and the size of a horse, if Rafael's painting is any gauge. This is the archetype of the Western dragon of lore throughout Europe, tinted by oral permutation of the legends of gods and monsters from ancient India and Mesopotamia and Egypt.[4]

Of course, somewhere along the line, the impression of size magnified to something elephantine or even more colossal. The myth became that these giants are ferociously protective of their eggs and exceptionally long-lived. There are dragons of land and air and sea. The sea serpents are of especially ancient lore, including the Babylonian *Tiamat* and *mushussu*, the Judaic Leviathan, and the Māori *Taniwha*. Some stories have dragons discharging bolts of lightning or belching poison gas or acid. But, at some stage in their mythological development, fire-breathing arrived on the scene as a dragon's weapon of choice, perhaps inspired by the *Iliad*'s Chimaera, or her brother the Colchian dragon, and subsequently rendered iconic in Beowulf's final and mortal battle.

What foul breath does nature already have to offer? Skunks, spitting cobras, bombardier beetles: they all spray their foes in defense. Their mechanisms, however, all differ. The skunk's noxious chemicals erupt directly from their anus-adjacent scent glands that can store enough stink for up to six sprays, courtesy of honed muscles in their rear end that can direct the foul sulfur-rich thiol chemical irritants in a stream up to 3 m (10 feet) away.[5] Despite the horror of volatilized skunk musk, Thomas Aldrich waxed almost poetic of its aesthetics in 1896[6]: "Obtained from the sacs a few hours after the death of the animal, the secretion is a clear, limpid fluid, of a golden-yellow or light-amber colour, of a characteristic, penetrating, and most powerful odour, and having a specific gravity, at ordinary temperature, less than water." Making use of the opposite end of the gastrointestinal tract, spitting cobras can squeeze their venom glands to eject a 2.5-m (8-foot) stream of the especially painful 3FTX and PLA2 toxins directed at the eyes of an assailant.[7] The fangs of spitting cobras like *Naja nubiae* have evolved unusual front-facing openings near the tips to allow such projectile toxicity, in addition to the usual bite-derived venom delivery.

Despite its diminutive size, the bombardier beetle's means of long-range weaponry may be the most sophisticated. When perturbed, a *Stenaptinus insignis* beetle initiates a controlled explosive chemical reaction inside a special chamber within the tail-end of its body. With a rapid-pulsed injection of a reactant solution into a pool of enzymes, they produce a series of audible detonations, creating a 100 °C (212 °F) boiling hot and toxic fluid that jets out of a miniature rear-end firehose.[8] This apparatus is reinforced with a distinctive composite of hard cuticle chitin, proteins, and waxes that protect the black-and-red-bodied bombardier from itself. When the reactant solution of hydrogen peroxide and hydroquinone make contact with the pool of peroxidase and catalase enzymes, the rapid chemical reaction explosively produces steam, oxygen gas, and noxious benzoquinone irritants with such enormous pressure that it ejects in a spray in any direction aimed by the tip of the tail.[9] This makes an effective defensive weapon against targets up to 15 body lengths away.

Projectile poisons and chemical explosions and electrical shocks, as impressive as they may be, are not fire. No animals produce fire, save humans, and we need tools to do it. Some plants depend on fire for seed germination. These pyrophytes, in some cases, even have evolved to promote flammable conditions to encourage fire to erupt from lightning strikes.[10] Such bystanders to conflagration, however, are not the same thing as sparking ignition.

There is no shortage of biological sources for combustible material. Most habitats abound with thin and dry plant leaves and bark, at least some times of the year. There also are high-surface-area combustible powders like fern spores or flaky skin, as well as creatures that produce flammable gases like volatilized eucalyptus oil (eucalyptus trees), oxygen (bombardier beetles), hydrogen (*Bacillus* bacteria), and methane (*Methanobrevibacter* archaeal microbes in cow guts).[11] One could even imagine mechanisms to ignite such kindling that could, in principle, evolve. Keratin and chitin could be produced to make clear, hard lens-like structures to concentrate sunlight, much like Boy Scouts might do to earn a Primitive Fire Making Merit Badge. Structures capable of dispensing electrical discharges, like the sparker on your propane barbecue grill, also seem technically feasible. Biomineralization of calcium or iron is common in the formation of bones, shells, teeth, and exoskeletons which suggests the possibility of biomineralization of compounds capable of high-friction scraping, like a fire-bow or hand drill, or sparking, like with a flintstone. Some chemical mixtures can even lead to spontaneous combustion, deriving in part from biological sources: ammonium nitrate in feces combined with salt, zinc, and water.

Evolution, apparently, has not engineered these possibilities.[12]

★

In South and East Asia, the dragon depictions are decidedly more serpentine. The enormous cobra-like form of the Indian Naga is legless, whereas the sinuous shape of Chinese and Japanese and Korean dragons sport a quartet of short, muscular, claw-tipped legs. They are mustachioed with sensory whiskers and have a sharp-fanged foxlike maw, stiff horns, elaborate head decor reminiscent of some crenulated chicken combs and gobbles, and fuzzy tail-tips. In these traditions, dragons frequently express a benevolent temperament and intelligence and are often associate with the onset of rain. The Aztec Quetzalcoatl—literally, the precious-feathered serpent—also was a rain god of a dragon. Its sinuous snake of a body was coated in exquisite feathers, more brilliant even than the iridescent green and red sheen coloring the elongated tail feathers and body of the resplendent quetzal bird *Pharomachrus mocinno* of Central America that lends its name.

From the stories of St. George to Quetzalcoatl, their features take real-world characteristics and combine them in a mishmash, tinged with the supernatural or with nonliving physical phenomena: mouths and limbs, or lack thereof, from snakes, crocodiles, lions, and eels; horns from goats; wings and feathers of birds. The enormous fossil bones of dinosaur megafauna and whale skeletons washed upon a beach inspire fables of wyrms of gargantuan size. These features were glued together with the artistry of the brush and the spoken word, stitched with the fine thread of myth into chimeras of form.

In myth, the extinction of dragons often coincides with the evaporation of magic from the world. If we were to evolve a dragon, it would mark, in a sense, the birth of a new magic. The magic of genetic welding spliced into the fabric of nature. The first question before us—in advance of wading into the sticky ethical and environmental consequences—is how might one apply the artistry, sorcery, and science of genetic engineering and the ancient forces of evolution to constitute such distinctive combinations in the flesh.

11.2   Decisions

To evolve a new kind of animal, an essential early decision we must make is which animals to build from. What founding stock to use for the subsequent steps of our evolutionary algorithm? To proceed efficiently, this stock must satisfy several practical criteria. The founding stock should develop quickly to reproductive maturity and produce many off-spring when reproducing. These features ensure reliable propagation from generation to generation. The founding stock also should be hardy and have only modest demands for space and food. Finally, the founding stock should be amenable to genetic manipulation and not derive from a rare or endangered species. Fortunately, many of these characteristics are correlated with one another. More specifically, they are most prevalent among creatures that are physically small.

The next set of criteria define the physical features to prioritize, accounting for constraints that each feature may impose on others. Foremost is body plan. A key decision here is whether to prioritize powered flight: given the rarity and complexity of wing evolution, tweaking a flight-capable founding stock ought to prove much more feasible than developing flight ability from scratch. Actively flying vertebrates thus restrict our founding stock to bats or birds. Bats typically reproduce on an annual cycle with very few pups per litter; some birds can cycle through two or more generations per year with large clutches of eggs. The rearing and feeding of birds is common practice, whereas bats tend toward finicky and sensitive living conditions. In this matchup, birds seem the inevitable choice.

If we consider passive gliding to be sufficient, then *Draco* gliding lizards and flying squirrels enter the mix. The *Draco* genus contains some 40 or so species found throughout southeast Asia.[13] As compelling an attribute is their Latin name, much of *Draco*'s reproductive biology remains unknown, though lizards commonly grow for multiple years prior to sexual maturity. The southern flying squirrel, *Glaucomys volans*, by contrast, can reproduce within six months of birth. Some folks even keep them as pets. Reptile as well as avian reproductive physiology, however, may make genome editing a greater logistical hurdle for lizards and birds than for squirrels and bats. In this matchup, *Glaucomys* takes the lead. The trouble with flying squirrels, thinking ahead, is that they seem so very mammalian, perhaps too distant from a vision of what an extraordinary species might be.

We might also consider aquatic forms for evolving a dragon, using fish or eels as a founding stock. The tantalizing transparent leptocephalus larval stage of eels, however, is offset by the logistical reality of a complex migratory life cycle that often requires a decade or longer to reach sexual maturity followed by a single reproductive episode. Reproductive maturity in sharks is also protracted. Guppies grow fast, but, alas, they are guppies. Some plants and microbes also would make outstanding metaphorical dragons, but for brevity and focus we will stick to land animals.

There are creatures that we already call dragons of one sort or another. In that minimalist sense, dragons already exist. Might they make the job of evolving a dragon that much easier? There are the 3-m (10-foot) Komodo dragons, a kind of monitor lizard whose nearly 80 kg (175 pounds) bulk inhabits a slew of Indonesian islands.[14] There are dragonflies and dragonfish and dragon fruits; there are snapdragons (a flower you may have in your garden), and bearded dragons (a lizard you may have as a pet), and leafy seadragons (a kind of seahorse pipefish found off the south coast of Australia). There are even the beautifully toxic shocking pink dragon millipedes and blue dragon nudibranchs.

These creatures all are dazzling and exotic, gorgeous and fascinating. Some are even frightful and deadly. But none fully captures our mind's eye of a dragon of myth and story.

★

Sex. Birds do it. Bees do it. But . . . dragons? Should we even contemplate the possibility? If we aim to evolve a new and extraordinary species, then, like it or not, the answer has got to be "yes." We have got to think about how they will reproduce, both to guide or manipulate their genetics and to get a viable population that can sustain itself. How an organism goes about reproducing its own kind defines one of its most intrinsic properties. Reproduction is, after all, the key factor in evolutionary fitness, what one survives to do. For all of life, reproduction is *la raison d'être*. And there are so very many ways to go about it.

There are the parthenogens, species comprised of a single sex. Their reproduction is asexual, with embryos developing from an unfertilized egg, a genetic clone of their single parent. A permutation of this style of natural cloning is gynogenesis, in which mating still takes place, but the sperm of a mate simply provides a chemical trigger to start the egg's embryogenesis without actually fertilizing it or contributing any DNA—the gratification without the genes. The cuckold male, if *cuckold* is the right term for such sly circumstances, is usually a member of a different species altogether.[15] This makes interspecies sexual parasites of gynogens like the Amazon molly fish *Poecilia formosa* and the silvery salamander *Ambystoma platineum*. Aphids are more calculating still in their asexuality. In the pleasant summer months, mothers clone daughters in a rapid succession of live virgin births for 10 generations or so. The onset of autumn then triggers the production of two sexes that copulate to produce an overwintering clutch of eggs that lay quiescent until spring, proving that their parthenogenesis is not truly obligatory.

Other species absolutely require sexual interaction between two individuals to reproduce, but they do so with just a single sex: hermaphrodites. Hermaphrodite individuals make both types of gametes, sperm and eggs, in some cases simultaneously and in some cases changing which gamete type they produce over the course of adulthood: for example, clown fish. The largest clown fish in a group acts as a female by making only eggs, but if she dies, then the fish that had been functioning as male by providing sperm will transition to take on the egg-producing female role, giving up sperm production altogether. One of the smaller, juvenile fish in such sequentially hermaphroditic species will then sexually mature to initiate sperm production. More commonly in other sequentially hermaphroditic fish, like wrasse, individuals initially develop female reproductive organs before transitioning to male reproductive function later in life.

Hermaphrodites of some species, however, eschew mating with one another altogether. They simply fertilize themselves. You can find such self-fertilization in nematode roundworms, like the elegant roundworm *C. elegans*,[16] and some snails, but it is especially common in plants.[17] While self-fertilization has some genetic drawbacks—it is, after all, the most extreme possible form of inbreeding—it nonetheless gives an organism a means of assuring reproduction when populations are at low density or pollinators are rare, as at the edge of a species range or upon colonizing a new habitat.[18]

★

The sexual mode that most often tantalizes our imaginations, of course, is gonochorism. You might not know it by that bit of biological jargon, but this is the kind found in all mammals and all birds. This is the kind with female and male biological sexes. Those two sexes get defined by having reproductive organs with the propensity to make eggs versus sperm, as well as distinct secondary sexual characteristics, though intersex individuals may also occur that have a mix of characteristics of both females and males.[19] The distinct ways that different species manage to get the sexes to come together to put their gametes (sperm and egg) in sufficiently close proximity for fertilization is as varied as the distinct shapes of leaves among trees. Just as all shapes of leaves are successful in photosynthesizing, all species' styles of sexual interactions have proved successful in propagating the next generation.

Among those animals that fertilize one another externally, like in salmon and many other fish, the union of cells to make a baby is not a particularly intimate affair. The two sexes sidle up near to one another, then males squirt their sperm over the eggs getting squirted out by females. In scorpions, it is more personal, but only slightly. A male will deposit a globule of sperm—a spermatophore—onto the ground, then lead the female scorpion to walk over it to squat down and pick it up with her genital operculum. Subsequent to this genteel arrangement, the female scorpion's fertilized eggs develop inside her body until she births a brood of live baby scorpions that she'll carry around on her back for the duration of their first juvenile stage. In both salmon and scorpions, the release of gametes at a distance, however, follows what counts as an elaborate dancing routine of courtship between the sexes.

Things get a bit more touchy-feely for many birds, like chickens and osprey. They use their cloacas to transfer gametes. The cloaca is a multipurpose urogenital opening that connects to either testes or ovaries, depending on the sex of the osprey, as well as connecting to the colon. Males, though penis-less, transfer sperm to females through what is affectionately termed a *cloacal kiss*, as each sex scissors their tail feathers in a mutually dependent cloaca-accessible orientation.

Duck copulation, however, is decidedly less dainty.[20] Unlike a male osprey, a male duck does have a penis.[21] Normally, they will keep it tucked away discretely inside their body cavity. But in one-third of a second, a male Muscovy duck *Cairina moschata* can explosively evert his penis into an erect 20-cm (8-inch) phallus, more than a quarter of the total body length of the animal.[22] Such alarming erectile speed arises from use of the lymphatic system rather than from blood flow of the vascular system. This elaborate ballistic genital structure spirals in a counterclockwise corkscrew shape with which he attempts copulation with a female duck, inserting it into her cloaca whether she is willing or not, the semen transferring along the external groove of the corkscrew rather than in an enclosed urethral canal.[23] In what reflects an evolutionary arms race over control of reproduction, Muscovy duck females have evolved sophisticated vaginal architecture comprised of extra pouches with dead-ends or with corkscrew chirality in the opposite clockwise direction to stymie the reproductive will of the males. These counter-adaptations ensure greater female control over fertilization in the face of sexual conflict. This sexual conflict can be extreme, with male ducks imposing forced copulations that sometimes result in the female's death by drowning.

Rather than an inflatable balloon of an intromittent organ, the penis of many mammals is rigid all the time. This permanent hardness comes courtesy of a literal bone in their penis known as a baculum (Figure 11.2). In evolutionary terms, mammalian baculum bones have come and gone nearly 20 times, being present in 90% of species from bats to wolves to walrus to primates, including chimpanzees.[24]

Human males, of course, are the odd man out. Perhaps this contributes to the enigma and multiple hypotheses that biologists have brainstormed to explain what exactly is the function of a baculum? They vary in shape and size, features that don't correlate much with body size, as it turns out, though the 60-cm (2-foot) baculum of a walrus is nearly as long as one of their tusks. One way to help figure out what they do is to create X-ray 3D scans. Not long ago, a research team did just that with a collection of bat penises, each inflated with formalin from little bat cadavers and scanned with a miniature version of a hospital CAT scanner. From those X-rated X-ray scans, it has become clear that the baculum of *Pipistrellus* bats acts to protect the tip of the urethra, as it does also in primates.[25]

Male dogs have a baculum, an *os penis*, and some female dogs develop a homologous bone, the baubellum or *os clitoris*, as an intersex trait. The function, if any, of the baubellum, which is found as a typical feature of female walrus and squirrel anatomy, remains to be demonstrated.[26] The perks of a penis bone, however, differ across distinct kinds

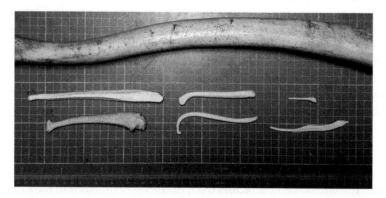

Figure 11.2 *The skeletons for males of many species of mammal include a baculum, also known as an* os penis, *a bone in the penis that can serve to protect the urethra and promote paternity. This assortment of bacula, used for teaching vertebrate anatomy and evolution in the Department of Ecology & Evolutionary Biology at the University of Toronto, includes specimens from (top to bottom, left to right): walrus, brown bear, river otter, beaver, sea lion, raccoon, and domestic dog. The walrus baculum specimen extends over 60 cm in total; grid squares show 1-cm spacing.*
Photo by the author.

of mammals, acting as genital protection, or to promote paternity in mice, or to signal male vigor in seals.

You might think that penetrative copulation is as intimate as sex can get, slipping one animal's protruding body part inside of another animal's body cavity. But you'd be wrong. Deep-sea anglerfish achieve the apogee of sexual intimacy in whole-body fusion of tissue. The minute males of *Ceratias holboelli* seek out females that may be 60-times larger than themselves, the males using their big eyes and, in other species, big nostrils to home in on potential mates by their glow and their pheromone scent. In the dark of the ocean 300 m (1000 feet) below the surface, a male can see a female by the shine of her bioluminescent lure, the lure being a glowing organ derived from evolution of a spine, a lure perched above her snout and her prodigious jaws—a lure that also attracts prey.

When he finds a mate, the anglerfish male bites her belly with special teeth to hang on. Then, ending the free-living phase of his life, the skin of the male's head fuses with the skin of his mate, followed by degeneration of a variety of his bodily structures. In this "anatomical joining" that is permanent in more than 20 species of ceratioid fish, the cellular structure of the male intercalates with that of the female so that he becomes a sort of nutritional and physiological parasite, paying his rent in sperm. It is hard to beat William Beebe's 1938 description of what happens, "driven by impelling odor headlong upon a mate so gigantic ... in such Stygian darkness, and willfully to eat a hole in her barbed-wire skin, to feel the gradually increasing transfusion of her blood through one's veins, to lose everything that marked one as other than a worm, to become a brainless, senseless thing that was a fish" as "the muscles of the male, its sense organs, digestive tract, the very skeleton itself gradually disappear."[27]

To permit this two-body chimera to persist, the genomes of anglerfish have undergone radical disruption to the genes responsible for immune reactions. Different species have lost distinct sets of genes that would normally allow their immune systems to detect and fight off potential foreign pathogens—the genes that help an animal identify "self" from "nonself"—a genetic version of the immune-suppressing medications administered to organ transplant recipients.[28] In fact, these true-life chimeras may be made of many biological roommates: in species like *Cryptopsaras couesii*, up to eight males may permanently fuse their bodies to a single female in this most intimate polyamorous embrace.

★

To be able to recombine genetic factors from distinct individuals into the same genome, reproduction must involve biparental sexual reproduction. Genetic welding similarly requires biparental sexual reproduction. Self-fertilization and the various incarnations of asexual parthenogenesis and vegetative budding simply won't do. The two parents could be female and male, or they both could be hermaphrodites. Interfertile hermaphrodites, however, are not known to occur among mammals, birds, or reptiles, our primary candidates in sourcing a founding stock. Despite the diversity of reproductive modes across the tree of life, the path of least resistance in coaxing evolutionary change—given that recombination is indispensable—will be to settle for one of the many commonplace bisexual gonochorist options, for a founding stock comprised of females and males.

★

Humans have sought out dragons for eons in our exploration of the world and to understand how the world works. Evolutionary engineering offers a means to create dragons of our own making, with myth, nature, whimsy, and practical needs as inspiration. The logistics of evolving a dragon in the flesh demand consideration of factors that will allow evolution and genetic engineering to be practicable and efficient. Founding stocks most amenable to dramatic evolutionary engineering will be hardy with short generation times and simple propagation requirements, readily available, have demonstrated tractability of genetic engineering techniques, and reproduce biparentally.

Notes

1. A sample of Boulenger's taxonomic efforts: https://en.wikipedia.org/wiki/Category:Taxa_named_by_George_Albert_Boulenger. Accessed October 13, 2021.

2. A wyrm is the name for a dragon-like creature, usually limbless in mythological tails of sea serpents.

3. With stories about Saint George dating back to the third century of the Common Era, it should be no surprise that they are often fragmented and inconsistent. Saint George may be the Christian rendition of an amalgam of Cadmus, Hercules, and Perseus, so considerable is the overlap of their dragon-slaying legends. Many places other than England also share claims to Saint George as a patron saint. George himself, though, appears to have been Greek, with his supposed feat of dragon-slaying having taken place in what is now Libya; certainly, he never set foot in the British Isles. Lingersoll, E. 1928. *Dragons and Dragon Lore* (2014 edition). New York: Cosimo Classics.

4. When depicted with two legs, such a dragon often is referred to as a *wyvern*; when legless and perhaps wingless, sometimes called a *wyrm*. On occasion, they are shown with more than one head. In the extreme, there is the nine-headed Lernaean Hydra slain by Hercules. Developmental anomalies certainly can lead to multiple heads, like conjoined twinning. We might choose, however, to interpret the notion of multiple heads as decoy or warning displays, or sexy costumery, to dazzle an onlooker, much like the vivid extra eye-like spots on a peacock tail or the hindwings of an otherwise-drab moth.

5. The most abundant compound, comprising about 40% of skunk spray, is (E)-2-butene-1-thiol. Wood, W.F. 1999. The history of skunk defensive secretion research. *Chemical Educator.* 4: 44–50. https://doi.org/10.1007/s00897990286a.

6. Thomas Aldrich was, indeed, an intrepid scientist, also writing, "The exceedingly disagreeable odour possessed by the secretion of the Mephitinae, and the difficulty attending its collection in sufficient quantity for an exhaustive study have deterred investigators in the

past from studying this most singular and interesting fluid … I have been more fortunate than my predecessors in being surrounded by those who, for the cause of science, would endure even the odour of a skunk in close proximity" (pp. 327–328); p. 329 in Aldrich, T. B. 1896. A chemical study of the secretion of the anal glands of *Mephitis mephitica* (Common Skunk), with remarks on the physiological properties of this secretion. *Journal of Experimental Medicine.* 1: 323–340. https://dx.doi.org/10.1084%2Fjem.1.2.323.

7. *3FTX* is the venomologist's abbreviation for three-finger toxins that often act to disrupt cell membranes, making them cytotoxic CTX 3FTXs, the most common kind of venom protein found in all types of cobras. PLA2 stands for phospholipase A2, a kind of enzymatic protein that cuts lipid fat molecules to result in the production of acid, which accelerates extreme pain from cytotoxic 3FTXs. PLA2s are especially abundant in spitting cobra venom. Kazandjian, T.D., et al. 2021. Convergent evolution of pain-inducing defensive venom components in spitting cobras. *Science.* 371: 386–390. https://dx.doi.org/10.1126/science.abb 9303.

8. Arndt, E.M., et al. 2015. Mechanistic origins of bombardier beetle (Brachinini) explosion-induced defensive spray pulsation. *Science.* 348: 563–567. https://dx.doi.org/10.1126/science. 1261166.

9. Eisner, T., and D.J. Aneshansley. 1999. Spray aiming in the bombardier beetle: photographic evidence. *Proceedings of the National Academy of Sciences USA.* 96: 9705–9709. https://dx.doi. org/10.1073/pnas.96.17.9705.

10. Mutch, R.W. 1970. Wildland fires and ecosystems: a hypothesis. *Ecology.* 51: 1046–1051. https://doi.org/10.2307/1933631.

11. The spores of common club moss *Lycopodium clavatum* were used historically as flash powder for theatrical and magician pyrotechnic effects when the spores are dispersed as a fine mist of powder. Skunk spray is also highly flammable and can "burn with a luminous flame" (p. 330), according to the 1896 report of Thomas Aldrich cited previously.

12. In addition to fire as a weapon or germination signal, with even more speculative whimsy, one could even envision fire-breathing as a mechanism for thermo-regulation.

13. McGuire, J.A., and R. Dudley. 2011. The biology of gliding in flying lizards (Genus *Draco*) and their fossil and extant analogs. *Integrative and Comparative Biology.* 51: 983–990. https:// doi.org/10.1093/icb/icr090.

14. Jessop, T.S., et al. 2006. Maximum body size among insular Komodo dragon populations covaries with large prey density. *Oikos.* 112: 422–429. https://doi.org/10.1111/j.0030-1299. 2006.14371.x.

15. Perhaps a spinoff of the interspecies romantic exploits of Captain James T. Kirk from the original *Star Trek* would do well to explore this concept. *Star Trek*'s literal next generation.

16. The full story, however, is more nuanced. Populations of *C. elegans* nematode roundworms may contain both hermaphrodites and males, a rare reproductive mode known as androdioecy. Interestingly, the hermaphrodites cannot inseminate one another, so reproduction occurs either by self-fertilization of hermaphrodites or cross-fertilization between a male and a hermaphrodite. Corsi, A.K., et al. 2015. A transparent window into biology: a primer on

Caenorhabditis elegans. 2015. *Genetics*. 200: 387–407. https://doi.org/10.1534/genetics.115. 176099.

17. In addition to simple hermaphroditism, some species show androdioecy with males, as well as hermaphrodites, akin to *C. elegans* roundworms. Many plants also have the reverse sexual system: gynodioecy, in which the two sexes in populations are hermaphrodites and females. Less common is trioecy, with all three sexes: males, females, and hermaphrodites. Plants add layers of reproductive complexity by virtue of whether both kinds of gametes get produced within the same or different flowers of a given plant, whether hermaphrodites may self-fertilize, and whether sexual and asexual budding reproduction may co-occur. Some plants, nonetheless, are dioecious—what animal biologists call *gonochorism*, and what most people call having just two sexes, male and female—which seems downright vanilla with so many alternatives. Vanilla a.k.a. *Vanilla planifolia*, it turns out, are hermaphrodites that can fertilize themselves.

18. This virtue of reproductive assurance through self-fertilization, in many species, is employed only as a last resort, after prior attempts at cross-breeding have failed.

19. Intersex refers to individuals with either a discordance between phenotypic and chromosomal sex, mosaicism in sex chromosome composition across their bodies, or reproductive structures of both males and females. In humans, approximately one in 4500 babies (0.02%) are born intersex, or in clinical nomenclature, born with disorders of sex development. Note, again, the distinction from gender identity and sexual orientation. Hughes, I.A., et al. 2006. Consensus statement on management of intersex disorders. *Archives of Disease in Childhood*. 91: 554–563. http://dx.doi.org/10.1136/adc.2006.098319; Sax, L. 2002. How common is intersex? A response to Anne Fausto-Sterling. *The Journal of Sex Research*. 39: 174–178. https://dx.doi.org/10.1080/00224490209552139.

20. Isabella Rossellini's Green Porno series of video vignettes about animal reproduction are stylized, bizarre, and surprisingly biologically accurate. Many of them are viewable on YouTube.

21. Male ducks are unusual among birds in having a penis. Roughly 97% of bird species lack a penis. Chickens, it turns out, have "evolved a highly reduced, nonintromittent phallic rudiment." Herrera, A.M., et al. 2013. Developmental basis of phallus reduction during bird evolution. *Current Biology*. 23: 1065–1074. https://doi.org/10.1016/j.cub.2013.04.062.

22. Brennan, P.L.R., et al. 2010. Explosive eversion and functional morphology of the duck penis supports sexual conflict in waterfowl genitalia. *Proceedings of the Royal Society B*. 277: 1309–1314. https://doi.org/10.1098/rspb.2009.2139.

23. This depiction of waterfowl mating contrasts with a serene 1937 description of mating behavior among British smew ducks, which begins, "January 10th, 1937. A cloudless sky, warm sun and a light cool S. wind. About 11 a.m. a pair were together, the duck with her tail up at angle of 30° and neck stretched forward at the same angle. ..." Hollom, P.A.D. 1937. Observations on the courtship and mating of the smew. *British Birds*. 31: 106–111.

24. Schultz, N.G., et al. 2016. The baculum was gained and lost multiple times during mammalian evolution. *Integrative and Comparative Biology*. 56: 644–656. https://doi.org/10.1093/icb/icw034.

25. Herdina, A.N., et al. 2015. Testing hypotheses of bat baculum function with 3D models derived from microCT. *Journal of Anatomy*. 226: 229–235. https://doi.org/10.1111/joa.12274.

26. Sumner, S.M., et al. 2018. Os clitoris in dogs: 17 cases (2009–2017). *Canadian Veterinary Journal*. 59: 606–610. https://pubmed.ncbi.nlm.nih.gov/29910473; Lough-Stevens, M., et al. 2017. The baubellum is more developmentally and evolutionarily labile than the baculum. *Ecology and Evolution*. 8: 1073–1083. https://doi.org/10.1002/ece3.3634.

27. Quotation from Beebe, W. 1938. *Ceratias*: siren of the deep. Through perpetual darkness the tiny male seeks out his mate and attaches himself to her spiny skin for life. *Bulletin of the New York Zoological Society*. 41: 50–53; Pietsch, T. 2005. Dimorphism, parasitism, and sex revisited: modes of reproduction among deep-sea ceratioid anglerfishes (Teleostei: Lophiiformes). *Ichthyological Research*. 52: 207–236. https://doi.org/10.1007/s10228-005-0286-2.

28. As Jeremy Swann and his colleagues summarized reproduction between anglerfish males and females, "attachment is followed by fusion of epidermal and dermal tissues and, eventually, by connection of the circulatory systems so that the male becomes permanently dependent on the female for nutrients, with the pair becoming a kind of self-fertilizing chimera." Swann, J.B., et al. 2020. The immunogenetics of sexual parasitism. *Science*. 369: 1608–1615. https://dx.doi.org/10.1126/science.aaz9445.

12

Evolutionary engineering in the flesh

Let us settle on a flock of birds as our founding stock for a first foray in evolutionary engineering. They have many practical benefits. They also have another evocative virtue. Birds are, after all, the living descendants of dinosaurs.

But which birds? A couple of standouts emerge. *Gallus*: you know these avian beasts as junglefowl, more colloquially and less imposingly called "chickens." And *Coturnix japonica*, quail.[1] Let's peruse their pros and cons.

12.1 Fleshing out a founding stock

The chicken might seem a goofy option as the basis for evolving a dragon. But they have many perks. Chickens are robust and easy enough to care for that the poultry industry raises more than 68 billion of them each year. The American Poultry Association officially recognizes 461 distinct breeds and the United Nations Domestic Animal Diversity Information System lists 1824 chicken breeds around the world, each with an amazing array of distinct features.[2] With its well-understood genome sequence, the size and shape and color and behavior for many of these traits have known causes in terms of specific gene variants. Experimental protocols are well-established and growing year-over-year in innovativeness for performing genome editing. In addition to the spectrum of features among domesticated chickens and the potential to create new genetic variants with genome editing, chickens also have the capacity to access genetic variants from the four wild species of *Gallus* junglefowl through hybridization.[3] Through selective breeding following such hybridization, genetic introgression could further augment the novel combinations of features possible.

Quail. Quail lack the full-throated industrial backing that chickens enjoy, but the ace up the sleeve of this fellow galliform bird is the speed with which they can propagate. The Japanese quail *Coturnix japonica* can reproduce within just seven weeks after hatching from its 16-day incubation inside the egg, each female laying about one egg per day, yielding up to five generations every year.[4] Quail, similar to chickens, are quite hardy and easy to handle, occupy little space and are catholic in their food requirements.

Evolving Tomorrow. Asher D. Cutter, Oxford University Press. © Asher D. Cutter (2023). DOI: 10.1093/oso/9780198874522.003.0012

Quail share another perk with chickens: the ability to hybridize with other species. In particular, *Coturnix japonica* can hybridize with the common quail *Coturnix coturnix*, providing a simple means to source additional genetic variants. As petite as the quail might be, analysis of its genome has helped link gene variants to the characteristics that they influence and recent advances in avian stem cell biology now permits genome editing in quail as well as chicken.[5] Our understanding of its genes and mutations, however, lags far behind chickens, and the barnyards of the world showcase fewer distinct breeds of quail with fewer conspicuous differences in outward appearance. As compelling as quail may be for speeding evolution along due to their fast generation time, the genetic and trait limitations seem too overwhelming to prioritize *Coturnix* over *Gallus* as a founding stock for a first foray.

★

Chicken geneticists have zeroed in on 14,058 parts of the genome that contribute to differences in 439 different measurable characteristics.[6] The characteristics range from things like tail feather length to earlobe color to leg bowing to aggressive behavior; from contrafreeloading (a preference for food that requires effort to obtain) to polydactyly (extra toe digits) to excreta weight (how much poop they make); testes weight, foot feathering, wattle length, comb shape, feather pigmentation, skin color, and eggshell strength. The list goes on and on. In addition to external features, it includes an impressive array of physiological characteristics, disease susceptibilities, and details of vital organs and bones, such as gizzard weight, jejunum length, and tibia strength.

In the domestication of chicken breeds, the head comb was a common target of artificial selection. You may be familiar with the stereotypical carnation style of comb, a showy red display of fleshy skin in a single spiky crescent, arcing back atop a rooster's head, akin to a red punk-rock hairdo, found on breeds originating in Spain like Empordanesa. But chicken breeds have a wide range of comb shapes determined by their genetic composition, including the double-row of a buttercup comb, the devilish two-pronged V-comb, and the bulbous and bumpy mound of flesh that defines the walnut comb.[7] The rose comb forms a fleshy cushion with a flat top and, genetically, results from chickens having a type of genetic variant known as a chromosomal inversion.[8]

The beauty of all these links between measurable characteristics (phenotype) and their encoding in the genome (genotype), what geneticists call the genotype-phenotype map, is that it allows one to selectively breed interesting and distinct combinations of features. While not as simple as rearranging the features on Mr. Potatohead, the distinct genetic causes of distinct chicken characteristics let you conduct selective breeding to combine them into defined collections.

Let's focus on a few (Figure 12.1). First, the Dong Tao breed is famous for its striking legs: thick and scaly, like the foot of an ostrich's fat uncle transplanted onto a chicken. La Fleche chickens, in turn, have a big and bright-red V-shaped comb on their heads, like the horns on a child's devil costume at Halloween. Smooth-skinned featherless chickens would provide an especially arresting look to the body,[9] especially if combined with the gorgeous bluish-black hue of the skin found on the Silkie breed. Blue eggshells also are

(A) (B) (C) (D) (E)

(F) (G) (H)

Figure 12.1 *Diverse features among chicken breeds are genetically heritable. (A) Featherless chicken breed, owing to alteration of the FGF20 gene. (B) Laced feathers of the Silver Sebright chicken breed have black outlining around otherwise white feathers. (C) This rumpless chicken lacks a pygostyle and tail feathers. (D) The Saipan Jungle Fowl breed is among the tallest chicken breeds, approaching 90 cm in height among roosters. (E) A variety of eggshell colors from different chicken breeds, including white, brown, green, and blue; the large grayish egg in the middle is a duck egg. (F) Silkie and Ayam Cemani chickens produce dark melanin pigment in their skin, muscles, bones, and other tissues. (G) The fleshy red V-comb on this Appenzeller Spitzhauben chicken gives an arresting look to its head. (H) The Dong Tao breed of chickens from Vietnam are characterized by exceptionally thick and scaly legs. Image credits: (A) reproduced under the CC BY 2.0 license from Wells, K.L., et al. 2012. Genome-wide SNP scan of pooled DNA reveals nonsense mutation in FGF20 in the scaleless line of featherless chickens.* BMC Genomics. *13: 257. https://doi.org/10.1186/1471-2164-13-257; (B) Latropox, reproduced under the CC BY-SA 3.0 license; (C) reproduced under the CC0 license; (D) Amuseofpc, reproduced from the public domain; (E) photo courtesy Beatrix Cutter; (F) Shubert Ciencia, reproduced under the CC BY 2.0 license; (G) Nienetwiler, reproduced under the CC BY 2.5 CH license; (H) Hoang Quoc Phuong, Shutterstock ID#1658260363.*

a nice treat, as found in the Araucana. Some of this selective breeding to combine these traits together is already underway.[10]

To permit flight, however, the animal's stubby forelimbs need feathers. Perhaps proper expression of the *FGF20* gene—the gene disrupted to confer featherlessness—could be localized only to the wings and tail to allow growth of flight feathers despite an otherwise smooth body. A similar expression trick for *FGF20* might be needed for the legs, as *FGF20* affects the development of skin scales and so might adversely influence the growth of thick legs from Dong Tao. For feather pattern, I am partial to the Silver Sebright's so-called laced feathers, which look as if each white feather were outlined with black ink in preparation for a paint-by-numbers project. Alternatively, the yellowish hue inside the black rims of each wing feather on a Golden Laced Wyandotte—the black-laced feather edges often glinting an iridescent blue—against a backdrop of black bare skin of the body would cut a striking form reminiscent of black-and-yellow warning coloration. Conveniently, many individual gene variants controlling chicken feather color have been characterized,[11] opening the door to using genetic engineering and genetic welding in addition to selective breeding in creating specific trait combinations.

As for body stature, Modern Game chickens offer a look of powerful athleticism and confidence, a Roman statue of a velociraptor carved in bird flesh. Modern Game birds also are more flight capable than the meaty breeds, offering a source of genetic variants that could be targeted by selective breeding for improved flying ability. Junglefowl, after all, typically are not visions of airborne grace, in contrast to pigeons and geese and swallows. What about temperament? There is the aggression of the Shamo fighting birds or, perhaps more wisely, the cuddliness of a Silkie.

These conspicuous chicken features already exist and are available to arrange in this novel combination. To really push the envelope—to tweak traits to the extreme, to experiment with hypothetical changes and whimsy, to produce novelty—we must turn to artificial selection and genetic engineering.

★

Some chicken characteristics are genetically variable, but not in a way that arises from the influence of a single natural genetic variant with a large impact. These characteristics depend on the genetically variable influence of many genes. These are the so-called quantitative traits. Beak shape is one such quantitative trait. Surprisingly, however, unlike virtually every other aspect of chicken morphology, there appears to have been very little attention to modifying beak size and shape. Nonetheless, the length of chicken beaks is genetically heritable such that about 20% of the variability in beak length among a given set of birds can be explained by the genetic composition of their parents.[12] This substantial genetic variability means that artificial selection will be extremely effective in changing chicken beaks.

By selectively breeding just those birds with the largest beaks or just those birds with the smallest beaks, we can drive the evolution of beak size and shape to be bigger or smaller, broader or narrower. Depending on how selective that selection criterion is—the S term in the breeder's equation $R=h^2 S$—the evolutionary response can be rapid. In the case of the chicken beak, we could expect up to a 20% change in the average size each generation. This kind of response compounds like interest on a financial investment: after just five years, 10 chicken generations, their beaks could evolve to be nearly five times bigger or smaller than they started out.[13]

This single-trait selective breeding works just fine, but we might care about many traits, not just one. If there are many traits that we are interested in promoting simultaneously, then it would be helpful to create and use a multitrait selection index. This index would be a weighted average of all the features we care about, incorporating information about their variability and cross-correlations because of both genetic and environmental factors. In applying that breeder's equation to artificial selection, the typical presumption is that variability in the observed trait among individuals is due to the combined effects of many, perhaps thousands, of genes. If genome-wide association studies allow us to predict trait values from individually tiny effects of many locations across the genome, then we could create a genomic selection index to use in guiding a program of selective breeding. Depending on the feature, it might even be quicker and easier to apply the genomic selection index than to measure the trait of every individual.

For help in guiding a strategy for genome editing to alter beak size and shape, however, we might turn to genetic insights from the analysis of beak evolution in Darwin's finches. By comparing the expression of genes in tissues of developing finch beaks, Arhat Abzhanov and his team were able to identify, and then confirm in experiments with chicken beaks, important differences in how a collection of at least five genes influence beak size and shape.[14] One set of genes control the development of the premaxillary bone, the major structural component of the upper jaw, genes with the acronym-laden names of *TGFβIIr*, *β-catenin*, and *Dkk-3*. Differences in expression of another set of genes, *Bmp4* and *CaM*, influence the development of the other major determinant of beak morphology: prenasal cartilage. By using CRISPR-Cas9 genome editing to tweak the expression of these genes, we have another inroad to manipulating the size and shape of the mouthparts of the founding stock of our creature on its way to becoming a new and extraordinary species. Whether by artificial selection or genetic welding, we might drive the evolution of a short, wide beak as yet another distinctive trait, akin to the short-faced characteristic of fancy pigeon breeds like the so-called African Owl Pigeons.[15]

As we explored earlier in this book, developmental experiments have already demonstrated the potential to induce tooth expression in chicken jaws. Exploratory experiments in this vein could refine our ability to genetically engineer and evolve our novel creature to possess a toothed mouth rather than a beaked mouth. Similarly, the rudimentary relics of claws on some chicken's wing tips raise the possibility of elaborating the size and functionality of wing claws.[16] Birds like the Venezuelan hoatzin *Opisthocomus hoazin*, however, develop true claws on their wings, from the tips of their relatively long forelimb finger bones. Hoatzin chicks use their wing claws for alternating-limb quadrupedal walking and climbing and even swimming, though the wing digits don't continue to grow to have similarly large proportions in adults.[17] Developmental analysis of hoatzin chicks could point out which genes to target for altered expression in chickens to enhance the growth of wing claws and multifunctional digits. One can't help but wonder what genetic variability there might be in claw development, from one animal to the next across the world's barnyards, that natural selection or selective breeding might then drive for the evolutionary exaggeration of wing claws into fully functional digits.

★

Genetic engineering can assist not only in the modification of features like beak size, but in the exploration of extraordinary characteristics. We could, for example, consider engineering extraordinary sensory perception.

We could, for instance, upgrade the chicken's innate sense of smell. Their genome contains only 266 functional odorant receptors,[18] a far cry from the mammals you know best: with about 400 we've got half-again as many, and elephant genomes have nearly ten-fold more of them. We could expand the odorant receptor repertoire of the standard chicken genome with a constellation of distinct versions of these GPCR proteins (as introduced in Chapter 10). Because we might like to insert hundreds of types of cargo with our CRISPR-Cas9 genome editing, inserting them one-by-one would be extremely onerous. We'd want to adjust our strategy. One way to do this is to take advantage of the genomic graveyard of nonfunctional DNA in the form of repetitive sequences, the remnants of jumping gene transposable elements. The genomes of all species, including chickens, are littered with these DNA sequences. Design of guide RNAs that target repetitive sequence elements of the genome would allow, potentially, simultaneous editing of thousands of positions in the genome at once with just a single guide RNA.[19]

We also could upgrade their senses with tweaks to that other type of sensory protein: opsin light receptors. Chicken retinas express four types of cone opsin proteins to give them tetrachromatic vision, like many birds, providing a detectable span of wavelengths of light from 370 nm to 700 nm.[20] This contrasts with the less-sophisticated trichromatic vision we have as humans. Chickens, unlike starlings, however, can't see ultraviolet light. To expand their visual perception, we could edit the chicken short-wavelength-sensitive 1 opsin protein, known as SWS1, to mimic the starling SWS1 protein, shifting its peak detectable wavelength by about 50 nm toward UV-sensitivity. Proof of concept of this idea was demonstrated over 20 years ago in cell culture by changing a single amino acid encoded by the chicken SWS1 protein.[21] By placing such a modified gene into a CRISPR-Cas9 gene drive cassette, we could then drive it through our founding stock population without depleting natural genetic variation: genetic welding in a controlled setting.

More radical, one could express in chicken retina cells the NeoR opsin protein found in fungus or the far-red-sensitive opsin from *Malacosteus* stoplight loosejaw dragonfish to then test whether it permits detection of near infrared wavelengths of light.[22] If so, we'd have a means of shifting the detectable wavelengths by up to 120 nm longer than the standard chicken LWS long-wavelength-sensitive opsin protein can do. But why stop with tweaking just the ends of spectral sensitivity? By inserting another fistful of opsins with varying sensitivities into the genome, we might enhance their visual acuity throughout the visible range.

Speaking of wavelengths of light, how about a chicken that emits its own light? In truth, they already exist. The University of Edinburgh's Roslin Institute has generated genetically modified chickens, for research purposes, that glow green or red—the so-called Roslin Green and Flamingo lines of chickens.[23] Their tissues emit green or red light when exposed to a UV lamp, like the black lights at a dance club or in a crime scene investigation. This basic idea could be extended in many creative ways. We could, for instance, restrict red fluorescence to cells in the comb on the chicken's head, a punk-rock Rudolph. These fluorescent birds use a Nobel Prize–awarded technique pioneered with the humble nematode roundworm *C. elegans* to express, in a controllable way, the green fluorescent protein (GFP)—a protein sourced from the genome of the jellyfish *Aequorea victoria*. Biotechnologists have now tweaked GFP and identified other fluorescent proteins that can emit red, yellow, blue, and other wavelengths of light.[24]

Glowing creatures may seem synonymous with genetic modification run amok and cartoon visions of radioactivity, but nature abounds with species that already produce fluorescent proteins and bioluminescence.[25] You know the blinking lights of fireflies in summer, the flashes of these lusty beetles powered by their luciferase enzyme; perhaps you've witnessed the eerie incandescence of green-glowing fungi on a forest floor at night; or deep-sea documentary footage of the ominous lure, dangling and shining above the frightful gape of a toothy anglerfish. Literally thousands of different species can emit their own light.[26]

★

One feature that chickens lack, that all birds are missing, in fact, is a sinuous and muscular tail. They have tail feathers, sure, and a fleshy rump beneath, but no proper tail. Instead, birds have a pygostyle: a stumpy skeletal structure of fused vertebrae at the posterior end of the animal surrounded by fatty tissue.

The smaller number of distinct vertebrae in the tail-end of birds gets determined in embryogenesis. As the embryonic form takes shape, tissue regions called *somites* define the segments that govern body elongation. In vertebrates, many mutations can act to disrupt this elongation to truncate the tail. Only a couple of mutations are known to increase features associated with tail length, mutations to genes like *Hoxb13* and *Mesogenin-1*.[27] In nature, however, the evolution of lizard tail elongation involves developmental acquisition of additional vertebrae, and some populations of North American deer mice *Peromyscus maniculatus* recently experienced adaptive evolution for longer tails, arising through both more vertebrae and longer individual vertebrae.[28] Domestication of pigs also has led to the evolution by artificial selection of longer animals with up to four more vertebrae than their wild boar progenitors because of alterations of the function or expression of specific proteins, including those called NR6A1 and VRTN.[29]

By thoroughly investigating the genetic causes of these kinds of cases, we could decipher the developmental genetic pathways that produce more or longer vertebral bones. By tweaking the regulation of the resulting gene candidates, one could explore how genetic variants could elongate bird butts to turn a pygostyle into a tail. If genetic manipulations prove too difficult to reconstitute a windy tail complete with vertebrae and muscles in our founding stock, then we might choose to eschew a tail altogether and give it a truncated backside. To do so, we could incorporate the rumpless characteristic of chicken breeds like the Rumpless Araucana that sports no protruding pygostyle whatsoever.[30]

Brain size and behavior also exhibit genetically controlled differences among chickens. With a multitude of regions of the genome associated with brain size,[31] we might devise a genomic selection index to favor the evolution of animals with enlarged nidopallium and mesopallium portions of the brain. These regions form the so-called avian prefrontal cortex that is especially big in crows and parrots, the smartest of nonhuman animals.

How social might our new species be? Birds often flock together in groups. A *mob* of emu, a *murder* of crows, a *squabble* of gulls, a *congregation* of plovers. Among predatory mammals, we have *prides* of lions, *pods* of orcas, *packs* of wolves. Grazing animals also often congregate in groups, more uniformly termed *herds*. If we were to evolve a dragon that ends up clustering into social groups, what name would we give the collective? Perhaps a term would spring from the lips after seeing a quirk of their behavior, wrought as an unintentional byproduct of selection and engineering of other characteristics. But there is no cost to ruminating about premeditated possibilities of group names: a *squadron*, a *shriek*, a *squall* of dragons? Might we, one day, stumble across a *huddle* of herbivorous hyperexotics chewing their way across a field of kudzu?

★

So far, we've focused on evolving outwardly visible characteristics like coloration and feather production. These features are easy to understand and easy to explain. These features often have a well-understood genetic basis but come at the expense of appearing frivolous. A more compelling case for evolving a new and hyperexotic species, from a societal standpoint, could seek to produce features that serve a desirable ecological function, to help resolve some of the grand problems of the Anthropocene. Bioremediation, carbon sequestration, ocean deacidification. Perhaps a special predilection for consuming purple loosestrife or kudzu, two notoriously invasive plants in North America, the bane of northern wetlands and southern woodlands? Perhaps evolutionary tolerance of a gut microbiome capable of breaking down plastics or spilled crude oil?

Of course, it is debatable whether a big land animal is the logical choice for any such roles. It may be the case that evolving smaller critters—microbes, insects, snails, corals, and plants—with the tools of genetic welding to engineer bespoke species would yield a more effective and palatable, if less flashy, means to an end.

★

Among the outward traits to evolve, size comes last. To evolve an extraordinary creature toward gigantism, an animal founded with avian genetic stock, it will inevitably drag along with it a slower development and a longer time to reach reproductive maturity. Slower turnover of generations impedes the pace of change in other traits, so we should save for last any selective breeding and genome editing that aims to exaggerate the mass of the creature toward something truly extravagant.

Artificial selection and selective breeding, of course, depend on preexisting genetic variants. They can only realize evolutionary change because of those genetic variants that already exist in the founding stock population. In a given amount of time, they can only take evolution so far. Genetic engineering offers one path to circumvent the limits of preexisting variants, to push evolution in a particular direction, such as toward larger size.

Genetic engineering, however, suffers its own limitations, with two being most crucial. First is the limitation of existing knowledge: we first have to know what to modify in order to engineer a desired outcome. Sometimes we can take an educated guess, for example, by creating a modification based on knowledge gleaned from the genetics and development of other kinds of organisms—say, from studies in pigeons or mice or dogs or humans. Using this logic, CRISPR-Cas9 disruption of the chicken version of the myostatin protein yielded birds with more muscle mass and less fat, similar to what happens in "double-muscled" mammals.[32] But only in male birds. We know a lot, for sure, but what we don't know adds up to a lot more.

Second is the limitation of pleiotropy: a genetic modification that creates one desired outcome may also exert unacceptable additional effects on other characteristics because of the multiple effects that a single genetic variant can have on physiology and development. One example of big pleiotropic effects involves genetic variants of the KITLG gene. Most conspicuously, variants of KITLG influence the intensity of coloration of dog fur. Dogs with more copies of the KITLG gene—up to ten copies instead of two—have hair cells that make more eumelanin or pheomelanin, the pigments causing black or red coloration, and, therefore, a deeper color to their coat. You can see this genetic effect at work in breeds like the deeply melanic Standard Poodle and Rottweiler and in the rich red of Irish Setters and Brittany Spaniels. Unfortunately, this same genetic influence gives them a higher risk for squamous cell carcinoma, an aggressive cancer that afflicts their toe bones.[33] We expect multiple impacts to arise more commonly when genetic changes have a big effect, and big effects usually are the goal when editing particular spots in the genome.

Taking a longer-term view, another option is to resupply the catalog of genetic variants that are available for selective breeding by creating new variability. For example, by lightly mutagenizing the reproductive tissues or temporarily disabling DNA repair capabilities of reproductive tissues, we could accelerate the arrival of novel genotypes into the population. By mimicking the natural mutation process, simply ramping up its rate, these random changes into the genome would occur at unknown locations and with unpredictable effects. Of course, many measurable effects would be detrimental, and this could trigger concerns over animal welfare. But some mutations would be beneficial with respect to the goals of selective breeding. If it were untenable to accelerate rates of mutational input, we could, instead, expand the population to enormous numbers and wait for natural mutations to arise. A bigger census means more opportunities for DNA replication errors during the cell divisions that make sperm and eggs. More errors will introduce more novel genetic variants through mutation to any given gene—provided that we could identify it and select it for breeding.

In this way, a continuous program of selective breeding for 100 or 1000 generations—50 or 500 years with chickens, 20 or 200 years with quail—could bypass the restrictions of the necessarily limited assortment of genetic variability within the founding stock population. Genome editing or mutagenesis may hold the power to producing a descendant of the founder stock taller than the tallest Malay or Indio Gigante rooster (approximately 90 cm (3 feet)) to set it on an evolutionary course parallel to the sizes achieved by ostriches and moa.

12.2 Mary Poppins's umbrella

One thing that we absolutely cannot expect in evolving a dragon is perfection. Physics and genetic networks and variable environments and natural selection conspire to ensure that no so-called Darwinian demons will exist.[34] Darwinian demons are those hypothetical creatures with maximal fitness no matter the circumstances; seemingly, a characteristic of the cinematic sci-fi classic aliens of *Alien*. In our case, the measure of evolutionary success is biological novelty, perhaps aligned with features inspired by fantasy or environmental functions. Even with this narrowed view, those same constraints keep the evolution of a dragon from being, as Mary Poppins might say, practically perfect in every way. Mary Poppins may be practically perfect in every way, but the parrot-headed top to her umbrella speaks with the voice of common sense. No Darwinian demons and no Mary Poppins. Instead, we'll have to set our sights on what we might call a minimally acceptable dragon.[35]

What we can do is take a practical mindset to pursue genetic criteria. That is, try to avoid, as much as possible, all the kinds of evolutionary speed bumps. For instance, in conducting artificial selection and genetic engineering, we'd be wise to prioritize those traits and those genes that we expect to cause the fewest knock-on effects to other features. We'd also be wise to avoid butting heads with the laws of physics, to use preexisting genetic variants to our advantage, and to take steps to avoid the unintentional loss of genetic variability.

Having a very large number of individuals in the population will benefit selective breeding. A large population allows artificial selection to cull the breeding group down to a small subset of them without too badly depleting the overall genetic variability available for future selection. It also avoids creating excessively long strings of DNA along chromosomes that are devoid of variability; in the jargon of geneticists, we'd like to avoid creating large deserts of polymorphism with haplotype homozygosity. Failure to manage these potential problems can constrain future evolution. Not managing inbreeding also can produce the evolution of unintended and unwanted characteristics, as we discussed in Chapter 3, such as health problems and susceptibility to infectious disease. See domestic dogs, again, for their many examples of detrimental features dragged along with traits that are otherwise desirable, due to multiple effects of pleiotropy and to linkage effects of genetic hitchhiking. Cattle and horse farmers typically take pains to prevent inbreeding for exactly these reasons, despite the seeming merits of using just a single prize stud with all of the dams. A large population that is genetically well-mixed helps avoid these problems of inbreeding.

A large population, however, comes with a cost, a cost beyond the basic costs of extra food and space. A large population makes genome editing to modify the entire population more difficult. When we make a genome edit, the genetic modification occurs in just a single version of a genome. That single version, initially inside of just a single individual, would then need to be bred into the rest of the stock, while, at the same time, trying to avoid those problems of inbreeding in the procedure of selective breeding.

To spread genome edits throughout the founding stock population without this snag, we can turn to that evolutionary tool of biotechnology: we could tether them to a gene drive. That is, we can harness genetic welding to help evolve the population in a particular direction. CRISPR-Cas9 genetic welding with gene drives does not suffer these problems of inbreeding. With genetic welding, the engineered variant leapfrogs through the population in a way that does not depend on recombination or intentional pairing of particular parents. This means there will be no reduced genetic variability elsewhere in the genome when a gene drive is used to spread a genetic modification in our nascent species.

The concern, however, transforms in kind. Like how the pink stain comes back in a new guise to haunt Sally and her brother as they spar with the Cat in the Hat,[36] genetic welding shifts the challenge to one that asks: How can we contain the gene drive?

★

The Voom under the hat of Little Cat Z solved all of the Cat in the Hat's problems. We are not so lucky. The problem is that gene drives are designed to propagate themselves in DNA from one generation to the next at a pace faster than the normal Mendelian rules of inheritance allow. If we weren't sufficiently careful, the drive could spread to all of the birds in the species, not just the intended derivatives of our founding stock. The genetic modification might even jump into the genomes of other junglefowl species, their genomes bridged through occasional hybridization events. Global spread could arise from just a single DNA copy of a gene drive cassette in a single genome as the gametes of that individual mix with naive gametes of other individuals. Accidental mating with members of another population or species could be the spark to light genetic infection beyond the intended group of individuals.

This possibility is most worrisome when gene drives are designed for population suppression—as intended for the control of mosquito vectors of malaria. Genetic welding, however, exploits modification gene drives and not such suppression gene drives. This distinction may nonetheless provide cold comfort among the folks most alarmed by potential spread of any human-mediated genetic modification.

Laboratory research into gene drives, however, has already defined a suite of confinement strategies.[37] In the laboratory, for example, one can first engineer a synthetic DNA sequence that is not found in the wild; gene drives could then be designed only to integrate into the novel synthetic sequence, making wild populations immune. This trick, however, is infeasible when trying to maintain high genetic variability in a founding stock population to be subject to artificial selection, genetic introgression through selective breeding, as well as genetic welding. If the goal is to evolve a novel species that could find a home in nature, then standard lab biosafety guidelines for confinement are moot. We'll need to articulate some back-up plans.

To restrict gene drives from spreading beyond an intended population, we would want to engineer them to be self-limiting. The toxin-antidote ClvR (pronounced "cleaver") and TARE techniques for implementing CRISPR-Cas9 gene drives have this property.[38] Another way to induce self-regulation is to split up the different components of the gene drive cassette into interdependent separate pieces. The result is a so-called split drive or daisy chain gene drive.[39] The idea is to make multiple transgenes that get inserted into different spots in the genome, such that only one encodes the gene drive component that you intend to spread. For three pieces C-B-A intended to spread the effects of the edit to the genome caused by piece A, the C piece drives increase in prevalence of B, and B drives increase in prevalence of A—but, crucially, A can't drive its own increase like a standard global gene drive could. The capacity for this three-part CRISPR-Cas9 daisy-drive to spread requires that all three elements occur together in the same individual. Because there is no selective pressure to retain piece C, it will tend to decline in abundance over time and, consequently, put a break on the ability of the genetic modifications to invade, and will prevent genetic invasion of nontarget organisms.[40]

More ruthless: Engineer the gene drive to encode an inducible genetic kill switch. If the gene drive seems to be getting out of hand, flick the switch to disrupt CRISPR-Cas9 functionality or even to exterminate all individuals that have copies of the gene drive cassette. This permutation of a so-called sensitizing drive would provide a built-in genetic dragon slayer.[41] The trigger could be an external cue, such as a pheromone or diet change or a synthetic compound that is otherwise innocuous.[42] Similarly, the gene drive could be positively inducible, meaning that, by default, the Cas9 protein is inactive. With positive inducibility, the Cas9 and guide RNA only get activated in the presence of an appropriate external cue, requiring, say, a synthetic compound not present in the wild that gets added to the feedstock to make the gene drive functional only in a captive population.

A less scorched-earth tactic to prevent a gene drive from spreading beyond a target population is to genetically engineer the population to be reproductively isolated. In other words, engineer a species boundary.[43] We saw earlier how "dead" dCas9 can be co-opted to produce reproductively isolated populations, genetically separated by so-called underdominant effects. In evolving a new species, this kind of thing is something that we definitely want to do.

Of course, the walls that define species boundaries are tallest when they keep species from mating in the first place, using genetic enforcers only to keep out the few curious lurkers over the fence. If we could evolve species-specific mating tendencies—seeking one's own kind and rejecting the advances of others—then that would help to prevent hybridization in the first place. Because such behaviors are most stable in maintaining separate gene pools when they evolve from new genetic variants,[44] genetic welding with novel variants that create fixed differences relative to wild populations would be an especially potent force in fostering sexual selection in species discrimination. We might use the peculiar plumage and extended spectrum of visible wavelengths and smellable smells, courtesy of an expanded opsin and odorant receptor repertoire, to assist in mate recognition. We could engineer genetic elements with sexually antagonistic properties to foster rapid divergence in genes expressed by females versus males, further sending their behavior and physiology on an evolutionary trajectory distinct from the fowl of the jungle and barnyard.

12.3   A recipe

Let's sum up. We are here for an algorithm, after all. Here is a six-step recipe.

First, establish one or more large founding stocks of individuals with a convenient suite of traits that ease the logistics. The domestic chicken, *Gallus*, provides a compelling candidate.

Second, apply selective breeding to combine together those genetic variants controlling preexisting desired characteristics. Complement selective breeding with genetic introgression of distinctive features from other species capable of hybridizing with the founding stock, such as from other junglefowl species. Prioritize characteristics that are least likely to impact multiple aspects of development. Artificial selection on characteristics with complex or unknown genetic inheritance, such as beak size, will require large populations of genetically diverse interbreeding individuals. Apply techniques that allow evolution of many traits in parallel, rather than one after the other in series. A genomic selection index that uses DNA testing can help to avoid adverse effects of inbreeding and also to propagate efficiently a large constellation of features simultaneously.

Third, perform CRISPR-Cas9 genome editing to modify gene expression to elaborate existing features and to introduce novel genetic elements that confer novel properties. For example, to induce tail elongation and to expand the visibly detectable wavelengths of light. This step may require cloning via somatic cell nuclear transfer and use of surrogate eggs and surrogate wombs if a mammal is chosen as the founding stock. After establishing the desired effect, use CRISPR-Cas9 gene drives to spread DNA modifications throughout the genetically diverse population stock that is, simultaneously, being subjected to artificial selection.[45] This procedure merges genetic welding with the other forces of evolution.

Fourth, genetically engineer the captive stock population to be reproductively isolated from its progenitor. This step occurs after having completed all introgression of genetic material from distinctive breeds or hybridization with other species. Establish reproductive isolation through Dobzhansky-Muller incompatibilities, toxin-antidote systems, or underdominance, or perhaps employ synthetic biology techniques that involve incorporation of unnatural amino acids. This step establishes the population stock as an evolutionarily distinct entity, what we might name *Gallus novus* or *Gallus syntheticus* or *Gallus hyperexoticus*.[46] If this new creature is to be a dragon, then perhaps we must settle on *Gallus draco*.

Fifth, expand the gene pool with genetic variants induced through mutagenesis to lay the groundwork for long-term evolution by artificial selection. Using artificial selection in combination with further genome editing propagated with genetic welding, we evolve body size toward gigantism. This stage of change may require concurrent modification of diet, with appropriate genetic alterations, as well as adjustment to the gut microbiome.

Sixth, conduct controlled tests to quantify the effects on, and by, *Gallus novus* when exposed to a diverse set of environmental conditions and species interactions. Identify an appropriate habitat that would allow it to persist by maintaining a viable and genetically diverse population. Perform test introductions with sterile individuals to assess any immediate ecological effects.

★

This recipe, of course, could be cooked up again and again. One could experiment with different evolutionary trajectories from *Gallus* founding stock. One could experiment with those alternative founding stock species that dropped from the initial running—the quail and pigeons and lizards, bats and flying squirrels, the guppies and eels—to see how far selective breeding and genetic welding could go in your lifetime, or in your grandchildren's lifetime. One might instead prioritize dragons of the plant kingdom or insects or fungi or microbes, and traits that serve-up environmental benefits.

By following this recipe, the Anthropocene could produce a stable of distinct species, a collection of minimally acceptable dragons—*mynnacks*. An entire novafauna of creatures with unique combinations of features and abilities to allow them to thrive as mynnacks of land and sea, of prairies and swamps, of mountains and cities, of sewage-treatment plants and clearcuts, of landfills and oil spills and the Great Pacific Garbage Patch. We have an updated peek at what Henry Bates reported to Darwin in an 1861 letter about his study of Amazonian butterflies:[47] "*I think I have got a glimpse into the laboratory where Nature manufactures her new species.*" In this way, evolutionary engineering of new and extraordinary species could generate ecosystem engineers to play key roles in restoring human-perturbed ecological systems.[48]

Evolving new species like these might also be symbolic for the generation of new bio-diversity. But make no mistake, it would not even make a dent in offsetting the extreme extinction rates the world is experiencing today. The most expedient way to have a lot of species in the world is to prevent their undue extinction in the first place.

★

We can now see a path for how to evolve a dragon, but what capital costs would it impose? What would it cost in money, space, infrastructure, personnel, and time, if built from scratch? Let's acquire 100 acres of farmland, which currently averages $4400 per acre in the United States. We'll build a few poultry barns and a molecular biology facility, along with their relevant equipment, a facility that we might entitle the Artemis Institute. Instead, you might set up a partnership with an ag-school or private industry. Depending on the location, this startup cost might, conservatively, set us back $5 million. Now, ongoing costs. Let's envision a modest staff of 20, from farm hands to gene jocks to paperwork pushers. The per-animal cost of feed, using chickens as a guide, might run $50 per year, making a huddle of 1000 run to $50,000 annually. We might donate excess eggs to a local food bank, if the law allows, perhaps offering a modest tax break. Personnel will represent about 75% of the ongoing expenditures, which, at $2 million per year for round numbers, implies an overall annual outflow of about $2.7 million. To get 20 or 30 generations of evolution through selective breeding and genetic welding, at 2 generations per year, we need to plan for at least 15 years of effort; but let's call it an even two decades. Twenty years would add up to about $60 million in today's dollars.

Is a dragon—or, as naysayers might shout, a very fancy chicken—worth $60 million? In today's market, an all-black inside-and-out Ayam Cemani chicken might fetch $2500, and a single fancy racing pigeon could set you back $1.9 million.[49] Private investors just injected $15 million to start up a de-extinction company, which as we'll see in the next chapter, explores a similar niche to evolving new and extraordinary species. Suddenly, $60 million for a population of very fancy chickens doesn't sound so fanciful. What intellectual property might be devised along the way? What new scientific knowledge and educational opportunities might the project produce? What bioremediation solution might result? What public-private partnerships might emerge as business spin-offs? Ask your local venture capitalist.

★

Biotechnology has the capacity to help produce novel forms of life. Genetic welding can hijack the genetic composition of nature in what was the wild, and suppression gene drives are capable even of driving species extinct through intentional genetic interventions. The full consequences for the ecology of organisms that lie beyond our direct management are only partially predictable. We must ask ourselves, when we lay our moral compass on the map of the world, does the needle point toward the realm that declares "here be dragons"? And, if we embark on a quest to bring about new creatures, will we sleep soundly?

Should we succeed in incorporating a cornucopia of genetic modifications into a founding stock—evolving a creature through selective breeding and artificial selection and genetic welding, producing a population of individuals with a unique combination of characteristics never before witnessed by the wilds of the Earth, that includes newly engineered features and capabilities, reproductively isolated from other biological entities as a distinct species—will what we have created be a beautiful beast or an abomination? How much does the answer lie in the eye of the beholder, and do we weigh the judgments of some beholders in higher regard than others? So far, we've set to the side the ecological consequences and ethical implications as we've simply considered the what and the how of technical possibility. The answers to these ecological and ethical questions are not written in the tech specs of our algorithm. But we must write about them, nonetheless. These are the dragons of the next chapters.

★

It is possible to evolve new species through a combination of selective breeding and genetic welding. Genetic engineering permits the directed evolution of organisms as genomic chimeras, using genes sourced from distant species, as well as genetic properties not found in any existing organism. The chicken, *Gallus gallus*, provides an especially promising founding stock on which to base the evolution of new extraordinary animal species. Evolutionary engineering that uses genetic welding to form new species or to modify wild organisms, however, could in principle be applied to any sexually reproducing animal or plant. Evolving a species with visible novelties alone, however, does not in itself justify the creation or release of a new viable population.

Notes

1. You might have expected *Columba livia* pigeons on this short list, as pigeons share many of the virtues of chickens. They are easy to rear and capable of two generations per year. Their genome is well-characterized, and researchers have pinned down many of the gene variants that cause differences in their traits. The variation in features, however, is not as diverse as for chickens. Pigeons are smaller birds and stronger flyers, which may be viewed

both as advantages and disadvantages. Another limitation, relative to *Gallus*, is the less-mature foundation it provides that is due to the smaller historical investment in research and breeding techniques. This practical consideration, in the absence of a clear bonus feature that sets them apart, points to the chicken or quail as the most expeditious alternative to start with.

2. https://ourworldindata.org/meat-production#number-of-animals-slaughtered accessed June 3, 2021; http://amerpoultryassn.com/sample-page/apa-breeds-varieties/accepted-breeds-varieties and http://www.fao.org/dad-is/dataexport/en accessed June 3, 2021.

3. van Grouw, H., and W. Dekkers. 2019. Various *Gallus varius* hybrids: variation in jungle-fowl hybrids and Darwin's interest in them. *Bulletin of the British Ornithologists' Club.* 139: 355–371. https://doi.org/10.25226/bboc.v139i4.2019.a9.

4. The king quail, also known as the Chinese painted quail *Excalfactoria chinensis*, has similar reproductive capabilities to the Japanese quail. King quail have especially beautiful blue-hued feather coloration and are about half the size of Japanese quail, though less well-established in terms of genetic features. Tsudzuki, M. 1994. *Excalfactoria* quail as a new laboratory research animal. *Poultry Science.* 73: 763–768. https://doi.org/10.3382/ps.0730763.

5. Huss, D., et al. 2008. Japanese quail (*Coturnix japonica*) as a laboratory animal model. *Lab Animal.* 37: 513–519. https://dx.doi.org/10.1038/laban1108-513; Minvielle, F. 2009. What are quail good for in a chicken-focused world? *World's Poultry Science Journal.* 65: 601–608. https://dx.doi.org/10.1017/S0043933909000415; Morris, K.M., et al. 2020. The quail genome: insights into social behaviour, seasonal biology and infectious disease response. *BMC Biology.* 18: 14. https://doi.org/10.1186/s12915-020-0743-4; Lee, J., et al. 2019. Direct delivery of adenoviral CRISPR/Cas9 vector into the blastoderm for generation of targeted gene knockout in quail. *Proceedings of the National Academy of Sciences USA.* 116: 13288–13292. https://dx.doi.org/10.1073/pnas.1903230116.

6. https://www.animalgenome.org/cgi-bin/QTLdb/GG/summary and https://www.animalgenome.org/cgi-bin/QTLdb/GG/ontrait?trait_ID=2414. Accessed June 3, 2021.

7. Chicken combs provided a key point of evidence in William Bateson's arguments to resurrect Mendel's buried ideas about inheritance in the early 1900s. Bateson's description of the pea comb shape is especially evocative, "a comb consisting of three fairly regular longitudinal ridges, along each of which are several more or less lumpy tubercles. . . . Any one who desires to examine such combs can see them any day in a poulterer's shop." p. 87 of Bateson, W. 1902. Part II: experiments with poultry. *Reports to the Evolution Committee of the Royal Society.* 1: 87–124. https://wellcomecollection.org/works/jt2ejhx9.

8. An inversion mutation causes a segment of DNA to reverse its order, so instead of a sequence reading ACCGGTTA a four-nucleotide inversion might make it read ACTGGCTA. The inversion variant responsible for the rose comb trait involves an inversion spanning over 7 million nucleotides on chromosome 7 of chickens. Imsland, F., et al. 2012. The rose-comb mutation in chickens constitutes a structural rearrangement causing both altered comb morphology and defective sperm motility. *PLoS Genetics.* 8: e1002775. https://doi.org/10.1371/journal.pgen.1002775.

9. Wells, K.L., et al. 2012. Genome-wide SNP scan of pooled DNA reveals nonsense mutation in FGF20 in the scaleless line of featherless chickens. *BMC Genomics.* 13: 257. https://doi.org/

10.1186/1471-2164-13-257; https://www.amusingplanet.com/2015/06/the-dragon-like-legs-of-dong-tao-chicken.html. Accessed June 3, 2021.

10. Wang, H., et al. 2021. Genetics and breeding of a black-bone and blue eggshell chicken line. 1: body weight, skin color, and their combined selection. *Poultry Science*. 100: 101035. https://doi.org/10.1016/j.psj.2021.101035.

11. Hua, G., et al. 2021. Genetic basis of chicken plumage color in artificial population of complex epistasis. *Animal Genetics*. 52: 656–666. https://doi.org/10.1111/age.13094.

12. The crossed-beak deformity in chickens is proposed to be due to a genetic variant in the *LOC426217* gene that affects keratin, but I am unaware of other described genetic causes of natural genetic variability in beak shape of chickens. Icken, W., et al., 2017. Selection on beak shape to reduce feather pecking in laying hens. *Lohmann Information*. 51: 22–27. Joller, S., et al. 2018. Crossed beaks in a local Swiss chicken breed. *BMC Veterinary Research*. 14: 68. https://doi.org/10.1186/s12917-018-1398-z.

13. This calculation presumes $S = 0.75$, which corresponds to breeding each generation only among the 25% of birds with the most extreme beak sizes, and that the heritability of $h^2 = 0.2$ persists for the duration of artificial selection. Given an initial average beak length of 2 cm (4/5 of an inch), we would predict the evolution of beaks as small as 0.4 cm (3/16 of an inch) or as large as 81 mm long (3.25 inches) in 10 generations. Refer back to Chapter 2 for deeper discussion of the breeder's equation.

14. Mallarino, R., et al. 2011. Two developmental modules establish 3D beak-shape variation in Darwin's finches. *Proceedings of the National Academy of Sciences USA*. 108: 4057–4062. https://dx.doi.org/10.1073/pnas.1011480108.

15. The Parrot Beak Aseel breed of chicken might also provide a good source of genetic variants to explore this beak morphology.

16. Fisher, H. 1940. The occurrence of vestigial claws on the wings of birds. *American Midland Naturalist*. 23: 234–243. https://dx.doi.org/10.2307/2485270.

17. Abourachid, A., et al. 2019. Hoatzin nestling locomotion: acquisition of quadrupedal limb coordination in birds. *Science Advances*. 5: eaat0787. https://dx.doi.org/10.1126/sciadv.aat0787.

18. Vandewege, M.W., et al. 2016. Contrasting patterns of evolutionary diversification in the olfactory repertoires of reptile and bird genomes. *Genome Biology and Evolution*. 8: 470–480. https://doi.org/10.1093/gbe/evw013.

19. Smith, C.J., et al. 2020. Enabling large-scale genome editing at repetitive elements by reducing DNA nicking. *Nucleic Acids Research*. 48: 5183–5195. https://doi.org/10.1093/nar/gkaa239.

20. Seifert, M., et al. 2020. The retinal basis of vision in chicken. *Seminars in Cell & Developmental Biology*. 106: 106–115. https://doi.org/10.1016/j.semcdb.2020.03.011.

21. Yokoyama, S., et al. 2000. Ultraviolet pigments in birds evolved from violet pigments by a single amino acid change. *Proceedings of the National Academy of Sciences USA*. 97: 7366–7371. https://dx.doi.org/10.1073/pnas.97.13.7366.

22. Broser, M., et al. 2020. NeoR, a near-infrared absorbing rhodopsin. *Nature Communications*. 11: 5682. https://doi.org/10.1038/s41467-020-19375-8.

23. https://www.ed.ac.uk/roslin/national-avian-research-facility/avian-resources/genetically-alter ed-lines. Accessed June 3, 2021.

24. Rodriguez, E.A., et al. 2017. The growing and glowing toolbox of fluorescent and photoactive proteins. *Trends in Biochemical Sciences*. 42: 111–129. https://doi.org/10.1016/j.tibs.2016.09. 010.

25. Oba, Y., and D.T. Schultz. 2014. Eco-evo bioluminescence on land and in the sea. In G. Thouand and R. Marks (eds). *Bioluminescence: Fundamentals and Applications in Biotechnology-Volume 1. Advances in Biochemical Engineering/Biotechnology*, vol 144. Berlin: Springer, pp. 3–36. https://doi.org/10.1007/978-3-662-43385-0_1.

26. Some animals express their own proteins or biochemicals that bioluminesce, whereas other animals use the bioluminescent properties of symbiotic bacteria. My most intimate foray with bioluminescent creatures involves an evening skinny dip in my youth amidst the gentle waves of the Coral Sea, in Zoe Bay on Australia's Hinchinbrook Island National Park, the water glowing blue-white from the dinoflagellate marine microorganisms. Tracts of bioluminescent microorganisms in the ocean, so-called milky seas, can be so vast as to be visible from space. As stated eloquently in a recent study, "The massive bodies of glowing ocean, sometimes exceeding 100,000 km^2 in size, persist for days to weeks, drift within doldrums amidst the prevailing sea surface currents." Miller, S.D., et al. 2021. Honing in on bioluminescent milky seas from space. *Scientific Reports*. 11: 15443. https://doi.org/10.1038/s41598-021-94823-z; Davis, M.P., et al. 2016. Repeated and widespread evolution of bioluminescence in marine fishes. *PLoS One*. 11: e0155154. https://doi.org/10.1371/journal.pone.0155154.

27. Economides, K.D., et al. 2003. Hoxb13 mutations cause overgrowth of caudal spinal cord and tail vertebrae. *Developmental Biology*. 256: 317–330. https://doi.org/10.1016/S0012-1606(02)00137-9; Fior, R., et al. 2012. The differentiation and movement of presomitic mesoderm progenitor cells are controlled by Mesogenin 1. *Development*. 139: 4656–4665. https://doi.org/10.1242/dev.078923; Rashid, D.J., et al. 2014. From dinosaurs to birds: a tail of evolution. *EvoDevo*. 5: 25. https://doi.org/10.1186/2041-9139-5-25.

28. Bergmann, P.J., and G. Morinaga. 2019. The convergent evolution of snake-like forms by divergent evolutionary pathways in squamate reptiles. *Evolution*. 73: 481–496. https://doi.org/10.1111/evo.13651; Kingsley, E.P., et al. 2017. The ultimate and proximate mechanisms driving the evolution of long tails in forest deer mice. *Evolution*. 71: 261–273. https://doi.org/10.1111/evo.13150.

29. Charles Darwin even cited this anatomical fact in 1868, as reported in 1837 by Thomas Eyton, esquire. Eyton, T.C. 1837. *Proceedings of the Zoological Society of London*. 23. https://doi.org/10.1111/j.1469-7998.1837.tb06823.x; Mikawa, S., et al. 2007. Fine mapping of a swine quantitative trait locus for number of vertebrae and analysis of an orphan nuclear receptor, germ cell nuclear factor (NR6A1). *Genome Research*. 17: 586–593. https://dx.doi.org/10.1101/gr.6085507; Duan, Y., et al. 2018. VRTN is required for the development of thoracic vertebrae in mammals. *International Journal of Biological Sciences*. 14:667–681. https://dx.doi.

org/10.7150/ijbs.23815; Rubin, C.-J., et al. 2012. Strong signatures of selection in the domestic pig genome. *Proceedings of the National Academy of Sciences USA*. 109: 19529–19536. https://doi.org/10.1073/pnas.1217149109.

30. Freese, N.H., et al. 2014. A novel gain-of-function mutation of the proneural IRX1 and IRX2 genes disrupts axis elongation in the Araucana Rumpless chicken. *PLoS One*. 9: e112364. https://doi.org/10.1371/journal.pone.0112364.

31. Höglund, A., et al. 2020. The genetic regulation of size variation in the transcriptome of the cerebrum in the chicken and its role in domestication and brain size evolution. *BMC Genomics*. 21: 518. https://doi.org/10.1186/s12864-020-06908-0.

32. Kim, G.-D., et al. 2020. Generation of myostatin-knockout chickens mediated by D10A-Cas9 nickase. *FASEB Journal*. 34: 5688–5696. https://doi.org/10.1096/fj.201903035R.

33. Bannasch, D.L., et al. 2020. Pigment intensity in dogs is associated with a copy number variant upstream of *KITLG*. *Genes*. 11: 75. https://doi.org/10.3390/genes11010075; Karyadi, D.M., et al. 2013. A copy number variant at the KITLG locus likely confers risk for canine squamous cell carcinoma of the digit. *PLoS Genetics*. 9: e1003409. https://doi.org/10.1371/journal.pgen.1003409.

34. It has been argued that ant queens come closest to being Darwinian demons: "The unusual positive association between fecundity and longevity and their seclusion from external mortality risks in the shelter of the nest make social-insect queens highly efficient egg-laying machines that come closest to Law's (1979) hypothetical Darwinian demons, which simultaneously maximize all fitness components." Quotation from Schrempf, A., et al. 2017. Royal Darwinian demons: enforced changes in reproductive efforts do not affect the life expectancy of ant queens. *American Naturalist*. 189: 436–442. http://dx.doi.org/10.1086/691000; Law, R. 1979. Optimal life histories under age-specific predation. *American Naturalist*. 114: 399–417. https://doi.org/10.1086/283488.

35. To give such a *min*imally *acc*eptable dragon a name, inspired by genetic predilection for abbreviations and the whimsy of tall tales and Tolkien, one would be hard-pressed not to dub it a *Mynnack*.

36. I am, of course, referring to the storyline of the 1958 classic children's book by Dr. Seuss, *The Cat in the Hat Comes Back*.

37. Akbari, O.S., et al. 2015. Safeguarding gene drive experiments in the laboratory. *Science*. 349: 927–929. https://dx.doi.org/10.1126/science.aac7932.

38. Refer back to Chapter 7 for the details on how ClvR and TARE gene drive schemes operate.

39. Noble, C., et al. 2019. Daisy-chain gene drives for the alteration of local populations. *Proceedings of the National Academy of Sciences USA*. 116: 8275–8282. https://dx.doi.org/10.1073/pnas.1716358116.

40. This kind of C-B-A daisy chain gene drive would not become a fixed feature of the species' genome, instead introducing some feature as an evolutionarily transient characteristic.

41. Esvelt, K.M., et al. 2014. Emerging technology: Concerning RNA-guided gene drives for the alteration of wild populations. *eLife*. 3: e03401. https://dx.doi.org/10.7554/eLife.03401.

42. Heffel, M.G., and G.C. Finnigan. 2019. Mathematical modeling of self-contained CRISPR gene drive reversal systems. *Scientific Reports*. 9: 20050. https://doi.org/10.1038/s41598-019-54805-8.

43. Moreno, E. 2012. Design and construction of "synthetic species." *PLoS One*. 7(7): e39054. https://doi.org/10.1371/journal.pone.0039054; Maselko, M., et al. 2020. Engineering multiple species-like genetic incompatibilities in insects. *Nature Communications*. 11: 4468. https://doi.org/10.1038/s41467-020-18348-1.

44. Mendelson, T. C., et al. 2014. Mutation-order divergence by sexual selection: diversification of sexual signals in similar environments as a first step in speciation. *Ecology Letters*. 17: 1053–1066. https://doi.org/10.1111/ele.12313.

45. Note that this kind of gene drive would be termed a *modification drive* and not a *suppression drive*, where suppression drives are those biological bombs that can drive a population or species—such as *Anopheles gambiae* mosquito vectors of human malaria—to extinction.

46. While *gallus* means *rooster* in Latin, it also references the ancient Celtic people who inhabited Gaul prior to Roman domination of Europe. Interestingly, the religion of the Celts, led by druids, venerated animals as incarnations of gods and spirits, worshiping them and sacrificing them in equal measure.

47. Darwin Correspondence Project. Letter no. 3104. https://www.darwinproject.ac.uk/letter/?docId=letters/DCP-LETT-3104.xml. Accessed December 15, 2021.

48. Byers, J.E., et al. 2006. Using ecosystem engineers to restore ecological systems. *Trends in Ecology & Evolution*. 21: 493–500. https://doi.org/10.1016/j.tree.2006.06.002.

49. https://www.nytimes.com/2020/11/16/world/europe/racing-pigeon-auction-record.html. Accessed October 8, 2021.

13

Nature, rewilded

Your relationship with nature is complicated. And, by you, I mean me and you and everyone we know.

We depend on natural resources for our survival and livelihoods. In our dependence, we use and exploit the living world and the products of the living world. We must eat to survive, and so humanity has taken to mass agriculture as a logical strategy. In so doing, we have carved out 14.7% of the land surface across the world for croplands and have dedicated livestock grazing rangelands to another 26.5%.[1] Human curiosity and inventiveness is insatiable in developing new technologies. We have transitioned mining as an enterprise in digging up simple flints for spears to what now amounts to over 57,000 square kilometers (14 million acres)[2] of direct impact to extract the entirety of the periodic table of the elements.[3] To name just a few: gold, silver, copper, nickel, zinc, lead, iron, phosphorus, lithium, aluminum, uranium, and titanium, along with extraction of coal, oil, and methane, and marble, slate, granite, sand, sapphires, and diamonds. We also mine the living: wood from logging; biochemical extraction for drug discovery, both illicit and pharmaceutical; and harvesting of wild creatures as animal companions and to fill window sill flower boxes. We humans appropriate for our own use, in one way or another, nearly a quarter of all of the biomass that plants produce around the globe each year.[4] A recent World Economic Forum report put it bluntly: "more than half of the world's total [gross domestic product] is moderately or highly dependent on nature and its services."[5] As our own species has proliferated to top 8 billion people this year, the footprint of our homes and villages and cities stamp the Earth in ways that exclude other species from co-inhabiting the same space.

Woven in with these threads of human neediness, use, and abuse, are other colors: threads of caring, cherishing, and protection for the living world. In the United States, more than half of all households care for one or more pets, amounting to over 76 million pet dogs among the 251 million pet animals across the country.[6] Outside of our homes, the desire to spend time in and to experience exotic parts of the natural world spurs the $180 billion ecotourism industry as hundreds of millions of people seek out national parks and reserves each year in both their home countries and abroad.[7] Some places get special recognition: the United Nations Educational, Scientific, and Cultural Organization has sanctioned for preservation over 250 World Heritage sites for their natural

Evolving Tomorrow. Asher D. Cutter, Oxford University Press. © Asher D. Cutter (2023). DOI: 10.1093/oso/9780198874522.003.0013

wonders,[8] and the 1959 Antarctic Treaty established an entire continent to peaceful nonexploitative use through scientific exploration and ecotourism.[9]

But, for many of us, even these amazing areas aren't nearly enough to fully satisfy the value we hold for the living world. The "30 × 30" movement that aims to devote 30% of the world's land and sea to protected conservation areas by the year 2030 has gained over 100 nations as signatories.[10] One might find it surprising that the flipside of that percentage—more than 70% of the world's surface—could legitimately be at risk of adverse human influence. But that is a fact of the Anthropocene era. Less than 26% of the Earth's land area now qualifies as "wildlands," regions with minimal human disturbance.[11] When you further consider whether those areas are fully intact in their complement of species or ecosystem functions, this number drops below 3%.

As self-anointed stewards of the planet, we need to decide collectively what, exactly, stewardship entails. The key decision lies with how proactive, how interventionist to be. To what extent does humanity strive simply to step back, to set aside lands as untouchable, to let the laws of nature run their course? Is it enough to leave nature to pick up from the perturbations that we've made? Don't forget that we will continue to perturb, given our dependence on interacting with the natural world for our own survival—and the whims of political turnovers. How much patience do we have? To what extent do we intercede to accelerate the processes of ecological, environmental, and evolutionary restoration? And, if we choose to intercede, what kinds of extreme measures are we willing to take?

13.1   Recall of the wild

You can find the call of the wild by visiting a preexisting wild area deep in the heart of a nature reserve.[12] Another strategy is to rehabilitate a tract of nature out of a human-impacted landscape, a wilderness phoenix raised from the ashes of the Anthropocene. This is the strategy of rewilding.[13]

Rather than a focus on "saving" any particular species or even restoring a particular bundle of predefined species,[14] rewilding aims to reconstitute the fundamental ecological circumstances that promote biodiversity itself. Perhaps lightening the load that pandas bear as a conservation icon, rewilding, too, has its poster child. It is the wisent.

The wisent, *Bison bonasus*, less poetically known as the European bison, once roamed the European continent from the Spanish Pyrenees to Kazakhstan. Recent centuries, however, witnessed dwindling numbers until its extinction in the wild in 1927.[15] A dirty dozen captive wisent have since been bred to produce a constellation of at least 33 free-living herds, now numbering over 5000 animals in total (Figure 13.1). These days, this largest living European mammal represents a key element of schemes to establish rewilded habitats complete with the beefiest of herbivores.

The largest reestablished population of wisent, some 1200 bison, resides in the aptly named Białowieża Primeval Forest (BPF).[16] The BPF, as it is known among friends, is a swathe of 1500 square kilometers (370,000 acres) straddling the border of Poland

Figure 13.1 *The wisent, or European bison* Bison bonasus, *now numbers in the thousands in its populations across Europe that expanded from a founding dozen that narrowly escaped total extinction in the 1920s.*
Image credit: Michal Köpping, reproduced under the CC BY-SA 4.0 license.

and Belarus. Large tracts of the BPF provide never-cut primary forest—a true rarity in Europe's long history of dense human habitation. Each day inside the BPF, a single wisent eats nearly four pood of food, a *pood* being just over 16 kg (36 pounds), and the food being a combination of lichens, mosses, grasses, leaves, acorns, and bark. The massive creatures share these food resources in the primeval forest with competing moose, red and roe deer, and wild boar and, to complete the trinity of tritrophic food web interactions, with apex predators: wolves and lynxes.

The core goal of rewilding a landscape like this is to regenerate an ecosystem to persist in a self-sustaining fashion. "The restoration of wildness, rather than wilderness is thus the key goal of rewilding efforts," a team of advocates stated recently, where wildness is "the autonomy of natural processes" without ongoing management.[17] To do so, the idea is to intentionally bolster the complexity of food webs, enable dispersal of animals and plants, and both acknowledge and permit small-scale catastrophes in the form of so-called stochastic disturbance as a normal part of ecosystem functioning: tree-falls, fires, floods, and mudslides. Estimates project that the percentage of Earth's surface with intact communities of species could go from 2.8% to 20% simply by reintroducing one to five species in areas that otherwise have a low human footprint.[18]

Getting an ecosystem to be self-sustaining is the sticky trick, especially in Anthropocene landscapes that have witnessed the extinction or displacement of large herbivores and their predators. For example, look at the experiment in rewilding that is the Oostvaardersplassen nature reserve in the Netherlands. Oostvaardersplassen was rewilded to contain plenty of large herbivores, but neither predators of them nor easy means for dispersal beyond the reserve. An expanse of 56 square kilometers (nearly 14,000 acres) sounds like lots of space to roam but is small enough to make a megaherbivore claustrophobic when food gets scarce. Public outcries ensued when thousands of deer, horses, and cattle—up to 34% of the population in a given year—starved to death during lean winters or were euthanized.[19]

★

Biodiversity losses, whether of large herbivores or of large predators, also exact knock-on effects to the kinds of plants that can persist. In the Americas, plants with large or fleshy fruits, like the calabash *Crescentia alata* and pawpaw and persimmon, were termed "ghosts of evolution" by Connie Barlow, and evolutionary "anachronisms" by Daniel Janzen and Paul Martin.[20] Anachronisms and ghosts by virtue of having co-evolved for attraction, consumption, and dispersal by now-extinct mammalian megafauna. Without megaherbivores, the plant species were left to dwindle in abundance for 10,000 years toward their own extinction. The dispersal movements of large herbivores provides an important mechanism for seed dispersal, an ecosystem function that would offer a special virtue in the face of climate change for immobile plants with heavy seeds and fruits. Those heavy fruits that have proved tasty to humans and easy to grow, however, like avocados *Persea americana*, have found rejuvenated ranges through agriculture. For these select few plant species, humans have become the megafauna proxy for extinct glyptodonts and gomphotheres.[21]

The glyptodonts and gomphotheres are gone. The truck-sized giant ground sloths, the cow-sized giant armadillos, the mammoths, the rhinoceros-like toxodons have all met their fate in the state of extinction. Changes to prehistoric climates weren't the decisive factor: the disproportionate extinction of large-bodied animals from the Pleistocene through more recent times is a consequence of humans colonizing the world, due to combined effects of hunting "overkill" and ecosystem engineering.[22] In one view, the loss of these key elements of the food web, the trophic structure, has left ecosystems out of whack for thousands of years. What today looks like a balance of nature in seemingly pristine habitats is really just a partial reshuffling of species abundances and ranges in response to those catastrophic extinctions.

Our unfamiliarity with mourning these big losses of the Pleistocene as part of how we view biological conservation is a product of what Daniel Pauly calls the Shifting Baseline Syndrome.[23] The Shifting Baseline Syndrome is a kind of implicit bias that we have about the natural world: the innate tendency to redefine what is "natural" based on our own lived experience, in spite of real and accumulated compounding change caused by past generations. We reminisce about our own youth but have little nostalgia for the youth of our grandparents that we ourselves didn't experience.

★

Could we actively reconstitute a simulacrum of the Pleistocene—of ecosystems from 50 thousand years ago? Could we substitute some living species in place of glyptodonts and gomphotheres? Those substitutes could then rewild ecosystems in a way that mimics their composition prior to humanity's outsized impacts on megafauna species extinctions. Such rewilding by proxy interprets the key missing variable for self-sustaining wildness to be a set of keystone species, often apex predators or large herbivores, and so anything that can fulfill those ecological roles, native or not, will do.[24]

This perspective, sometimes termed *trophic rewilding* or *Pleistocene rewilding*, emphasizes the revival of apex herbivores and predators within ecosystems as a strategy for ecological restoration. The worry, of course, is one of hubris: that we do not sufficiently understand the knock-on effects of introducing an untested substitute species in any given environment. The geneticist in me can't help but draw analogy to Dobzhansky-Muller incompatibilities for what can happen when you put together untested pieces of a complex system.[25] This worry of unintended consequences is founded in our checkered history of species introductions intended for biocontrol. We shall see in a moment some cautionary tales.

More visceral reactions to trophic rewilding invoke rhetoric deriding such efforts as promoting "Frankenstein ecosystems."[26] There are practical complaints, too. Big animals require big tracts of land. Given the heavy human demands on the landscape that have fragmented much of what remains into small chunks, excess attention to and tinkering with more-or-less secure large reserves might seem a distraction from the urgent risks of biodiversity loss in smaller plots of rare habitats.

Proponents of trophic rewilding, however, point to successful ecological outcomes after wolf reintroduction into Yellowstone National Park, as well as after intentionally installing nonnative large tortoises on oceanic islands as substitutes for their extinct counterparts. Even the Dutch Oostvaardersplassen, despite its rounds of herbivore mass starvation, garnered success in many of its primary ecological goals: the activity of feral horses and cattle, introduced as mimics of their archaic ancestors, did indeed maintain grassland as habitat for greylag geese that, in turn, also enhanced the ecosystem complexity through their own effects on vegetation.[27] More eyebrow-raising, however, is the notion of introducing cheetahs and lions into the wild of North America as proxies for extinct big cats. Putting it mildly in their proposition for this repatriation idea: "Lions, which prey on wild equids and other large herbivores, offer a bold and exciting vision for Pleistocene rewilding."[28] If lions and cheetahs find themselves without an invitation, the extirpated jaguar, at least, has a welcome repatriated home in Arizona and New Mexico in addition to much of Central and South America.[29]

Big cats or no, the biggest argument for heavy-handed intervention lies with the inherent and imminent risks of inaction. Nonintervention does not eliminate ongoing human impacts on ecosystems. There are long-simmering impacts of our Anthropocene perturbations that reach every corner of the globe, many of which will not soon simmer down. Nonintervention alone means that, even if all of humanity blinked out of existence tomorrow, our imprint up through today would continue to ramify into the future.[30] There are species out there on the brink, doomed to extinction without proactive conservation. These are the species whose members have insufficient resources—habitat space or environmental conditions or mate availability—to adequately grow their population sufficiently to become sustainable. They are the dead species walking, even though they don't know it—a sticky situation that ecologists term the extinction debt.

13.2   Total recall of the wild: de-extinction

Oostvaardersplassen also gives us a lesson in applying selective breeding to produce the characteristics of one of Europe's most iconic extinct megaherbivores: the aurochs. Aurochs, with their dramatic forwardly curved 1-m-long (3-foot-long) horns, grace the cave walls in Lascaux, France, the artwork of prehistoric painters 17,000 years ago (Figure 13.2).

The diverse breeds of modern-day cattle are domesticated descendants of the aurochs, a bovine repeat of the relation between dog and wolf. It was two independent domestications, actually, leading to humped zebu cattle in India and taurine cattle from Mesopotamia, each dating to about 10,000 years ago. Aurochs, a name both singular and plural, were big: 25% taller than your typical cow, reaching 1.8 m (6 feet) at the shoulder.[31] Aurochs, unlike wolves, however, became extinct in 1627, with a geographic kinship to the wisent's extinction in the wild in where their last reserves were located: a forest in Poland. But the genome of the aurochs lives on, in modified form, in the genomes of cows.

In the 1920s, two German zookeepers, brothers, began the process of back-breeding cattle to derive a stock with characteristics most similar to aurochs. The result of intercrossing eight types of cattle was Heck cattle, which is the bovine breed that now

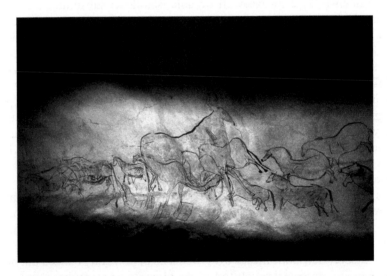

Figure 13.2 *The aurochs,* Bos primigenius, *known to humans since prehistory, as seen in the wall paintings inside of France's Lascaux Cave, went extinct despite being the ancestor of modern cattle. Modern breeding efforts aim to de-extinct the aurochs through back-breeding of cattle, guided by aurochs traits and genetic similarity to DNA that was sequenced from preserved ancient DNA in exhumed aurochs bones.*

populates Oostvaardersplassen. Heck cattle, of course, are not aurochs. They are, how-ever, hardy and "very large, very big-horned and unruly to the point of being downright dangerous."[32]

Others have since followed in the Heck brothers' footsteps, using selective breeding to try to chase stereotypical aurochs traits out of oblivion and into the here-and-now. This in-progress Tauros cattle breed project, in collaboration with the similar Auerrind Project, aims to make even bigger bovids with longer and forward-pointing horns, among other characteristics.[33] And its success can be measured in DNA. The aurochs genome was sequenced in 2015 from the ancient DNA locked inside of a preserved aurochs' leg bone found in a cave in Derbyshire, England.[34] Its genome provides a reference against which selectively bred cattle may be compared or used as a genetic guide for selective breeding itself by using a genomic selection index.[35]

This exact approach of a "genetically informed captive breeding program" is now underway, on the other side of the planet, for another kind of species. Breeders are using selection based on genomic features to reconstitute the extinct Floreana giant tortoise of the Galapagos Islands. The idea is to combine genome fragments within the "genomic archive" of rediscovered hybrid tortoises that were found on a remote volcano to recon-stitute individuals with nearly the complete genetic makeup of what seemed to be a lost species.[36]

★

Rewilding, in a sense, presents the ultimate irony of the Anthropocene: in trying to undo what we have done, we do more. This is, as Elizabeth Kolbert states it, "the recursive logic of the Anthropocene."[37] Then there is the most drastic means of doing more: de-extinction. In a clapback to the criticism that rewilded environments with proxy introductions of nonnative ecological analogs of extinct species will do little more than create "Frankenstein ecosystems," de-extinction shows a path that might avoid proxy organisms altogether. Why rewild with elephants when you could resurrect the woolly mammoth itself, using genetic engineering? If the wisent is the poster child of rewilding, then the woolly mammoth is the poster child of de-extinction.

On paper, de-extinction follows a straightforward five-step plan: (i) *in silico*, (ii) *in vitro*, (iii) *in vivo*, (iv) *ex situ*, and (v) *in situ*.[38] Translated from Latin, the de-extinction plan starts with determining the DNA sequence of the genome of the species to de-extinct. You extract that ancient DNA from preserved samples of some kind, whether intentionally cryopreserved, exhumed from permafrost, or isolated from an old bone found in a dry cave. For the woolly mammoth *Mammuthus primigenius*, this *in silico* step is complete several times over: most recently, mammoth genomes were sequenced from 1.6-million-year-old DNA extracted from permafrost-preserved molars.[39] The process then proceeds to genome editing to convert the DNA sequence of a living relative's genome to match the genome sequence of the extinct reference. This step might make use of laboratory cell lines. This is the step that has been in progress for the past nine years, the mammoth project led by George Church's team, using genomes inside of Petri dish cells from Asian elephants as the *in vitro* vessel.[40] Next, the genomes from these Petri dish cell lines are introduced into surrogate embryos, and those embryos into a surrogate mother, to allow full animal development *in vivo* with the edited genome. These animals would then be bred and assessed *ex situ* in captivity to form controlled populations prior to the final *in situ* phase: release into the wild. There is an unspoken sixth step. It might be called *in regimen*: the practicality of release and long-term monitoring in the face of laws, politics, and management.

Even before facing step six, mammoth de-extinction faces mammoth logistical challenges. The mammoth genome differs from the genome of its closest living relative, the Asian elephant *Elephas maximus*, by at least 1.4 million DNA nucleotides that affect the protein coding sequence of over 1600 genes.[41] An unknown portion of those 1.4 million differences, in turn, affect gene expression of the entire genome. Through clever bioinformatics analysis, this curse of riches can be narrowed down to some prime candidates to prioritize for genome editing, in hopes of regenerating key traits that distinguish mammoths from elephants. For example, we might prioritize those genes anticipated to play a role in growth of fur, in fat deposition, and in cold-temperature metabolism—cold tolerance may come from some genetic variants of hemoglobin, the protein that carries oxygen in red blood cells. Mammoths also had small ears, but I don't think that any geneticists have yet figured out the prime gene candidates for controlling the flap size of one's pinnae.

Elephant cell lines in Petri dishes now are accessible to researchers for genome editing,[42] but how many edits are enough? Are 50 edits sufficient to make an Asian elephant into a "mammophant": hairy, fatty, and cold-tolerant? Or do we need to go for the whole 1.4 million? How do we incorporate genetic variability in any resulting population of mammophant genomes? Asian elephants are an endangered species, making their wombs ill-suited to the risk of serving as surrogates for embryonic implantation. Supposedly, an artificial womb is in the works, albeit with no promise yet to endure the typical 18-to-22-month gestation of an elephant offspring that one might presume also would apply to a mammophant. And then, an elephant—and, presumably, a mammophant—grows for 10 to 14 years before reproductive maturity.

Not all species proposed as de-extinction candidates will have all these hurdles. Some species offer an easier go of it for technical and logistical reasons. Nevertheless, this approach to de-extinction with genetic engineering is a long-haul and substantially slower than the rates at which species are going extinct in the first place.[43]

★

The core biological rationale for de-extincting the mammoth is to restore its prehistorical role, a job currently vacant, as an ecological engineer in arctic landscapes. In the millennia before their total extinction about 3700 years ago, an extinction for which "humans were a synergistic cofactor," millions to hundreds of millions of mammoths traversed the broad circumpolar grasslands that predominated across the far north during the Pleistocene.[44]

One school of thought holds that the mammoths were responsible for maintaining those grasslands, a landscape that mostly disappeared along with their extinction. Now, this cold northern zone is made up mostly of tundra with its scrubby short shrubs, intermittent grass and sedge, moss, and lichens, and taiga, boreal forests of mostly pine and spruce. Re-establishing grasslands on a big scale, in place of these current biomes, is proposed as a climate change buffer: grassland habitat shows an outstanding ability to sequester carbon. Moreover, grazing animals in winter displace insulating snow to expose the ground to the frigid air, helping to retain permafrost in a deep freeze even in warmer seasons.[45] This scheme defines a biological means to geoengineering Earth's climate.[46] We now refer to these vast prehistoric grasslands as the "mammoth steppe." Don't we all wonder, though, what word our ancient ancestors uttered to describe this long-gone landscape when telling tales over a campfire to their children, tales of the day's travails with the local herd of woolly mammoths?

The latest analyses argue, however, that it was climatic forces that were most crucial in drastically diminishing appropriate mammoth habitat,[47] thus questioning the mammoth's ability to single-handedly reconstitute holarctic grasslands. In this view, no amount of tramping by de-extinct mammophants would terraform taiga into steppe. Regardless, de-extinct mammoths currently have an open invitation to populate Siberia's Pleistocene Park. Whether they would be welcome throughout the rest of the 29 million square kilometers (7 billion acres) of tundra and taiga, however, remains to be seen: this zone across Canada and Russia and Northern Europe represents nearly 20% of Earth's land surface. If mammophants and grassland species are winners with such rewilding, then the current occupants of tundra and taiga will be losers. With climate change already disrupting the prospects of those species, what responsibility do we have for their persistence?

Mammoth de-extinction at the scale imagined for ecological influence in the arctic has, realistically, a century-long or longer timescale. Is climate change too fast to reap their proposed benefits? Almost certainly. And these are just the technical challenges, separate from ethical and political considerations. Should steps ii, iii, iv, and v ever come to completion, they will usher in the full armament of issues around rewilding together with passions about genetically modified organisms as a cherry on top.

★

The mammoth isn't alone as an active de-extinction project that steeps itself in genetic engineering. It's joined by Australian gastric-brooding frogs and thylacines (Tasmanian tigers).[48] There is also the flock of birds sponsored by the Revive & Restore Foundation: the passenger pigeon, the heath hen, and perhaps soon the auk. The goal of the de-extinction strain of trophic rewilding, after all, is one of "intended consequences": the intervention into ecosystems to resuscitate key lost functions that are hypothesized to jump start ecosystem self-sustainability so as to foster biodiversity.[49]

The arctic tundra has been touted, in some quarters, to be a crippled ecosystem-in-waiting. It's waiting for mammoths to regenerate the lost Pleistocene mammoth steppe ecosystem, in so doing, perhaps preventing permafrost thaw and the accompanying sky-high CO_2 release that could further exacerbate global climate change.[50] What ecosystem awaits, however, for the passenger pigeon?

Passenger pigeons were a species that, in Darwin's day, numbered in the billions as its flocks carpeted the sky across eastern North America, alighting upon billions of American chestnut trees to fuel up on their bounty. The American chestnut *Castanea dentata*, after all, succumbed to blight, an invasive fungal infection. Wild stands of mature *C. dentata* that once accounted for 25% of trees along the Appalachian Mountain range were extirpated just 50 years after the passenger pigeon *Ectopistes migratorius* itself went extinct in the wild in 1901. Today's fragmented forests of eastern North America still have beechnuts and acorns, but no good chestnuts, and perhaps no good home for *E. migratorius* to roost.

Among the great optimists of the world are the chestnut aficionados. The chestnut aficionados of North America are giving full attention to the few remaining sturdy roots and their tenuous trunks with the aim of smiting the blight with genetics. Conveniently, Chinese chestnut trees *Castanea mollissima* are resistant to the blight, a fungus named, malevolently, *Cryphonectria parasitica*. Through selective breeding, the gene variants responsible for resistance in Chinese *C. mollissima* are being introgressed into the genomes of orchard collections of the American *C. dentata*. After six generations of such controlled breeding, the set of 7600 so-called BC3F3 trees have genomes with roughly 90% *dentata*-derived DNA and survival rates that look promising after eight years in forest replanting trials.[51] In parallel, another team has genetically engineered *C. dentata* to express a resistance gene that was snagged from the genome of wheat.[52] Both approaches offer viable extinction-avoidance strategies for the repopulation of North American deciduous forests with American chestnut, the high wild crowns of which are only distant memories of our oldest citizens.

★

Mammoth de-extinction depends on genome editing and genetic engineering. But there are two other ways to de-extinct a species.[53] The least technical approach is back-breeding. This is the scheme to resurrect the aurochs, or its verisimilitude of form, from the DNA that it squirreled away inside the genomes of the diverse breeds of modern-day cattle. Back-breeding can proceed purely from outward characteristics, the phenotype of the organisms. It also can incorporate DNA testing to aggregate together those particular segments of chromosomes identified as ancestral that have scattered representation among different individuals.[54] This DNA-guided approach is the tack taken for the Floreana Galapagos giant tortoise. Back-breeding to de-extinction uses selective breeding to concentrate into the same individuals those key characteristics of a now-extinct or soon-to-be-extinct organism that are deemed crucial for supporting ecosystem function in the wild.

The other de-extinction technique is cloning. Cloning requires an intact cell nucleus, which contains the full complement of DNA on chromosomes. This cell nucleus gets injected into a different cell, an egg cell from a donor species—an egg cell with its own nucleus removed—a procedure known as somatic cell nuclear transfer (SCNT). That cell nucleus put into the egg constrains the scope of de-extinction that can be possible with this technique: it must come from a living creature, or from cells cultured in Petri dishes, or from high-quality cryopreserved cells. By bathing the SCNT egg in just the right bath balm, it will begin developing as an embryo using the DNA instructions from the foreign nucleus. The developing embryo can be reared inside a surrogate host, either the womb of a related species, such as the species used to provide the donor egg that had its own nucleus sucked out, or potentially an artificial womb. Despite all of these surrogates and donors, it is the DNA from the creature being cloned that directs the developmental show.

A consequence of all these requirements is that cloning in itself can be a viable approach for species on the verge of extinction, but not for long-extinct species. And only some kinds of species: it doesn't yet work for egg-laying species like birds. Cloning briefly made the bucardo de-extinct, a kind of goat from Spain's Pyrenees mountains also known as the Iberian ibex.[55] Cloning is also an approach being taken in efforts to prevent extinction of black-footed ferrets—a ferret named Elizabeth Ann (without-an-E) being the first member of an endangered species to be cloned—as well as northern white rhinos, despite the death in 2018 of the white rhino's last male.[56]

You'll notice that the genetic engineering approach envisioned for mammoths incorporates SCNT at the *in vivo* step iii. And all approaches ought to incorporate some form of selective breeding at the *ex situ* step iv, in order to stock any newly de-extinct species with ample genetic variability prior to release into nature.[57] In this way, these three de-extinction techniques are like nesting Russian matryoshka dolls, each using what lies inside: selective breeding inside of cloning inside of genome engineering.

Technically speaking, none of these programs of de-extinction will resurrect an extinct species in a form truly indistinguishable from its prior occurrence in the wild.[58] A so-called de-extinct creature would be engineered through use of proxy eggs and proxy genomes from other living organisms, those proxy genomes used as a scaffold for genome editing, editing to produce genome sequences that match ancient DNA of extinct representatives as much as possible. The editing would not be 100% perfect. The editing also would not capture 100% of the differences between the ancient DNA genome and the modern-day scaffold genome. A population of such a creature would not contain the pre-extinction complement of genetic variability, of epigenomic marks, of microbial constituents in its gut, or of socially learned behaviors.

Of course, the same can be said for any ordinary present-day species: the genetic composition today is different from the genetic composition a generation ago, and even more different from a generation before that. This fact is the simple inevitable byproduct of evolutionary change, attributable to genetic drift, natural selection, and their comrades among the ancient forces of evolution. So, it is a truism that any technically-successful endeavor in de-extincting an old species actually creates something new. By design, that new biological entity ought to bear strong resemblance to most features of the extinct species and, perhaps most importantly from a rewilding perspective, ought to embody the essence of the functional traits intended for reconstitution of ecosystem function. But it would, at the end of the day, be a new and different biological entity.

New and only partly de-extinct, does it matter that we might, one day, come to peer through binoculars at a herd of "mammophants" rather than mammoths? As Rob Desalle and George Amato wrote, "The de-extinction movement is an example of an apparently well-meaning desperate measure ... But whether de-extinction can help make a real dent in the extinction crisis—whether it will help save species—is much less clear."[59]

Given the long-haul, in time as well as effort, to de-extinct a species, it is hard to escape the conclusion that de-extinction is not a viable path to offset the alarmingly high extinction rates in the world today. Rather than a mandate of reversing extinction to save species, however, the de-extinction mandate is perhaps best viewed as a special strand of rewilding, an extreme measure to create "a possibly useful proxy of an extinct species"; in this view, de-extinction is "any attempt to establish free-ranging populations of functional proxies of extinct species."[60]

★

Newspapers, magazines, blogs, scientific journals: all have expressed the technological hype and ethical gnashing of teeth in discussions about de-extinction. It almost feels anticlimactic to point out that the potential for de-extinction already exists in nature, in a sense, without any human genetic tinkering at all.

Natural de-extinction is practically a way of life in deserts. Erratic rains may lead to bumper crops of wildflowers in some years with little to no germination for many years in between. This was the case in the Sonoran Desert surrounding my time living in Tucson, Arizona, as a graduate student. In March of 2001, the flat playa around Kitt Peak just west of town erupted in astonishing multicolored fields carpeted with poppies and lupines and asters, emerging from the bank of dry seeds under the dust of prior years. It is a fine line between dormant and de-extinct.

Now let's look north. Take, for example, the arctic sedge plant, *Carex bigelowii*, found in Alaska.[61] Each summer, they drop their seeds to the ground to chill for the winter before sprouting the next year. Some of the sedge seeds continue to slumber, however, and instead skip a year before they wake up and germinate. Or skip another year, and another, nestled below the soil. This collection of seeds creates what is called a seedbank. If you dig into the soil about 17 m (56 feet) you can find seeds dating to 295 years ago. Some of these sleeping beauties are viable and can grow into healthy sedge plants. At the opposite end of the Earth, entire moss plants survived more than 600 years of burial beneath Antarctic glaciers. And viable plant embryos from seeds of arctic *Silene*—a kind of campion—have been awoken after 31,800 years in Siberian permafrost, buried in the fossil burrows of Pleistocene squirrels.[62]

Animals, as well as plants, exhibit such suspended animation known as cryptobiosis.[63] They survive long spells of harsh environmental conditions like aridity or cold in a state of metabolic quiescence. There is the famed tardigrade, renowned survivor of extreme conditions. Rehydrated after 30 years of having been frozen with a piece of Antarctic moss, *Acutuncus antarcticus* tardigrade water bears emerged seemingly unphased. In dry conditions, the egg cysts of *Artemia* brine shrimp persist for 15 years in a parched state of suspended animation termed anhydrobiosis, and the African midge *Polypedilum vander-planki* can experience such living desiccation for 17 years. Nematode roundworms also are stalwarts of long-term survival, with multiple species documented to have 30-year-plus revivals. I've revived such nematodes after a decade in my lab's deep freezer, though that hardly compares to Russian permafrost cryopreservation. Nematodes from arctic permafrost in Siberia are reported to have been thawed and revived to crawl on their own. No mean feat for the two lucky worms, after 32,000 and 41,000 years of natural cryogenic storage, one extracted from a derelict Siberian squirrel burrow near to those that housed the *Silene* seeds.[64] And then there are Pleistocene permafrost extractions of viable amoebae, rotifers, ciliates, algae, and fungi.[65]

Bacteria, though, leave the longest legacy. Bacteria appear able to survive at least half a million years, perhaps even 3.5 million years, in suspended animation, 470 million such cells cryopreserved in each gram of permafrost.[66] Microbes might not be able to populate a Jurassic Park, but a microscopic Pleistocene or even Pliocene Park is in the cards without any genetic engineering whatsoever. This is all to say that genomic phoenixes arising from the Earth, their DNA seemingly having been lost to history, have ample precedent in nature.

13.3 Back to the future of the wild: novafauna

If de-extinction by selective breeding (aurochs) and synthetic biology (mammoths) is suitable for producing a megaherbivore for trophic rewilding, might genetic engineering of entirely novel creatures also do the trick? This truly would be the origin of extraordinary species in the wild, what we might call the novafauna, a biological community of literal and figurative dragons.

What goals could such a novafauna achieve? One obvious outcome is that introducing a newly evolved organism into an ecosystem would create biological novelty and distinctiveness—that is, increased biodiversity. But bespoke creation of species through directed evolution and genetic welding is a costly enterprise. In itself, certainly it is not an efficient means of offsetting extinction. Genetic engineering, whether to de-extinct a species or to evolve a new one, is a clumsy way to manage extinction. For the rewilding advocate, management is anathema; instead, we should be "Moving away from managing extinction and toward actively restoring ecological and evolutionary processes."[67]

Rather than evolving novel creatures for their own sake, we might instead evolve them to assume a particular ecological role. We might take an imperfect ecological proxy organism as founding stock and tweak their features with genetic welding in ways that make them a better match to the habitat that one intends to rewild. This would counter some of the criticisms lodged against proxy species that differ from their extinct analogs. It would remain to be seen, however, whether dropping a newly engineered proxy species into a landscape would cascade through the ecosystem to increase or stabilize biodiversity, or if it would exacerbate the existing problem. That is the risk. There is no way around some kind of risk. There is risk in inaction, after all, albeit the devil we know.

Michael Soulé and Reed Noss defined a rationale for introducing new species as an interventionist way for us to regain "human opportunities to attain humility." This idea is one of the original motivations for the rewilding movement, articulated in their classic 1998 manifesto in the *Wild Earth* magazine, which also could apply to novafauna. On its face, evolving a dragon might seem a direct adversary of this philosophy. After all, we've outlined a recipe for a kind of synthetic biology, infusing nature with the new evolutionary force of genetic welding, concocted by humanity, to direct evolution according to our will. Once an organism gets set loose into an uncontrolled environment, however, to witness the subsequent life-and-death interplay of individual organisms with one another in all of its complexity, *that* certainly can spark humility and awe. If applying the algorithm with a founding stock of carnivores or to evolve gigantism with new megafauna, then that humility would be physical in the way that reintroduced cheetahs in the American prairies would be.

Regardless of size and intimidation factor, when set out into an ecosystem, out of the controlled confines of an environment devised for selective breeding, a population of hyperexotic creatures would be free to evolve to the local conditions. They would experience new pressures of natural selection that were absent under controlled breeding; they also would experience relaxed selection on traits important in controlled breeding that, in the wild, no longer impact fitness in such a strong way. Such local adaptation defines a ready means of further diversification. Local adaptation is something readily seen in nature, the distinct pressures of natural selection that differ from one place to another to drive the evolution of novel features unique to those different places.

As Philip Seddon and Mike King wrote recently,[68] "If we want to sustain and enhance a biodiverse natural world we might have to be forward looking and embrace the notion of bio-novelty by focusing more on ecosystem stability and resilience, rather than backward looking and seeking to try and recreate lost worlds." This view underscores the fact that it is impossible to preserve the prior state of the world, or at least our idea of it. Conservation biology does not preserve nature like peaches in a jar or pelts in a museum. Ecological systems are dynamic; populations of organisms evolve. As Rosemary Grant, the doyenne of Darwin's finches, says,[69] "Species don't stand still. You can't 'preserve' a species."

★

There was an additional clause that I cut from the end of a quotation that I gave a moment ago, the quote about embracing bio-novelty and moving past lost worlds. I truncated the phrase "using Pleistocene history as a guide." That prehistorical guide is the mantra of the strain of conservationists beholden to Pleistocene trophic rewilding. Some critics balk at what they see as a boot-shined incarnation of Jurassic Park, rebranded with an unconvincing sheen of a more recent geological era. In Eastern Siberia, Russia even has that privately owned place dubbed Pleistocene Park. Private corporations have sprung up with the hopes of populating such a park, like Colossal Biosciences that rose out of George Church's academic laboratory tasked with de-extincting a mammoth and developing intellectual property. Announcing venture capitalists' initial investment of $15 million with a public relations push in September 2021, it remains to be seen whether society and hype-averse scientists will be persuaded by the testimonial on the Colossal webpage by celebrity life coach Tony Robbins. Naysayers aside, their aim to "jump-start nature's ancestral heartbeat" does not end with making a freakshow in a zoo, but with the evocative vision to "see the Woolly Mammoth thunder upon the tundra once again."[70]

Jurassic Park or Pleistocene Park—or perhaps Novopark as the home for newly evolved species—how to enforce the park's boundaries? Should transgressions beyond the park boundaries be enforced at all? The wolves of Wyoming's Yellowstone National Park are not fenced in. Wolf populations expanded throughout western states including Idaho and Montana following their reintroduction, reaching sufficient abundance to get delisted as endangered species in 2011. Had wolves been penned inside even the nearly 9000 square kilometers (2.2 million acres) of Yellowstone, their positive ecological influence would be unable to spread through the ecosystems of the broader western United States. This wolf expansion, however, came much to the chagrin of big-game hunters and ranchers nearby. So strong was some of the antipathy toward them, despite just 85 cows and calves getting killed by wolves out of Idaho's 2.7 million cattle in the 12 months prior to July 2020, that politicians recently enacted laws to cull wolf populations by up to 90%.[71] Understandably, environmental advocates have launched legal challenges, and scientists proposed an ecologically plausible, if politically dubious, rewilding network of reserves across 16% of 11 states in the American West (496,000 square kilometers) that features reintroduction of wolves and beavers.[72] The policies and criteria for range boundaries for introductions of hyperexotic species into a Novopark, or de-extinct species into a Pleistocene Park, will need such care of consideration to make wolves seem like teddy bears.

★

If humanity were to evolve a dragon, where would we put it? If you were to make a map of which countries create the world's breakthroughs and developments in genetic engineering, and gene drives in particular, and which countries are prioritized as potential candidates for deploying those biotechnologies, then you would notice a stark lack of overlap. This mismatch is most striking, perhaps, for the case of suppression drives intended as a form of biocontrol over malaria. Despite the prickles of potential exploitation you might feel, the African Union supports exploration of gene drives as a means to fighting the devastating toll of malaria on that continent.[73] Field tests to explore the potential for suppression gene drives in mosquitoes are in the works in Burkina Faso, based on biotechnology developed in the UK and United States. Should things go awry with unintended consequences, however, "the global South might have the most to lose."[74] Rewilding, de-extinction, and genetic welding not only impinge on ecology and evolution, but also on geopolitical justice.

★

Our relationship with nature is complicated: we depend on it, we exploit it, we cherish it. In compensating for the profound impacts of the Anthropocene, one option is passive rewilding: to leave nature alone in its current state to allow ecological and evolutionary forces to proceed with minimal human influence. Another rewilding option is to intervene proactively by introducing ecological proxies for extinct organisms, especially those with roles in ecosystem engineering. Ecological proxies may derive from existing species with similar ecological roles, selective breeding to generate features that mimic lost ecosystem functions, genetically engineered de-extinction, or creation of entirely novel species. De-extinction has a natural counterpart in cryptobiosis. Introduction, and reintroduction, of species will raise complex ecological and political considerations.

Notes

1. Ellis, E.C., et al. 2020. Anthropogenic biomes: 10,000 BCE to 2015 CE. *Land.* 9: 129. https://doi.org/10.3390/land9050129.

2. This combined area is bigger than Croatia or Costa Rica.

3. Maus, V., et al. 2020. A global-scale data set of mining areas. *Scientific Data.* 7: 289. https://doi.org/10.1038/s41597-020-00624-w.

4. Haberl, H., et al. 2007. Quantifying and mapping the human appropriation of net primary production in earth's terrestrial ecosystems. *Proceedings of the National Academy of Sciences USA.* 104: 12942–12947. https://dx.doi.org/10.1073/pnas.0704243104.

5. https://www.weforum.org/reports/nature-risk-rising-why-the-crisis-engulfing-nature-matters-for-business-and-the-economy. Accessed September 29, 2021.

6. According to 2016 data from the American Veterinary Medical Association, about 40% of pet owners keep more than one species of pet: https://www.avma.org/resources-tools/reports-statistics/us-pet-ownership-statistics. Accessed May 7, 2021.

7. https://www.alliedmarketresearch.com/eco-tourism-market-A06364. Accessed May 7, 2021.

8. Among the criteria for World Heritage status designation is criterion number nine, "to be outstanding examples representing significant on-going ecological and biological processes in the evolution and development of terrestrial, fresh water, coastal and marine ecosystems and communities of plants and animals." https://whc.unesco.org/en/list/stat and https://whc.unesco.org/en/146. Accessed May 7, 2021.

9. With the tense political relations among the diverse set of countries that have established scores of Antarctic bases, the spirit of the treaty ideal may not always be met in practice. https://www.bbc.com/news/magazine-27910375; https://www.ats.aq/e/antarctictreaty.html. Accessed May 7, 2021.

10. All 10 of the continental North American countries have signed on to the 30 × 30 idea. Unfortunately, the list of supporting countries has six notable absences among the 10 geographically most-expansive nations in the world: Algeria, Argentina, Brazil, China, Kazakhstan, and Russia. https://www.hacfornatureandpeople.org/hac-members. Accessed August 23, 2022; https://www.washingtonpost.com/climate-environment/2021/05/06/biden-conservation-30x30; https://www.weforum.org/agenda/2021/01/planet-coalition-environment-conservation-protection-france-climate-change/. Accessed May 7, 2021.

11. Ellis, E.C., et al. 2020. Anthropogenic biomes: 10,000 BCE to 2015 CE. *Land.* 9: 129. https://doi.org/10.3390/land9050129; Plumptre, A.J., et al. 2021. Where might we find ecologically intact communities? *Frontiers in Forests and Global Change.* 4: 26. https://doi.org/10.3389/ffgc.2021.626635.

12. My most romantic vision of experiencing wilderness was shaped by viewing the 1983 film *Never Cry Wolf* as a child on vacation in Ocean Park, Maine. Later, I developed a more alpine view, with dreams of through-hiking the Appalachian Trail. Alas, I have not yet fulfilled either of those aspirations of months of semiwild living in the Canadian Arctic or ensconced in the American mountains from Georgia to Maine.

13. Soule, M., and R. Noss. 1998. Rewilding and biodiversity: complementary goals for continental conservation. *Wild Earth*. 8: 1–11. http://www.environmentandsociety.org/mml/wild-earth-8-no-3.

14. Despite derision, in some quarters, of outsized emphasis on conservation efforts devoted to charismatic megafauna like pandas and elephants and wolves and owls, their benefit is that they "justify bigness" in acreage of ecological preservation. They do so as so-called keystone species "whose influence on ecosystem function and diversity are disproportionate to their numerical abundance" and/or as so-called umbrella species that offer "land protection under which many species that are more abundant but smaller and less charismatic find safety and resources" (Soule and Noss, 1998, *Wild Earth*).

15. Deinet, S., et al. 2013. *Wildlife Comeback in Europe: The Recovery of Selected Mammal and Bird Species*. Final report to Rewilding Europe by ZSL, BirdLife International and the European Bird Census Council. London, UK: ZSL. https://api.semanticscholar.org/CorpusID:114705002; Węcek, K., et al. 2017. Complex admixture preceded and followed the extinction of wisent in the wild. *Molecular Biology and Evolution*. 34: 598–612. https://doi.org/10.1093/molbev/msw254; Samojlik, T., et al. 2019. Historical data on European bison management in Białowieża Primeval Forest can contribute to a better contemporary conservation of the species. *Mammal Research*. 64: 543–557. https://doi.org/10.1007/s13364-019-00437-2; Kuijper, D.P.J., et al. 2013. Landscape of fear in Europe: wolves affect spatial patterns of ungulate browsing in Białowieża Primeval Forest, Poland. *Ecography*. 36: 1263–1275. https://doi.org/10.1111/j.1600-0587.2013.00266.x.

16. "*Białowieża*" comes from the Polish for *white tower*, supposedly in reference to a white lodge erected around the year 1400 by Jogaila a.k.a King Władysław II Jagiełło, Lithuania's last pagan ruler and victor over the Teutonic Knights in the near-mythical Battle of Grunwald.

17. Perino, A., et al. 2019. Rewilding complex ecosystems. *Science*. 364: eaav5570. https://dx.doi.org/10.1126/science.aav5570.

18. The primary candidates for such regions lie in the arctic and subarctic portions of Canada, Russia, and the Alaskan United States, along with parts of the Sahara and the Amazon and Congo Basins. Plumptre, A.J., et al. 2021. Where might we find ecologically intact communities? *Frontiers in Forests and Global Change*. 4: 26. https://doi.org/10.3389/ffgc.2021.626635.

19. As Frans Vera notes, insufficient food is a natural check on population growth in the wild and animal welfare in the wild park was, perhaps unfairly, "compared against an agricultural benchmark." Vera, F.W.M. 2009. Large-scale nature development: the Oostvaardersplassen. *British Wildlife*. 20: 28–36. https://www.researchgate.net/publication/235249399.

20. Janzen, D.H., and P.S. Martin. 1982. Neotropical anachronisms: the fruits the gomphotheres ate. *Science*. 215: 19–27. http://www.jstor.org/stable/1688516; Barlow, C. 2001. Anachronistic fruits and the ghosts who haunt them. *Arnoldia*. 61: 14–21. http://arnoldia.arboretum.harvard.edu/issues/172.

21. Glyptodonts were those magnificent automobile-sized herbivorous mammals of South America of the Pleistocene, related to modern-day armadillos, with a shell like an enormous lobster's carapace and with an intimidating spiky club of a tail. The gomphotheres of the Pleistocene were elephant-like creatures, but from a separate evolutionary lineage from their fellow proboscideans that are modern-day elephants. The Pleistocene is that 2.5-million-year span of time leading up to its end 11,700 years ago, which marked the onset of the Holocene, in which the present-day Anthropocene resides.

22. Hunting, however, appears not to have been the leading driver of megafauna extinctions in arctic latitudes. Wang, Y., et al. 2021. Late Quaternary dynamics of Arctic biota from ancient environmental genomics. *Nature*. 600: 86–92. https://doi.org/10.1038/s41586-021-04016-x; Andermann, T., et al. 2020. The past and future human impact on mammalian diversity. *Science Advances*. 6: eabb2313. https://doi.org/10.1126/sciadv.abb2313; Araujo, B.B.A., et al. 2017. Bigger kill than chill: the uneven roles of humans and climate on late Quaternary megafaunal extinctions. *Quaternary International*. 431: 216–222. https://doi.org/10.1016/j.quaint.2015.10.045.

23. Pauly, D. 1995. Anecdotes and the shifting baseline syndrome of fisheries. *Trends in Ecology & Evolution*. 10: 430. https://doi.org/10.1016/S0169-5347(00)89171-5.

24. Svenning, J.-C., et al. 2016. Science for a wilder Anthropocene. *Proceedings of the National Academy of Sciences USA*. 113: 898–906. https://dx.doi.org/10.1073/pnas.1502556112.

25. Recall from Chapter 8 that Dobzhansky-Muller incompatibilities are those devastating interactions between genes that occur in interspecies hybrids.

26. Nogués-Bravo, D., et al. 2016. Rewilding is the new Pandora's box in conservation. *Current Biology*. 26: R87–R91. https://doi.org/10.1016/j.cub.2015.12.044; Rubenstein, D.R., and D.I. Rubenstein. 2016. From Pleistocene to trophic rewilding: a wolf in sheep's clothing. *Proceedings of the National Academy of Sciences USA*. 113: E1. https://doi.org/10.1073/pnas.1521757113; Oliveira-Santos, L.G.R., and F.A.S. Fernandez. 2010. Pleistocene rewilding, Frankenstein ecosystems, and an alternative conservation agenda. *Conservation Biology*. 24: 4–5. https://doi.org/10.1111/j.1523-1739.2009.01379.x.

27. Recent years have seen numerous experiments and versions of Pleistocene rewilding, including adding Przewalski's horses and aurochs-like cattle into Hungary's 25-square-km (6000-acre) Hortobágy National Park and a slew of animals into Russia's 160-square-km (40,000-acres) privately owned nonprofit Pleistocene Park in northeastern Siberia. Today, the Rewilding Europe organization lists 66 distinct rewilding efforts of various sizes and scopes across 27 European countries, https://rewildingeurope.com/european-rewilding-network. Accessed May 14, 2021;Kerekes, V., et al. 2019. Analysis of habitat use, activity, and body condition scores of Przewalski's horses in Hortobagy National Park, Hungary. *Nature Conservation Research*. 4: 31–40. https://dx.doi.org/10.24189/ncr.2019.029; Macias-Fauria, M., et al. 2020. Pleistocene Arctic megafaunal ecological engineering as a

natural climate solution? *Philosophical Transactions of the Royal Society B*. 375: 20190122. https://doi.org/10.1098/rstb.2019.0122; Vera, F.W.M. 2009. Large-scale nature development: The Oostvaardersplassen. *British Wildlife*. 20: 28–36. https://www.researchgate.net/publication/235249399.

28. Quotation from p. 670 in Donlan, C.J., et al. 2006. Pleistocene rewilding: an optimistic agenda for twenty-first century conservation. *American Naturalist*. 168: 660–681. https://doi.org/10.1086/508027. This view echoes one rationale in early perspectives on rewilding strategies, "Wilderness is hardly 'wild' where top carnivores ... have been extirpated ... nature seems somehow incomplete, truncated, overly tame. Human opportunities to attain humility are reduced" (p. 24 in Soule and Noss, 1998, *Wild Earth*). In line with this vision, Kuno National Park, a nature reserve in India, is scheduled to reintroduce cheetahs for the first time since their extirpation from the country around 1950. https://nymag.com/intelligencer/2021/05/indias-effort-to-fight-climate-change-involves-cheetahs.html. Accessed May 25, 2021.

29. Sanderson, E., et al. 2021. A systematic review of potential habitat suitability for the jaguar Panthera onca in central Arizona and New Mexico, USA. Oryx. 56(1): 116–127. https://dx.doi.org/10.1017/S0030605320000459.

30. Randers, J., and U. Goluke. 2020. An earth system model shows self-sustained thawing of permafrost even if all man-made GHG emissions stop in 2020. *Scientific Reports*. 10: 18456. https://doi.org/10.1038/s41598-020-75481-z.

31. Stokstad, E. 2015. Bringing back the aurochs. *Science*. 350: 1144–1147. https://dx.doi.org/10.1126/science.350.6265.1144.

32. A dark piece of the Heck cattle history is the role of Lutz Heck in Nazi administration when installed as head of the German Nature Protection Authority, because of his close ties to Hermann Goring. Heck's breeding rationale unfortunately meshed with the perversion of genetics co-opted by Nazi politics and propaganda. None of Lutz Heck's cattle survived the Second World War, but his brother Heinz's breed did. Quotation from p. 31 in Campbell, D.I., and P.M. Whittle (eds). 2017. Three case studies: aurochs, mammoths and passenger pigeons. In *Resurrecting Extinct Species*. Cham: Palgrave Macmillan, pp. 29–48. https://doi.org/10.1007/978-3-319-69578-5_2.

33. https://rewildingeurope.com/rew-project/auerrind-project/;   https://rewildingeurope.com/news/new-tauros-release-boosts-natural-grazing-in-the-velebit-mountains. Accessed August 18, 2022.

34. Park, S.D.E., et al. 2015. Genome sequencing of the extinct Eurasian wild aurochs, *Bos primigenius*, illuminates the phylogeography and evolution of cattle. *Genome Biology*. 16: 234. https://doi.org/10.1186/s13059-015-0790-2.

35. Recall from Chapter 5 that a breeder can use a combination of DNA markers in the genome as an "index" to conduct artificial selection and selective breeding due to how they correlate with a broad set of desired characteristics.

36. Miller, J.M., et al. 2017. Identification of genetically important individuals of the rediscovered Floreana Galápagos Giant Tortoise (*Chelonoidis elephantopus*) provides founders for species restoration program. *Scientific Reports*. 7: 11471. https://doi.org/10.1038/s41598-017-11516-2.

37. Quotation from p. 117 in Elizabeth Kolbert's 2021 book *Under a White Sky*.

38. *in silico* (i.e., *in a computer*); *in vitro* (i.e., *in a test tube*); *in vivo* (i.e., *in a live organism*); *ex situ* (i.e., *in captivity outside the laboratory*; and *in situ* (i.e., *in nature*); Novak, B.J. 2018. De-extinction. *Genes*. 9:548. https://doi.org/10.3390/genes9110548.

39. van der Valk, T., et al. 2021. Million-year-old DNA sheds light on the genomic history of mammoths. *Nature*. 591: 265–269. https://doi.org/10.1038/s41586-021-03224-9.

40. Although popular press articles reported as early as 2014 that Church's team has successfully edited elephant genomes in cell cultures to contain mammoth sequences, I am unaware of published scientific studies documenting such findings to date. http://dx.doi.org/10.1038/nature.2015.17462; https://www.vice.com/en/article/5393vn/the-plan-to-turn-elephants-into-woolly-mammoths-is-already-underway. Accessed May 19, 2021. Preface to the 2020 edition of Shapiro, B. 2015. *How to Clone a Mammoth: The Science of De-Extinction*. Princeton, NJ: Princeton University Press.

41. Lynch, V.J., et al. 2015. Elephantid genomes reveal the molecular bases of woolly mammoth adaptations to the arctic. *Cell Reports*. 12: 217–228. https://doi.org/10.1016/j.celrep.2015.06.027.

42. African elephant *Loxodonta africana* cell lines were established first, followed more recently by Asian elephant *Elephas maximus* cell lines. Siengdee, P., et al. 2018. Isolation and culture of primary adult skin fibroblasts from the Asian elephant (*Elephas maximus*). *PeerJ*. 6:e4302. https://doi.org/10.7717/peerj.4302.

43. Fortunately for the mammoth, its backers have a long-haul view. The Revive & Restore and Long Now Foundations explicitly point to acceptable time horizons of 10,000 years. https://longnow.org/about. Accessed May 19, 2021.

44. MacDonald, G., et al. 2012. Pattern of extinction of the woolly mammoth in Beringia. *Nature Communications*. 3: 893. https://doi.org/10.1038/ncomms1881; Vartanyan, S.L., et al. 2008. Collection of radiocarbon dates on the mammoths (*Mammuthus primigenius*) and other genera of Wrangel Island, northeast Siberia, Russia. *Quaternary Research*. 70: 51–59. https://doi.org/10.1016/j.yqres.2008.03.005; Cantalapiedra, J.L., et al. 2021. The rise and fall of proboscidean ecological diversity. *Nature Ecology and Evolution*. 5: 1266–1272. https://doi.org/10.1038/s41559-021-01498-w.

45. Novak, B.J. 2018. De-extinction. *Genes*. 9:548. https://doi.org/10.3390/genes9110548; https://reviverestore.org/projects/woolly-mammoth. Accessed May 19, 2021.

46. https://www.nature.com/articles/d41586-021-02844-5. Accessed October 20, 2021.

47. Nogués-Bravo, D., et al. 2008. Climate change, humans, and the extinction of the woolly mammoth. *PLoS Biology*. 6: e79. https://doi.org/10.1371/journal.pbio.0060079; Wang, Y., et al. 2021. Late Quaternary dynamics of Arctic biota from ancient environmental genomics. *Nature*. 600: 86–92. https://doi.org/10.1038/s41586-021-04016-x; Monteath, A.J., et al. 2021. Late Pleistocene shrub expansion preceded megafauna turnover and extinctions in eastern Beringia. *Proceedings of the National Academy of Sciences USA*. 118: e2107977118. https://dx.doi.org/10.1073/pnas.2107977118.

48. https://tigrrlab.science.unimelb.edu.au/. Accessed August 18, 2022.

49. Phelan, R., et al. 2021. Why intended consequences? *Conservation Science and Practice*. 3:e408. https://doi.org/10.1111/csp2.408.

50. https://www.theatlantic.com/magazine/archive/2017/04/pleistocene-park/517779. Accessed May 17, 2021.

51. In 1983, the American Chestnut Foundation initiated their program of intercrossing between American and Chinese chestnut to introgress blight-resistant gene variants into the American chestnut genome. Westbrook, J.W., et al. 2019. Optimizing genomic selection for blight resistance in American chestnut backcross populations: a trade-off with American chestnut ancestry implies resistance is polygenic. *Evolutionary Applications*. 13: 31–47. https://doi.org/10.1111/eva.12886; Clark, S. L., et al. 2019. Eight-year blight (*Cryphonectria parasitica*) resistance of backcross-generation American chestnuts (*Castanea dentata*) planted in the southeastern United States. *Forest Ecology and Management*. 433: 153–161. https://doi.org/10.1016/j.foreco.2018.10.060.

52. Zhang, B., et al. 2013. A threshold level of oxalate oxidase transgene expression reduces *Cryphonectria parasitica*-induced necrosis in a transgenic American chestnut (*Castanea dentata*) leaf bioassay. *Transgenic Research*. 22: 973–982. https://doi.org/10.1007/s11248-013-9708-5; Steiner, K.C., et al. 2017. Rescue of American chestnut with extraspecific genes following its destruction by a naturalized pathogen. *New Forests*. 48: 317–336. https://doi.org/10.1007/s11056-016-9561-5.

53. Shapiro, B. 2017. Pathways to de-extinction: how close can we get to resurrection of an extinct species? *Functional Ecology*. 31: 996–1002. https://doi.org/10.1111/1365-2435.12705.

54. In principle, a eugenicist dictator could apply such DNA testing each generation to guide human breeding to collect the fragments of Neanderthal DNA found across humanity's genomes. In so doing, one could assemble individuals with a genome that contained approximately 50% *Homo neanderthalensis* DNA, rather than the roughly 2% that each of us typically have embedded in our chromosomes. Vernot, B., et al. 2016. Excavating Neandertal and Denisovan DNA from the genomes of Melanesian individuals. *Science*. 352: 235–239. https://dx.doi.org/10.1126/science.aad9416.

55. Derived from frozen cells from the last living specimen of *Capra pyrenaica*, the single cloned bucardo that survived embryonic development in the uterus of a captive ibex surrogate, however, died soon after birth because of a lung abnormality. Folch, J., et al. 2009. First birth of an animal from an extinct subspecies (*Capra pyrenaica pyrenaica*) by cloning. *Theriogenology*. 71: 1026–1034. https://doi.org/10.1016/j.theriogenology.2008.11.005.

56. https://www.nytimes.com/2021/02/18/science/black-footed-ferret-clone.html. Accessed October 18, 2021. https://www.theverge.com/2018/4/6/17175936/northern-white-rhino-de-extinction-stem-cells-sudan. Accessed September 21, 2021.

57. Steeves, T.E., et al. 2017. Maximising evolutionary potential in functional proxies for extinct species: a conservation genetic perspective on de-extinction. *Functional Ecology*. 31: 1032–1040. https://doi.org/10.1111/1365-2435.12843.

58. Shapiro, B. 2017. Pathways to de-extinction: how close can we get to resurrection of an extinct species? *Functional Ecology*. 31: 996–1002. https://doi.org/10.1111/1365-2435.12705.

59. There are other criticisms of de-extinction, as well, which we will explore in greater depth in later chapters. Quotation from p. S19 in Desalle, R., and G. Amato. 2017. Conservation genetics, precision conservation, and de-extinction. *Hastings Center Report.* 47: S18–S23. https://doi.org/10.1002/hast.747.

60. Quotations from p. 732 in Seddon, P.J., and M. King. 2019. Creating proxies of extinct species: the bioethics of de-extinction. *Emerging Topics in Life Sciences.* 3: 731–735. https://doi.org/10.1042/ETLS20190109.

61. Sedges look a bit like grasses, but they map to a distinct plant family that is also distinct from rushes. My college botany professor, Dr. Nick, taught me to distinguish them by their stems with this mnemonic limerick: "sedges have edges; rushes are round; grasses have joints when the cops aren't around."

62. McGraw, J., et al. 1991. Ecological genetic variation in seed banks I: establishment of a time transect. *Journal of Ecology.* 79: 617–625. https://dx.doi.org/10.2307/2260657; Cannone, N., et al. 2017. Moss survival through in situ cryptobiosis after six centuries of glacier burial. *Scientific Reports.* 7: 4438. https://doi.org/10.1038/s41598-017-04848-6; Yashina, S., et al. 2012. Regeneration of whole fertile plants from 30,000-y-old fruit tissue buried in Siberian permafrost. *Proceedings of the National Academy of Sciences USA.* 109: 4008–4013. https://dx.doi.org/10.1073/pnas.1118386109.

63. Cryptobiosis is not to be confused with cryptozoologist. The latter term refers to a believer in pseudoscience, intent on discovering evidence in nature for folktales, such as the Loch Ness Monster or Bigfoot. My sole encounter with a dyed-in-the-wool cryptozoologist was with no other than the late noted cryptid-seeker J. Richard Greenwell, a founder of the International Society of Cryptozoology. It was on October 16, 2002, and he was delivering a lecture at the International Wildlife Museum in Tucson, Arizona, entitled, "Sasquatch: Does a Large Unknown Primate Exist in North America?" This private museum is a peculiar institution, filled with trophy-hunter taxidermy that acts as a public-relations arm of the Safari Club International, which states its top purpose as: "To advocate, preserve and protect the rights of all hunters." Sneaking in just before the 7:00 p.m. start to avoid paying the $3 fee, with friend and fellow flabbergastee Barrett Klein, we witnessed an hour of entertainingly earnest tales of ambiguous booms in forest audio recordings (mating calls, perhaps?), false leads on Sasquatch scat (DNA analysis was inconclusive) and hopes for glimpses of sagittal crests through the too-dense foliage. Alas, each of his four expeditions to the Siskiyou Wilderness proved unsuccessful. Quotation from https://safariclub.org/about-us. Accessed May 20, 2021.

64. Rebecchi, L., et al. 2007. Anhydrobiosis: the extreme limit of desiccation tolerance. *Invertebrate Survival Journal.* 4: 65–81. https://www.isj.unimore.it/index.php/ISJ/article/view/144.; Tsujimoto, M., et al. 2016. Recovery and reproduction of an Antarctic tardigrade retrieved from a moss sample frozen for over 30 years. *Cryobiology.* 72: 78–81. https://doi.org/10.1016/j.cryobiol.2015.12.003; Shatilovich, A.V., et al. 2018. Viable nematodes from Late Pleistocene permafrost of the Kolyma River Lowland. *Doklady Biological Sciences.* 480: 100–102. https://doi.org/10.1134/S0012496618030079.

65. Shmakova, L.A., et al. 2020. Microeukaryotes in the metagenomes of late Pleistocene permafrost deposits. *Paleontological Journal*. 54: 913–921. https://doi.org/10.1134/S003103012008016X; Shmakova, L., et al. 2021. A living bdelloid rotifer from 24,000-year-old Arctic permafrost. *Current Biology*. 31: R712–R713. https://doi.org/10.1016/j.cub.2021.04.077.

66. There also are reports of live bacteria getting revived after being trapped for 25-plus million years in fossilized amber or for 250 million years in salt crystals, though these findings are treated with skepticism. Johnson, S.S., et al. 2007. Ancient bacteria show evidence of DNA repair. *Proceedings of the National Academy of Sciences USA*. 104: 14401–14405. https://dx.doi.org/10.1073/pnas.0706787104; Zhang, D.-C., et al. 2013. Isolation and characterization of bacteria from ancient Siberian permafrost sediment. *Biology*. 2: 85–106. https://doi.org/10.3390/biology2010085; Cano, R., and M. Borucki. 1995. Revival and identification of bacterial spores in 25- to 40-million-year-old Dominican amber. *Science*. 268: 1060–1064. http://www.jstor.org/stable/2888885.; Vreeland, R., et al. 2000. Isolation of a 250-million-year-old halotolerant bacterium from a primary salt crystal. *Nature*. 407: 897–900. https://doi.org/10.1038/35038060.

67. Quotation from p. 665 in Donlan, C.J., et al. 2006. Pleistocene rewilding: an optimistic agenda for twenty-first century conservation. *American Naturalist*. 168: 660–681. https://doi.org/10.1086/508027.

68. Quotation from p. 731 in Seddon, P.J., and M. King. 2019. Creating proxies of extinct species: the bioethics of de-extinction. *Emerging Topics in Life Sciences*. 3: 731–735. https://doi.org/10.1042/ETLS20190109.

69. Quotation from p. 250 in Weiner, J. 1994. *The Beak of the Finch*. New York: Vintage Books.

70. https://colossal.com/company. Accessed September 16, 2021.

71. Wolves also appear to have killed 64 sheep, five dogs, and one domestic bison during the same period. https://www.npr.org/2021/05/21/999084965/new-idaho-law-calls-for-killing-90-of-states-wolves. Accessed September 16, 2021.

72. Ripple, W.J., et al. 2022. Rewilding the American West. *BioScience*. biac069. https://doi.org/10.1093/biosci/biac069.

73. https://www.nepad.org/publication/gene-drives-malaria-control-and-elimination-africa; https://allianceforscience.cornell.edu/blog/2018/11/africa-kicks-proposed-gene-drive-moratorium-un-biodiversity-conference/. Accessed June 2, 2022.

74. Montenegro de Wit, M. 2019. Gene driving the farm: who decides, who owns, and who benefits? *Agroecology and Sustainable Food Systems*. 43: 1054–1074. https://dx.doi.org/10.1080/21683565.2019.1591566.

14

When nature comes to call

Intentional rewilding seems like the new kid on the conservation block. Humanity, however, has seen its fair share of unintentional rewilding. As civilizations collapse or disasters drive out human habitation, nature, in time, fills the void.

Nuclear disasters usually give little room for an upside. That certainly was the world's sentiment in the immediate aftermath of the events of April 26, 1986: the day the Chernobyl nuclear reactor melted down in what is now Ukraine near the Belarus border. The vicinity was evacuated to form a 30-km exclusion zone surrounding the reactor, which, along with abandonment of huge swathes of adjacent agricultural land, led to what is now "one of the most iconic natural experiments on rewilding in recent history."[1] Nature was left to its own devices, albeit with a heavy spritz of ionizing radiation. The passive rewilding around Chernobyl, excepting intentional reintroduction of wisent and Przewalski's wild horses, was subsequently formalized by establishing nearly 5000 square kilometers (1.2 million acres) of biological reserves in the form of the Belarusian Polesie State Radioecological Reserve and the Ukrainian Chornobyl Radiation and Ecological Biosphere Reserve. It's now a thriving home to the entire living repertoire of large European animals: wisent, wild boar, moose, red and roe deer; wolf, brown bear, and lynx; beaver, badger, fox, and the fox-relative known as the raccoon dog.[2]

At a smaller scale, we can see passive rewilding in ghost towns. The Mojave Desert of California and Nevada in the early 1900s was subjected to boom-and-bust mining operations, striking the ground hard under the heel of human activity. With township names like Kunze, Gold Valley, Furnace, and Skidoo, it almost seems a shame that no one calls them home any longer. When the ore ran out between 1906 and 1928, the rapidly abandoned ghost towns were left to the whims of the weather. Passive rewilding followed, from the surrounding arid-adapted plants and animals. Now, about a century later, the soil disturbance and vegetation is reaching more-or-less full recovery compared to adjacent plots.[3] The stark population declines in recent decades in rural Spain and Japan, as well as the exclusion zone around the Fukushima nuclear disaster site, similarly have given license for passive rewilding by plants and animals.[4]

On a longer timescale, we can look to the traces of collapsed civilizations in the rewilded ruins of their forsaken cities. Humans abandoned the grand Maya city of Tikal about 1200 years ago after a thousand years of habitation. They left, apparently, due to the combined effects on Mayan socioeconomic cohesion of drought, deforestation,

Evolving Tomorrow. Asher D. Cutter, Oxford University Press. © Asher D. Cutter (2023). DOI: 10.1093/oso/9780198874522.003.0014

and toxic drinking water of their reservoirs.[5] Within as little as 80 years, the landscape was overtaken entirely by tropical forest. Tropical forests that were deforested in modern times also appear capable of robust recovery. Nearly 80% of old-growth attributes return within 20 years of regrowth following abandonment of agricultural land, although recovery of species diversity and community composition takes many decades longer.[6]

The feral horses roaming the American west represent a curious case of unintentional rewilding, if actively instead of passively promulgated. Horses ran as natives in North America for millions of years, until close to the end of the last ice age. Their continental extinction tagged along as part of the great megafaunal extinction around the dawn of the Holocene. That extinction of wild stallions, however, was undone by their reintroduction in the form of domesticated horses starting in the late 1400s by European explorers and settlers. Are the 5000 or more mustangs that today roam the American west feral invasives, or are they righteous rewilded residents? Either way, ranchers seem to see them as trespassers, and so the US Bureau of Land Management intends to cut their population back by 70%.[7]

The bighorn sheep of Tiburón Island, the biggest island in the Gulf of California tucked just inside the elbow of Baja California, have a similar if more welcome story. Now comprising a healthy herd of 500 or more, a small flock of 20 *Ovis canadensis* was shepherded to the deserted island in 1975 as a conservation measure because of the species' rarity in Mexico.[8] Nearly 40 years later it was discovered that Tiburón was, in fact, an ancestral home for the species some 1500 years ago, based on analysis of ancient remains found within a cave. Native or nonnative often is all a matter of timescale, of how shifted is your baseline.

14.1   Manufacturing destiny

In the public relations rollout to encourage ecological restoration and rewilding, you would be forgiven for conjured visions of wilderness Pleistocene parks as living museums in rigid harmony with themselves. But this pictures a false stasis. The long generation times of oaks and tortoises, the serenity of a tall-grass meadow on a sunny day, the predictable seasonality of leaf-flush and butterfly migratory passages, all disguise the dynamism of evolutionary change. Evolution among interacting species, as we have seen, is inexorable and often poised for hurried change. This is to say, when set loose, the survival and reproduction of a species drives its own destiny as evolution takes on a life of its own.

Wherever there is a group of individuals, a population, inhabiting enough space with enough resources to grow their numbers, it implies the capacity for exponential growth. The capacity for exponential growth, in turn, implies an inevitable struggle for existence among members of that population because demand for food and space will eventually outstrip its availability. This is the basis for natural selection. The consequences go on full display among the large herbivores of Oostvaardersplassen in the chill of winter when food is scarce, cutting down the herds by 6% at the minimum.[9] The same principles apply to populations of organisms derived from genetic engineering: synthetic life evolves, too.

Today's workhorse of synthetic life is the *E. coli* strain C321.ΔA, a suitably *Blade Runner*-esque name. This bacterium has had its genome recoded. In encoding a protein, a gene uses triplets of nucleotide letters to designate particular amino acids within that protein, each triplet called a codon. With four nucleotide letters, that makes $4 \times 4 \times 4 = 64$ possible triplet combinations. This amounts to 43 more codons than are needed to map to the 20 typical amino acids plus one codon that tells a cell, when making a protein, where to stop. As a result, there is a lot of redundancy in this genetic code, which the creators of C321.ΔA exploited by reducing the redundancy ever so slightly in order to give it a new capability.

Normally, the cell's genetic code relies on three redundant codons to represent that stop signal (TAG, TGA, TAA). In C321.ΔA, however, the cell no longer reads the DNA nucleotide triplet TAG as "STOP." Instead, the genes in its genome have been edited to rely entirely on the two alternative stop signals. This feature frees up the TAG codon to do something else instead. It lets biotechnologists genetically engineer C321.ΔA even further. It provides the capability to use the TAG triplet to encode a different kind of information. In the C321.ΔA strain, TAG can encode *p*-acetyl-phenylalanine instead, one of a suite of possible nonstandard amino acids. It is nonstandard because it isn't found in any known life form—until now. By incorporating this nonstandard amino acid, microbiologists hope to expand the functional repertoire of possible biological chemistry that bacteria are capable of. The problem is that C321.ΔA is sick: it takes nearly twice as long to grow as other strains of *E. coli*. One remedy: evolution. After 1100 generations of experimental evolution, mutations arose and fixed in lab populations of C321.ΔA that recovered their poor fitness to normal levels.[10] Evolution did what genetic engineering didn't. Adaptation by natural selection yielded a healthy new bacterial workhorse for synthetic biology with evolution in its name: "C321.ΔA.M9adapted."

Evolution often is pitched in terms of history. It is true: genomes record a partial history of genetic change and fossil strata record a partial history of life on Earth. In this sense, what exists in the present depends on what happened in the past. Genome engineering of species put into the wild, however, would introduce a discontinuity into the living world with respect to the historical view of evolution. They jump life's queue, so to speak. Nonetheless, the way that the process of evolution works in any given moment is as a process of accretion. Evolution yields genetic additions and subtractions to whatever the biological state happens to be right now in the present. In this sense, evolution is memoryless, unaware of its own history, what a statistician might call Markovian. With respect to how evolution proceeds from here on out for a population of a new and extraordinary species that finds itself in the here-and-now, there is nothing special in the fundamentals of what evolution does next.[11]

★

For some species, human occupation of the landscape is a blessing, not a curse. They have proliferated right alongside people. You know many of these Anthropocene winners because they live with us and near us: they are human commensals. First, there are the domesticates. You could count for a long time enumerating the members of the select few species that we have cherry-picked for domestication and agriculture and horticulture, whether starting with chickens (26 billion), maize (1.1 billion metric tons), or tulips

(173 million sold). Despite these vast numbers of individuals, only one half of one percent of known species of plants and mammals were domesticated for human food consumption.[12] Then, there are the innocuous natural examples. You may not need to do more than look out a window to see some of them: the two most abundant undomesticated birds in the world are the House Sparrow (1.6 billion) and Common Starling (1.3 billion), with *Columba livia* pigeons rounding out the top 20 (285 million).[13] As for our furry relatives, it turns out that, on average, mammals are more abundant in areas with a stronger human footprint, excluding the densest urban spots.[14] Anecdotally, the trees of my Toronto postage stamp of a backyard host an exceptional density of squirrels compared to my experience in any patch of forest. Then, there are the pests and the weeds. The ship rat, the dandelion, and the constellation of species of mosquito.

In transforming the Earth's surface—and subsurface—we've created unanticipated habitats ripe for colonization by some preadapted species. Take, for instance, the London Underground. England's famed subway train system started operation in 1863, just in time for passengers to read Charles Darwin's letter to *The Gardeners' Chronicle* on the subject of "Vermin and Traps."[15] The subterranean tunnels of the Underground also served as bomb shelters during the Second World War, during which Londoners learned that they were not alone in the Tube. The tunnels might have made war-weary Tube-dwellers safe from explosions, but they were not safe, even in winter, from voracious Underground mosquitoes. The fittingly named *Culex pipiens molestus* had migrated from Mediterranean latitudes to colonize a new home in the year-round mild environs of the London Underground. *C. p. molestus* is behaviorally and reproductively distinct from local aboveground *Culex pipiens*, even if anatomically indistinguishable.[16] Subway systems from throughout Europe and the United States now have their own local populations of *C. p. molestus*, which, unfortunately for commuters, also happen to serve as a vector of West Nile Virus.

Anthropogenic influence over nature isn't just limited to ecological range expansions. Urbanization of environments also drives evolutionary change. I know white clover from eyeing patches of it amidst the grassy lawn in my parents' New England backyard, in hopes of spying the occasional developmental anomaly that gives an extra leaflet to *Trifolium repens* to form a four-leaf clover (Figure 14.1). Inside the leaf tissue, no matter if three or four leaflets, however, is a potent poison to deter herbivores: hydrogen cyanide (HCN). Clover growing in cities, it turns out, produce less cyanide than clover plucked from rural areas.[17] This pattern is repeated again and again from one city to the next when you march from an urban center to the countryside, reflecting repeated evolution of low cyanide production by the clover that inhabit cities.

The difference in cyanide for city and country clover is controlled primarily by the relative prevalence of genetic variants for two genes: a detoxifying enzyme known as the cytochrome P450 gene *CYP79D15* and the linamarase enzyme encoded by a gene known simply as *Li*. When a plant happens to be lacking a functioning copy of one or both of these two genes, then it doesn't make cyanide. While this may seem absolutely detrimental, making them susceptible to hungry leaf-eaters, there are other effects—another case of pleiotropy. Cyanide production also makes the plants more sensitive to

Figure 14.1 *White clover,* Trifolium repens, *grows commonly in association with humans, as is the case for these plants found in a small park in downtown Toronto, Ontario. Clover plants in settings that are more urban tend to have a genetic makeup that causes them to produce less cyanide than their rural relatives, making them more susceptible to herbivores but also more adapted to environmental conditions within cities.*

Photo by author.

cold temperatures and more resistant to drought. It appears that the balance of costs from herbivores and environmental stress vary consistently in such a way as to favor cyanide production most strongly in rural environments and its absence most strongly in urban environments.

Stands of clover do not stand alone in having evolved in response to human co-habitation. Dozens of species show documented evolutionary change due to city living. There is the peppered moth *Biston betularia*, of course, with its 1800s evolution of dark coloration making it inconspicuous to bird predation in sooty southern England, only to then evolve back to a preponderance of speckled white moth wings in the 1900s when air quality improved. Researchers also have documented urban-associated evolution for populations of bobcats and blackbirds, mice and salamanders, water dragons, and, yes, even us.[18]

Humans directly mediate some selection pressures on wild organisms, even if not through premeditated artificial selection. Nonetheless, those pressures can yield an entirely predictable, albeit unintended and undesired, outcome. Hunting for trophies from the biggest-horned bighorn sheep led to the evolution of smaller horn size in rams.[19] This hunting as a selective pressure in the mountains of Alberta, Canada through the mid-1990s butted heads with sexual selection pressure. Ordinarily, sexual selection in bighorn sheep confers greater reproductive fitness on those male rams with larger headgear. Because the trophy hunting typically occurred prior to breeding, it exacted a disproportionate influence on the evolutionary fate of horn size as a heritable feature. Selective harvesting in the sea is similar. Fishing for the largest fish, whether Atlantic cod *Gadus morhua* or the flounder *Hippoglossoides platessoides*, has induced the evolution of smaller fish and younger ages of maturation.[20]

The biggest example of humans inadvertently driving evolutionary change comes out of Africa. Poaching of elephant tusks in Mozambique led to an extreme evolutionary outcome in just 15 years. During the Mozambican civil war throughout the 1980s, ivory poaching drove a >90% decline in the elephant population. African savannah elephants *Loxodonta africana* that lacked tusks, however, were more likely to survive, with the incidence of tuskless females increasing from 19% to 51%. Genetic variants associated with the amelogenin gene AMELX increased in abundance and appear to contribute to tusklessness in females. The same gene variants also appear responsible for causing male embryos to die *in utero*. As a result, a greater fraction of birthed offspring are spared a death at the hands of ivory poachers: fewer offspring grow up to develop tusks at all as a consequence of females being more likely to be tuskless and males, which are tusked, being underrepresented at birth.[21] Given the pivotal role of tusks in how elephants enact their role as ecosystem engineers, it remains to be seen whether downstream impacts on Mozambique's savannah ecosystem will follow suit.

★

Species commensal to humans have successfully exploited the opportunities of humanity's ecological engineering. Other species, however, take ecosystem engineering into their own hands. Nowhere is this role more obvious than with the beaver. Thanks to their powerful gliriform teeth, *Castor canadensis* can cut down a 15-cm-diameter (6-inch-diameter) aspen in under an hour, drag the cut logs up to 200 m (650 feet) to a stream, and use a collection of such logs, trimmed to 2-m (6-foot) lengths, to construct a dam up to 3 m (10 feet) tall and 850 m (nearly half a mile) wide.[22] Stuffing the cracks with grass, mud, and rocks, these dams create new habitats. The dam locations, however, change over time, getting established and abandoned, producing dynamic patterns of ecological succession across the landscape through the turnover of plants, herbivores, and large carnivores.

In the oceans, corals are the builders, creating reefs as habitat for exceptional diversity of life. In these newest days of the Anthropocene, however, rising carbon dioxide in the atmosphere percolates into the oceans to make them more acidic. Combined with the stress of unusual events of warm water temperatures, coral reef construction has waned around the world.[23] With ocean acidification expected to halve the concentration of carbonate ions in seawater by the year 2100—the key building block of coral skeletons— the prognosis for coral ecosystem engineering is not looking good.

Because essentially all organisms modify their environment to some extent—digging, secreting concentrated chemical compounds, blocking wind—essentially all organisms are, to some extent, ecosystem engineers. It's a matter of degree: the direct habitat creation by the beaver versus the byproduct of the earthworm's tunneling. Even the very existence of an animal's body represents an unexploited habitat for evolution by parasites and pathogens. Organisms adapt to their environment in ways that modify it, that then influence the abundance and distribution of other species, which drives further evolution of their own and other species. This ongoing recursive biological feedback is sometimes termed eco-evolutionary dynamics.[24]

That is to say, organisms affect one another. They affect one another in the outward way that we can see, by which species are present or absent in a given place. They also affect one another at the most fundamental level, shifting the direction and magnitude of natural selection and genetic drift, altering the course that evolution is most likely to take their DNA from one generation to the next.

★

Rewilding. De-extinction. Hyperexotic species introduction. What is the worst that could happen?

The planet is riddled with intentional and unintentional species introductions into the wild. The astounding scale of human movement across the globe has led to the astounding realized potential for movement of other species, as well. Some are simply hitchhikers, and some were moved with benignly ill-conceived intentions, as for the nineteenth century's American Acclimatization Society aiming to populate North America with a panoply of European birds.[25] During the era of seafaring global exploration, canny sailors would convert Caribbean and South Pacific islands into living grocery stores by depositing domestic animals on their shores. These newcomer populations of pigs and sheep and goats, and rats, were left to run feral to provide food on subsequent stopovers. In the process, these creatures decimated the local flora and fauna. Among known extinctions, more than half have resulted in large part from the influence of invasive species.[26] The costs of invasive species accumulate from alterations to soils and water quality, erosion, cycling of nutrients through ecosystems, vectoring of diseases, and changes to ecological communities, including extinction of local species.

In a fight-fire-with-fire tactic, classical biological control takes the strategy of bringing in yet another nonnative species with the goal of managing the preceding invasive pest. Despite the alarm that this idea might trigger, it often works out reasonably well, though most cases target weedy plants or insects.[27] The trouble is that, when it goes wrong, it can prove disastrous. Australia has a litany of cautionary tales.[28] It's not all bad news, though. Among 8628 nonnative terrestrial species of plants and animals documented throughout Europe, 358 are vertebrates, and just about one third of those vertebrates exact an ecological or economic toll.[29] Clearly, many of the nonnative organisms are not quantifiably damaging or invasive at all. The vast majority of reports—86%—on programs to eradicate problematic invasive species have proven successful, 95% of which targeted vertebrates.[30] Perhaps such success in eradicating large animals isn't so very unexpected. After all, we humans have proved exceptionally capable of inducing extinction in large animals in the lead up to and throughout the Anthropocene.

Biological control is a paradigm of intended consequences. Sometimes—as when the introduced agent can't establish itself—it fails to achieve those intentions. Sometimes, instead, it achieves unanticipated terrible consequences, and sometimes it even works as advertised. It is a form of ecological restoration, just like rewilding. Rewilding is also a paradigm of intended consequences, and so we should anticipate cases for each of these three categories of outcome. Provided that a newly introduced species can persist in the new environment, however, with near certainty, we can expect that it will have an effect.

★

The literally biggest invasive animals, I'd hazard to guess, are the hippos in Colombia (Figure 14.2). The introduction of the hippopotamus to South America was not due to inconspicuous seeds stuck to the sole of a flight attendant's shoe. No, it was a drug lord's intercontinental whimsy that trafficked them to South America from Africa by way of shady dealings in the United States. A quartet of smuggled *Hippopotamus amphibius* from an American zoo arrived as pets for notorious leader of the Medellín drug cartel, Pablo Escobar, the wealthiest criminal in the world. After Escobar's dramatic death in 1993 after a rooftop chase, his 20-square-kilometer (5000-acre) Hacienda Nápoles estate became a tourist attraction—including the menagerie of exotic animals that accompanied the expanding, and now feral, population of hippo descendants. Native to sub-Saharan Africa, today they number perhaps 80 animals spread along 150 km (90 miles) of tropical tributaries and ponds in the Magdalena River basin. If left unchecked, the hippos of Colombia, however, could hit 1000 or more individuals in another 20 years.[31]

Figure 14.2 *The hippopotamus evolved in Africa, but humans transplanted these aquatic animals to South America, where they escaped to become feral in a river basin of Colombia. These ecological engineers have the capacity to alter ecosystems and, in some but not all ways, to mimic features of large mammals that went extinct in South America during the Pleistocene.*

Image credit: Paul Maritz, reproduced under the CC BY-SA 3.0 license.

Hippos are physically dangerous to people, with hippo attacks being more deadly than attacks by sharks, crocodiles, or grizzly bears.[32] Hippos also threaten South American water quality by perturbing sediments and depositing excessive nutrients through feces, altering microbial communities in ways that promote toxic algal blooms. And yet, the people love them. They are big and charismatic and darlings of animal rights activists; they are on the International Union for Conservation of Nature's (IUCN) "Red List" of vulnerable species in their native range; they accentuate the mystique and lore of the region around Medellín; they are a tourism magnet. In trying to thread this needle of competing interests, the Colombian government has elected to sterilize rather than cull the feral animals.[33]

Could Colombian hippos as intercontinental aliens be viewed yet another way: rewilded megafauna to reoccupy ecological roles of long-lost South American extinctions? Rather than sterilizing or culling—what about helping hippos to take hold? At a global level, some 42% of nonnative introductions of large herbivores have acted to offset the lost functions of extinct megafauna.[34] How imperfect can a Pleistocene mimic be to recoup lost ecosystem functions and thus define the strategy for the future of Colombian hippos? As it turns out, they are not perfect mimics of extinct megafauna, instead presenting "a chimera of multiple extinct species' trait combinations," including those of an extinct giant llama and of an extinct semiaquatic creature known as *Trigonodops*.[35] As ecological engineers, however, hippos certainly can make a mark. From one vantage, they are an ecosystem disruption; from another, some of that disruption might fill underserved functions of now-extinct South American megaherbivores.

In a region with just a quarter of its historical tropical forest remaining, the added stress of an imperfect ecosystem engineer might be more than we're willing to risk, however, given all the other sources of stress felt by South American tropical ecosystems owing to climate change and human land use. If we were to take that risk, then we'd also need to think about how to augment their inbred genetic variability. After all, the current population was spawned from just three females and a single male. Perhaps the hippo is too imperfect of a proxy, or just too vigorous of an ecosystem engineer. It remains an open question, however, whether the mark that hippos might leave on South America would provide a missing ecosystem service as a Pleistocene proxy or impose a lasting ecosystem disservice as an invasive species.

★

Nonnative and *invasive* are biological curse words, "culturally and emotionally freighted terms," as Lesley Head says, on many conservationist's lips.[36] How righteous is the biological xenophobia against nonnative species? As global climate change shifts the habitable ranges of species, perhaps we should welcome the arrival of newcomers as part of a range shift that enables species persistence, in many cases, rather than sounding the invasive species alarm.[37] So-called assisted migration is the proactive face of restoration ecology, after all, where conservationists intentionally move individuals to populate a new place to encourage the global persistence of species and ecosystems.

The case of large mammals, in particular, can present a so-called conservation paradox.[38] If a keystone species is threatened with extinction in its native range, might a nonnative introduction of it elsewhere provide a desirable refuge to retain in the world today its distinctive evolutionary lineage along with its potential to influence ecosystem function? This is the logic for a global benefit offsetting a local cost, given a prevailing view of nonnative harms to local ecosystems. And, if introduced in a region that has lost species that previously delivered that ecosystem function, might such trophic rewilding by proxy even provide a local benefit? If so, it would be the unicorn of conservation: a win-win.

The shifting baseline syndrome that we all hold as one manifestation of implicit bias is difficult to shake. We've all internalized from our youthful experience a memory of what we perceive as the norm of nature. But what is native versus nonnative often depends decisively on the timescale of reference. We saw this with feral horses in North America and bighorn sheep on Tiburón Island. As Erick Lundgren and his colleagues wrote recently, "Introduced species have been primarily studied in the context of recent historic states under the premise that their ecological functions are novel."[39] And "novel" in the context of species introductions often implies "*unnatural*" or *destructive*" or "*undesirable*." They go on to point out that, when using Pleistocene ecosystems as a reference point, "presumed novelty yields to functional similarity."

After all, North America was overrun by camels, lions, cheetahs, and mammoths not so very long ago. The Pleistocene ended only about 10,000 years ago, some 5% of the timespan that modern *Homo sapiens* has walked the Earth. But camels recolonizing California or Kansas would seem a surprising nonnative introduction on a recent timescale, often measured since European colonization. On the flipside, we often take the humble earthworm to be a normal and valued inhabitant of our North American garden and forest soils. That was the view I grew up with, at any rate, with further inspiration from Thomas Rockwell's classic kids book *How to Eat Fried Worms*. And yet, the *Lumbricus terrestris* nightcrawlers of North America are nonnative imports from Europe, dating only to the 1700s. Their spread, however, is persistent, inducing profound alterations to soil biodiversity and, perhaps, to the structure of our forests' futures.[40]

As Josh Donlan and colleagues pointed out, "Earth is now nowhere pristine . . . we will decide, by default or design, on the extent to which humanity tolerates other species."[41] The corollary to this statement is, of course, that the resulting biodiversity also depends on how those species tolerate one another. Sometimes, it must be admitted, nonnative species negatively influence the abundance and diversity of their new neighbors.[42]

14.2   What is nature?

In the next chapter of the Anthropocene, if the world finds itself with hyperexotic species roaming the plains and with gene drives infiltrating genomes, seemingly with minds of their own as new ecological and evolutionary forces—what is nature? Even now, in the stage of the Anthropocene without such interventions running rampant just yet, there is no place on Earth unaffected by human activity. Our sense of what is "authentically natural" is no doubt shaped by shifting baseline syndrome. The norm of what we call natural amounts to whatever it is that was most wild in our own experience, with that baseline of experience shifting to the less and less wild from generation to generation. Part of the common view that nonnative species are undesirable or that genetically modified organisms are abominable is that they are inauthentic, that they just ain't natural. Why do we think that way?

If you were a psychoanalyst and you were to psychoanalyze our connection with nature, you might view people as nature tourists. You might say, "nature tourists … desire corporeal engagement with wilderness as a way to (re)connect to their biological selves." Seeking an authentic sense of wilderness is a potent desire. And yet, "authenticity is a fantasy, yet a crucial fantasy to humans' successful functioning."[43] In a related take, Nigel Dudley points out that truly pristine environments are a fiction, and life is dynamic, making some fixed ideal of "authentic" nature an untenable concept. He rejects the notion of a baseline reference for what is authentically natural. Instead, Dudley defined an "authentic ecosystem" much in the way that rewilding proponents do, despite his 2011 book *The Authenticity of Nature* only twice mentioning the term *rewilding*.[44] The key feature for us to internalize is a sense of self-sustaining resilience of biodiversity. In this view, it doesn't matter how much or what kinds of human influence the habitat has experienced, so long as the ecological interactions necessary for self-sustaining ecosystem function are in place.

The entire notion of authentic nature, still others would argue, depends intrinsically on humans interacting with it, that we are an indispensable part of nature and not something to separate out from it. For nature to be authentic, that is, we must be interacting with it. The separateness of nature from humanity appears to be a distinct feature of cultures influenced by the Abrahamic religions of Judaism, Islam, and Christianity. The languages of many other cultures, however, don't even have a separate word for nature as a distinct entity. The idea of "nature" as conceived in much of Europe and North America perhaps did not even exist prior to the urbanization of civilizations.[45] Being a part of nature and interacting with other organisms in this integrative kind of way, however, means having deeply rooted respect for the lives of other creatures and for our interactions with them.

⋆

For here-and-now worries about species introductions—whether with simple rewilding, de-extinction, or de novo additions—our concerns center on disruption to current ecosystems in a way that would exacerbate Anthropocene extinction rates, diminish helpful ecosystem services, or alter the wild aesthetic that we have become used to. We worry about gene drives taking on lives of their own, spreading synthetic genetic material without bound. We worry about unintended consequences that make the world worse, in perception or experience. We worry about the human footprint on nature expanding rather than shrinking. The short-term consequences necessarily involve change, and any change evokes anxiety.

Concerns about consequences are all a matter of timescale. We are most sensitive to individually lived human timescales: the consequences for tomorrow and next year and a decade from now. Perhaps, in quiet moments, we feel glimmers of apprehension extending a century. In that longest time horizon, we ponder the state of the world for our grandchildren, the deepest personal touch in our lifetimes that we will have to the future. It is hard to imagine how changes today might ramify for 1000 years, or 10,000, or 10 million. But they do.

Not one of us alive today will live long enough to experience the long-term consequences of species introductions. In the long term, the pace of change will settle down to become the new normal. That longer term meshes with what we might consider to be the integration of nature. Populations of species will adapt by natural selection to their new environs and diverge into distinct forms. For example, the evolutionary origin of mammalian species might require 4 million years or more to attain a distribution of body sizes like that found prior to the Anthropocene.[46] And other populations will go extinct because of lack of reproductive persistence or by fusion, the gene pools of two becoming one. This long-term integration with the normal dynamics of natural ecological systems elides with rates of speciation and extinction reaching their pre-Anthropocene dynamic balance. The forces of evolution will continue on their merry way in ways both predictable and unpredictable.

Then there is the extreme term: everything will be okay. Whatever de-extinct or novel species that we might genetically engineer will, itself, eventually, go permanently extinct. Humans will become extinct. The planet will mark the extinction of dragons, whether the dragons are us or them. The Sun will snuff out; the universe will go cold; entropy will see to it that this existence's experiments with life will end. As I said: everything will be okay. No amount of short-term consequences to biological intervention, or lack thereof, will alter this ultimate outcome. But there is a long span of time between now and then—perhaps 1.6 billion years[47]—and we'd be wise to figure out how to make the most of it.

★

Humanity has profoundly altered Earth's living systems. Recovery of highly impacted areas takes decades to millions of years, depending on how recovery gets measured. Evolution takes on a life of its own for species associated with dense human habitation and when species get introduced to rewilding conditions. Value judgments on species introductions depend on timescales of consideration and baseline expectations about nature. Our cultural traditions bias our views on what makes nature "authentic," and whether humans are separate from or intrinsic to nature. The time horizon of human concerns over potential negative unintended consequences primarily spans one century, though our Anthropocene impacts on the living world will extend for millions of years.

Notes

1. Perino, A., et al. 2019. Rewilding complex ecosystems. *Science*. 364: eaav5570. https://dx.doi.org/10.1126/science.aav5570.

2. Ecological research on the reserves has come to a standstill since the invasion of Ukraine by Russia in February 2022, with land mines likely to hinder future research assessments for years to come.

3. Webb, R.H. 2002. Recovery of severely compacted soils in the Mojave Desert, California, USA. *Arid Land Research and Management.* 16: 291–305. https://dx.doi.org/10.1080/153249802760284829.

4. Spain is notable but isn't alone in contributing to the projections of huge areas across Europe becoming depopulated in the coming decades due to time catching up with human demographic shifts. https://www.theguardian.com/world/2021/jan/24/as-birth-rates-fall-animals-prowl-in-our-abandoned-ghost-villages accessed January 25, 2021; Perpiña Castillo, C., et al. 2020. An assessment and spatial modelling of agricultural land abandonment in Spain (2015–2030). *Sustainability.* 12:560. https://doi.org/10.3390/su12020560; Lyons, P.C., et al. 2020. Rewilding of Fukushima's human evacuation zone. *Frontiers in Ecology and the Environment.* 18: 127–134. https://doi.org/10.1002/fee.2149.

5. A city of 60,000 inhabitants at its peak, the vestiges of Tikal are located in what is now northern Guatemala near its borders with Belize and Mexico, at the southern end of the elevated interior region of the Yucatan Peninsula. Lentz, D.L., et al. 2020. Molecular genetic and geochemical assays reveal severe contamination of drinking water reservoirs at the ancient Maya city of Tikal. *Scientific Reports.* 10: 10316. https://doi.org/10.1038/s41598-020-67044-z; Turner, B.L., and J.A. Sabloff. 2012. Classic Maya collapse in the Central Lowlands. *Proceedings of the National Academy of Sciences USA.* 109: 13908–13914. https://dx.doi.org/10.1073/pnas.1210106109; Mueller, A.D., et al. 2010. Recovery of the forest ecosystem in the tropical lowlands of northern Guatemala after disintegration of Classic Maya polities. *Geology.* 38: 523–526. https://doi.org/10.1130/G30797.1.

6. Poorter, L., et al. 2021. Multidimensional tropical forest recovery. *Science.* 374: 1370–1376. https://doi.org/10.1126/science.abh3629.

7. https://wyofile.com/wild-horse-advocates-cry-foul-on-cusp-of-enormous-roundup-removal accessed October 5, 2021.

8. Wilder, B.T., et al. 2014. Local extinction and unintentional rewilding of bighorn sheep (Ovis canadensis) on a desert island. PLoS One. 9: e91358. https://doi.org/10.1371/journal.pone.0091358.

9. I look forward to learning the results of future studies that analyze genomic evolutionary responses of the populations of Heck cattle, Konik ponies, and red deer in Oostvaardersplassen.

10. Wannier, T.M., et al. 2018. Adaptive evolution of genomically recoded *E. coli. Proceedings of the National Academy of Sciences USA.* 115: 3090–3095. https://dx.doi.org/10.1073/pnas.1715530115.

11. Gene drives, however, present a key exception to there being "nothing special."

12. Milla, R., et al. 2018. Phylogenetic patterns and phenotypic profiles of the species of plants and mammals farmed for food. *Nature Ecology and Evolution.* 2: 1808–1817. https://doi.org/10.1038/s41559-018-0690-4.

13. Callaghan, C. T., et al. 2021. Global abundance estimates for 9,700 bird species. *Proceedings of the National Academy of Sciences USA.* 118: e2023170118. https://doi.org/10.1073/pnas.2023170118.

14. Tucker, M.A., et al. 2021. Mammal population densities at a global scale are higher in human-modified areas. *Ecography*. 44: 1–13. https://doi.org/10.1111/ecog.05126; Daskalova, G.N., et al. 2020. Rare and common vertebrates span a wide spectrum of population trends. *Nature Communications*. 11: 4394. https://doi.org/10.1038/s41467-020-17779-0.

15. http://darwin-online.org.uk/converted/pdf/1863_vermin_F1728.pdf. Accessed May 25, 2021.

16. Recent genomic analysis upended the initial hypothesis that *C. p. molestus* speciated after colonizing subway systems, instead confirming the idea of colonization by a southern subspecies. Yurchenko, A.A., et al. 2020. Genomic differentiation and intercontinental population structure of mosquito vectors *Culex pipiens pipiens* and *Culex pipiens molestus*. *Scientific Reports*. 10: 7504. https://doi.org/10.1038/s41598-020-63305-z.; Byrne, K., and R.A. Nichols, R.A. 1999. *Culex pipiens* in London Underground tunnels: differentiation between surface and subterranean populations. *Heredity*. 82: 7–15. https://doi.org/10.1046/j.1365-2540.1999.00412.x.

17. Johnson, M.T.J., et al. 2018. Contrasting the effects of natural selection, genetic drift and gene flow on urban evolution in white clover (*Trifolium repens*). *Proceedings of the Royal Society B*. 285: 20181019. https://doi.org/10.1098/rspb.2018.1019; Santangelo, J.S., et al. 2022. Global urban environmental change drives adaptation in white clover. *Science*. 375: 1275–1281. https://doi.org/10.1126/science.abk0989.

18. Johnson, M.T.J., and J. Munshi-South. 2017. Evolution of life in urban environments. *Science*. 358: eaam8327. https://dx.doi.org/10.1126/science.aam8327.

19. Pigeon, G., et al. 2016. Intense selective hunting leads to artificial evolution in horn size. *Evolutionary Applications*. 9: 521–530. https://doi.org/10.1111/eva.12358.

20. Kuparinen, A., and M. Festa-Bianchet. 2017. Harvest-induced evolution: insights from aquatic and terrestrial systems. *Philosophical Transactions of the Royal Society B*. 372: 20160036. https://doi.org/10.1098/rstb.2016.0036.

21. Campbell-Staton, S.C., et al. 2021. Ivory poaching and the rapid evolution of tusklessness in African elephants. *Science*. 374: 483–487. https://doi.org/10.1126/science.abe7389.

22. Müller-Schwarze, D. 2011. *The Beaver: Natural History of a Wetlands Engineer*. Ithaca, NY: Cornell University Press.

23. Guo, W., et al. 2020. Ocean acidification has impacted coral growth on the Great Barrier Reef. *Geophysical Research Letters*. 47: e2019GL086761. https://doi.org/10.1029/2019GL086761.

24. Fussmann, et al. 2007. Eco-evolutionary dynamics of communities and ecosystems. *Functional Ecology*. 21: 465–477. https://doi.org/10.1111/j.1365-2435.2007.01275.x; Hendry, AP. 2019. A critique for eco-evolutionary dynamics. *Functional Ecology*. 33: 84–94. https://doi.org/10.1111/1365-2435.13244.

25. The apocryphal story holds that they aimed to bring the birds of Shakespeare to North America, including the invasive starling, though this literary detail appears to have been a colorful fiction. Fugate, L., and J.M. Miller. 2021. Shakespeare's starlings: literary history and the fictions of invasiveness. *Environmental Humanities*. 13: 301–322. https://doi.org/10.1215/22011919-9320167.

26. Clavero, M., and E. García-Berthou. 2005. Invasive species are a leading cause of animal extinctions. *Trends in Ecology & Evolution.* 20: 110. https://doi.org/10.1016/j.tree.2005.01. 003.

27. Van Driesche, R.G., et al. 2010. Classical biological control for the protection of natural ecosystems. *Biological Control.* 54: S2–S33. https://doi.org/10.1016/j.biocontrol.2010.03.003.

28. The 1988 film *Cane Toads: An Unnatural History* perhaps tells the early tale most evocatively about this new species introduction gone awry.

29. Among the top-10 invasive species in Europe are the Canada goose, sika deer, and the coypu, a semiaquatic rodent superficially similar to a muskrat or small beaver. You yourself can read the biographies-in-brief for what the IUCN (International Union for Conservation of Nature) currently considers to be the 100-worst invasive species around the world: http://www.iucngisd.org/gisd/100_worst.php. Accessed May 28, 2021; Vilà, M., et al. 2010. How well do we understand the impacts of alien species on ecosystem services? A pan-European, cross-taxa assessment. *Frontiers in Ecology and the Environment.* 8: 135–144. https://doi.org/10.1890/080083.

30. Genovesi, P. 2011. Are we turning the tide? Eradications in times of crisis: how the global community is responding to biological invasions. In: C.R. Veitch, et al. (eds). *Island Invasives: Eradication and Management.* Gland, Switzerland: IUCN. pp. 5–8. https://www.researchgate.net/publication/233755935.

31. Shurin, J.B., et al. 2020. Ecosystem effects of the world's largest invasive animal. *Ecology.* 101: e02991. https://dx.doi.org/10.1002/ecy.2991; Subalusky, A., et al. 2021. Potential ecological and socio-economic effects of a novel megaherbivore introduction: the hippopotamus in Colombia. *Oryx.* 55: 105–113. https://dx.doi.org/10.1017/S0030605318001588.

32. Haddara, M.M., et al. 2020. Hippopotamus bite morbidity: a report of 11 cases from Burundi. Oxford *Medical Case Reports.* 8: omaa061. https://doi.org/10.1093/omcr/omaa061.

33. https://www.bbc.com/news/world-latin-america-58937415. Accessed October 18, 2021.

34. With the expansive loss of megafauna, domesticated animals represent an important biological refuge for large mammals. As Erick Lundgren and colleagues stated, "it appears that domestication has provided a crucial bridge for certain species from the pre-pastoral wild landscapes of the early Holocene to the post-industrial wild landscapes of the Anthropocene." Lundgren, E.J., et al. 2018. Introduced megafauna are rewilding the Anthropocene. *Ecography.* 41: 857–866. https://doi.org/10.1111/ecog.03430.

35. Quotation from Lundgren, E.J., et al. 2020. Introduced herbivores restore late Pleistocene ecological functions. *Proceedings of the National Academy of Sciences USA.* 117: 7871–7878. https://dx.doi.org/10.1073/pnas.1915769117; Dembitzer, J. 2017. The case for hippos in Colombia. *Israel Journal of Ecology and Evolution.* 63: 5–8. https://doi.org/10.1163/22244662-06303002.

36. Head, L. 2017. The social dimensions of invasive plants. *Nature Plants.* 3: 17075. https://doi.org/10.1038/nplants.2017.75.

37. Urban, M.C. 2020. Climate-tracking species are not invasive. *Nature Climate Change*. 10: 382–384. https://doi.org/10.1038/s41558-020-0770-8; https://www.vox.com/down-to-earth/ 22796160/invasive-species-climate-change-range-shifting. Accessed November 29, 2021.

38. Marchetti, M.P., and T. Engstrom. 2016. The conservation paradox of endangered and invasive species. *Conservation Biology*. 30: 434–437. https://doi.org/10.1111/cobi.12642; Lundgren, E.J., et al. 2018. Introduced megafauna are rewilding the Anthropocene. *Ecography*. 41: 857–866. https://doi.org/10.1111/ecog.03430.

39. Quotations from Lundgren, E.J., et al. 2020. Introduced herbivores restore late Pleistocene ecological functions. *Proceedings of the National Academy of Sciences USA*. 117: 7871–7878. https://dx.doi.org/10.1073/pnas.1915769117.

40. Frelich, L.E., et al. 2006. Earthworm invasion into previously earthworm-free temperate and boreal forests. *Biological Invasions*. 8: 1235–1245. https://doi.org/10.1007/s10530-006-9019-3; Ferlian, O., et al. 2018. Invasive earthworms erode soil biodiversity: a meta-analysis. *Journal of Animal Ecology*. 87: 162–172. https://doi.org/10.1111/1365-2656.12746.

41. Quotations from Donlan, C.J., et al. 2006. Pleistocene rewilding: an optimistic agenda for twenty-first century conservation. *American Naturalist*. 168: 660–681. https://doi.org/10. 1086/508027.

42. Vilà, M., et al. 2011. Ecological impacts of invasive alien plants: a meta-analysis of their effects on species, communities and ecosystems. *Ecology Letters*. 14: 702–708. https://doi. org/10.1111/j.1461-0248.2011.01628.x.

43. In other words, "The fantasy of authentic wilderness draws us, as alienated social subjects, to its landscape with the siren song of an experience of our whole selves," or, as the subject named Jeremy put it succinctly, "wilderness is in large part a state of mind." Quotations from Vidon, E.S. 2019. Why wilderness? Alienation, authenticity, and nature. *Tourist Studies*. 19: 3–22. https://doi.org/10.1177/1468797617723473.

44. In his words, "a resilient ecosystem with the level of biodiversity and range of ecological interactions that can be predicted as a result of the combination of historic, geographic and climatic conditions in a particular location." Dudley, N. 2011. *The Authenticity of Nature*. London: Routledge. https://dx.doi.org/10.1007/978-90-481-2611-8.

45. Bliege Bird, R., and D. Nimmo. 2018. Restore the lost ecological functions of people. *Nature Ecology and Evolution*. 2: 1050–1052. https://doi.org/10.1038/s41559-018-0576-5; Ducarme, F., et al. 2021. How the diversity of human concepts of nature affects conservation of biodiversity. *Conservation Biology*. 35: 1019–1028. https://doi.org/10.1111/cobi.13639.

46. Davis, M., et al. 2018. Mammal diversity will take millions of years to recover from the current biodiversity crisis. *Proceedings of the National Academy of Sciences USA*. 115: 11262–11267. https://doi.org/10.1073/pnas.1804906115.

47. Mello, F., and A. Friaça. 2020. The end of life on Earth is not the end of the world: converging to an estimate of life span of the biosphere? *International Journal of Astrobiology*. 19: 25–42. https://dx.doi.org/10.1017/S1473550419000120.

15

Inspection by the department of health, safety, and ethics

That particular Fourth of July was a typical Cutter family vacation. I was 15 years old and, like it or not, the five of us were going to spend some quality time together with our lanky black mutt cooped up on a boat. A few years prior, my parents had "invested" in a slightly beat up second-hand Parker Dawson 26 sailboat that we moored near Boston in Winthrop Harbor. At sea, sailing on that slow boat, when the wind whipped over the waves, the salt spray on your lips, if ever you had any doubts, you knew that you were fully alive. On this clear and calm evening, however, situated with a perfect view of the sky, we had grand plans to lie back with front row seats to the massive fireworks display to be shot off by the city from nearby Snake Island.

As darkness settled, a single rocket flared high and boomed into the night sky. After that solitary pronouncement, the explosive display shifted into high gear. Sparking trails shot up fast and furious, one after the other in a frenetic barrage. What we had failed to notice, however, was that the wind had shifted. That oversight was soon made vivid as we watched the trajectory of burning embers falling from the sky. Toward us. After each brilliant detonation, red cinder and ash and smoke descended upon us, raining down to paint a saltwater hellscape, the multicolored explosions lighting us all up in cartoon silhouette with each flash. Fiery shrapnel splashed and hissed into the sea, port and starboard, fore and aft, as the sailboat rocked in the sea's swells. The dog, a neurotic rescue, scampered and yapped around the slick deck in a terror-stricken frenzy. Captain Dad hollered a boatload of nautical orders as he scrambled from cleat to stay to rudder, and we battened down the hatches as best as a man, woman, three children, and a manic dog could do under the circumstances. The circumstances being, of course, that we were perfectly stuck, with the best of intentions, bobbing like a cork in a downwind zone of annihilation.

What can we do to minimize the possibility that disaster strikes? In this imperfect parable from a Fourth of July debacle, if the world is the boat, what is to keep it from going up in flames as we launch beautiful and newly wild genetically welded species into the night? There are, presumably, preemptive steps to help mitigate potentially negative consequences of our actions for health and safety—of humans and of natural ecosystems.

Evolving Tomorrow. Asher D. Cutter, Oxford University Press. © Asher D. Cutter (2023). DOI: 10.1093/oso/9780198874522.003.0015

What are some of the lessons from restoration ecology and the debates about genetically modified organisms (GMO)? Why do we feel about nature the way we do, and how do different cultures feel? How much agency do animals deserve? Let's think through the ethical entanglements that shape our choices and inform how we decide to allocate time and resources.

15.1   Restoration and reunion

In creating and releasing a new organism into nature, one must tussle with an ethical bureaucracy. One challenge is simply to disentangle and de-conflate the shopping cart of concerns that get raised with such an undertaking. We need to be appropriately thoughtful on each point. For example, we have practical fears of unintended detrimental consequences to ecosystem services and public health. We have ethical considerations about animal welfare, the value of nature, the morality of playing god. We have legal scrutiny over genetic modifications and liability. All of these issues bleed into one another and into the red meat of public perception. We must try to demarcate, as well as possible, what are universal obligations and what are split opinions, and the roots of each.

The most substantive worry is that a biological Artemis Earth mission would turn out to be less the twin of Apollo and instead making something, say, more like a Jersey City Project, a shadowy biological sibling to the unseemly parts of what atomic fission and fusion unleashed following the Manhattan Project of the 1940s. It is the fear of unintended consequences. The fear that intentional manipulations of nature could do more harm than good, expanding ecosystem disruption in ways that accelerate the pace of extinction and produce undesirable or impaired environmental services. As stated starkly by Matt Davis, Soren Faurby, and Jens-Christian Svenning,[1] "Extinction is part of evolution, but the unnatural rapidity of current species losses forces us to address whether we are cutting off twigs or whole branches from the tree of life." It would be a tragedy indeed if rewilding of a new or resurrected species—growing a thin twig— inadvertently, in the process, cleaved off a thick branch.

★

Why worry about retaining biodiversity and intact ecosystems? The perks of retaining biodiverse ecosystems extend beyond the aesthetic preferences of outdoorsy folks. Exposure to nature provides demonstrable benefits to mental health, physical health, and one's sense of wellbeing.[2] Reduced stress, improved mood, lower obesity, lower cardiovascular and cancer mortality, fewer respiratory ailments, just to name a few. The benefits differ with short-term and long-term exposures, but, interestingly, the personal interactions with nature generally need not be with truly wild habitats. Human-constructed green spaces also confer such dividends to our health.

If you were an economist, then you might be more persuaded of the merits of species richness by the analyst's conclusion that biodiversity has so-called insurance value.[3] Biodiversity acts as insurance in stabilizing ecosystem functions, even substituting for some types of financial insurance. The logic of biodiversity as insurance holds that a broad portfolio of species buffers an ecosystem against disruption of any one species caused by environmental variability, mitigating risk of potential negative economic consequences.

When we talk about ecosystem services, we mean a human-centric view of what we can "get" from nature. This includes directly mining living systems through biomass production and retaining consistent food supplies. It also means things like flood mitigation, high water quality and availability, pollination of crops, reduced pest and disease burdens in agriculture and public health, carbon storage as mitigation of climate change, production of fertilizer through nitrogen fixation, high soil quality through soil regeneration and as a source of undiscovered drugs and chemicals. Variability caused, for example, by drought or disease or even seasonality, can create mismatched supply and demand, an inefficiency in the market economy. Biodiverse ecosystems can offset such agricultural variability to mitigate risks to crops, farmers, and food availability. This, in turn, minimizes risks to consumers and financial markets.

★

The unease we have with introducing or transplanting any new species is that it might become an invasive species, negatively impacting the landscape beyond expectation. Rather than creating new habitat for more species, or sets of species that are more desirable, it might degrade the ecosystem. It might make the environment suitable only for itself and other invaders through some irreverent combination of predation, competition, toxicity, ecosystem engineering, and disease transmission.

Usually, we concern ourselves with the invasive potential of a species that is, today, resident in one locale but that has distinctive traits that might allow explosive growth elsewhere. Even though many new species introductions do not become invasive, we see the aura of risk like a sixth sense. De-extinction to rewild ecosystems suggests, in addition to today's invaders, the possibility of "invaders from the past," or, in the case of new species evolved through selective breeding and genetic welding, "invaders from the future." Even if they don't lead to a problem of population explosion, they might still contribute to population implosions of other species as a source of pathogens or as an incidental disease reservoir. The ensuing dynamics of disease that expedite transmission of what are currently low-grade infectious diseases could then increase the health burden to humans or vulnerable plants and other animals.

"Nature, in other words, is widely seen as either hopelessly fragile or already completely broken," lamented Stewart Brand, before declaring that "Nature is not broken."[4] While the world's habitats may not all be pristine, we can say that some habitats are rather unlikely to be hit with adverse hypothetical outcomes of intentional species introductions. Habitats that are already degraded, those most impacted by human activity, may have no place to go but up. As we've seen, the addition of just a few ecosystem engineers or keystone species to such destitute environments can starkly improve their prospects.[5]

Other habitats are on the brink, however, with scientists already taking desperate measures. These are desperate times for coral reefs, for example, ecosystems that support exceptional oceanic biodiversity. Ocean heating events and increased acidity present existential threats to coral life as the climatic fallouts of elevated atmospheric CO_2. Perhaps genetic welding of coral species, using gene drives to spread heat tolerance or broader pH tolerance, would prove an acceptable means to avoiding *The End* of tropical reef ecosystems (Figure 15.1). Genetic variants that confer thermal tolerance could spread especially fast when combined with gene drives, while also avoiding the undesirable diversity-reducing side-effect of genetic hitchhiking, though asexual phases of coral reproduction would constrain genetic welding's potential. The prospect of implementing genetic welding in this way grows more plausible with the recent demonstration of CRISPR-Cas9 genome editing in the coral *Acropora millepora*, and as researchers home in on the genetic variants that enable thermal tolerance in species like the brain coral *Platygyra daedalea*.[6] As it stands, coral biologists already have ramped up the forces of evolution using selective breeding with this goal in mind.[7] Genetic welding might find an unexpectedly welcome home inside of restoration ecology as a form of "genetic rescue" for these and other organisms.

If the right genetic defense can be worked out in time, then genetic welding could be deployed similarly as a conservation tool in bats to evade deadly white-nose syndrome or in frogs to resist chytrid fungus mortality.[8] Plans are afoot for even more extreme conservation measures. The GBIRd (Genetic Biocontrol of Invasive Rodents) group aims to introduce suppression gene drives to drive extinct invasive rodent populations

Figure 15.1 *CRISPR-Cas9 genome editing has proved feasible in the hard branching coral* Acropora millepora, *which can still be found growing in colonies on Australia's Great Barrier Reef. This coral species is also the subject of conservation research intended to increase heat tolerance through selective breeding and assisted migration.*

Image credit: Petra Lundgren, Juan C. Vera, Lesa Peplow, Stephanie Manel, and Madeleine J.H. van Oppen, reproduced under the CC BY 4.0 license.

on islands, with similar ideas percolating for a variety of organisms in Australia and for invasive sea lamprey in the Great Lakes.[9]

Biocontrol of invasive species, at its heart, is a form of ecological restoration to recover perturbed ecosystem functions or to mitigate threats to one region's biodiversity. Restoration ecology operates under the presumption or observation that nonnative and invasive species are damaging to ecosystems, and so require active intervention to resuscitate some biological golden era of the past. It seeks to make ecosystems great again.

Changes to species' geographic ranges through migration and colonization, however, are entirely natural features of biological communities, albeit accelerated by human activity. Consequently, some ethicists espouse such an extreme view of the "authenticity" of nature, even, that they reject the entire enterprise of restoration ecology. They see it as "an example of the human domination of nature" that acts to "subvert, and render meaningless, the environmentalist goals of the protection and preservation of natural systems and entities."[10] Regardless of how convincing or not such rhetoric might be, sometimes, counterintuitively, eliminating an invasive species can do more harm than good. For example, introduced filter-feeders may improve water quality; invasive shrubs may create nest sites favorable to certain native animal species; animal consumers may come to depend on introduced species as a food resource.[11] Rapid elimination of the introduced organisms may then create disequilibrium that puts the entire community at risk. Ecosystem function, it turns out, is complex and often difficult to boil down into what is good and what is bad in a simple sound bite.

How far are we willing to push restoration ecology? With suppression gene drives, we could eradicate mosquitoes in Africa, black rats in New Zealand, sea lampreys in the Great Lakes—maybe even ragweed.[12] We could bolster the genomes of endangered species through genetic welding to resist pathogens, tolerate insecticides intended for pests, or replace those deleterious gene variants made abundant by inbreeding in undersized endangered populations. How curated, though, do we want nature to be, even in heavily impacted ecosystems? Are we willing to trade our current ecosystems that are perturbed by extinctions and invasive species for future ecosystems as domesticated gardens, stamp collections manicured with genetic welding and de-extinction? Do we owe reparations to nature for all of our history of disruption, in the form of intervention? What do we owe our descendants—perhaps adopting the Haudenosaunee (Iroquois) Seventh Generation Principle: how do we be "good ancestors" to future generations of humanity?[13]

★

We've discussed human intervention with ecosystems, some might complain, from a very top-down, Eurocentric, colonialist perspective. This backdrop, in caricature, pits a strategy of complete exploitation of human-dominated landscapes against complete preservation of wilderness habitats that are best off with the complete absence of humans.

As argued by Rebecca Bliege Bird and Dale Nimmo, however, "discussions about ecological restoration rarely acknowledge the missing ecological functions once performed by people" in preindustrial history.[14] They reject the idea that wilderness should be off-limits to people, that there is even such a thing as nature separate from people.[15] This view considers *Homo sapiens* as one among many keystone species that interact with other organisms within ecosystems. People are, and should be, participants in natural processes. After all, humans interacted with ecosystems for tens of thousands of years prior to industrialization.

Ecosystem functions like fire regimes, soil turnover, and seed dispersal that were carried out by prehistorical *Homo sapiens* and have since been lost, Bliege Bird and Nimmo go on to say, result from "what industrialized societies do when they displace indigenous peoples." Such displacement, they argue, leads to ecosystem consequences of the elimination of human communities being "comparable to the extinction of species." Erle Ellis and colleagues concluded recently, "The primary cause of declining biodiversity, at least in recent times, is the appropriation, colonization, and intensifying use of lands already inhabited, used, and reshaped by current and prior societies."[16] And yet, we also know that prehistoric megafauna extinctions arose in large part from the seemingly low-intensity land use of our ancient ancestors. Despite such consequences, this perspective considers the ecosystem states that result from traditional practices of Indigenous peoples to represent a desired set of states for nature. It remains to be seen whether the argument that "wilderness is an inappropriate and dehumanizing construct" is a winning outlook that will lead the world "to embrace situated Indigenous and local knowledge systems in ... landscape management" in order to foster biodiversity in the face of global-scale ecological challenges.[17]

Perhaps the biggest implication of the view that humans are an essential component of ecosystems, rather than simply a saboteur, is in how we set a baseline. Proponents of Pleistocene rewilding reference a baseline prior to the evolution of humans, or at least prior to substantial human impacts. This ad hoc baseline is convenient and high-minded, in a sense, but, at the end of the day, arbitrary. An alternative is to define the baseline for ecological restoration as the era of human-inhabited preindustrial ecosystems. This alternative baseline is also arbitrary. There's no intrinsic reason why either of these baselines is inherently "good." In recognizing the distinction in outlook, however, we can then conscientiously debate worldviews about what we think *is* good and about the place of *Homo sapiens* in nature.

Caretaker, curator, manager, gardener, sculptor. You can utter these descriptions of our interaction with nature with the curled lip of derision, as if commenting on a hubristic fascist, or with the soft-chinned pride of someone who attends to their duties and responsibilities. The domination of nature by segregating it as "other" or by domesticating it or by extracting whatever uses benefit humans; the caring for nature by shaping it to be part of one's life. Leave it alone, be part of it, engineer it; all provide a defensible rationale to help get natural systems back on the right track.

But which is best? It is indeed a tall order to reconcile between the belief that "The idea of wilderness is destructive, and must be abandoned," as Indigenous environmental scholar Michael-Shawn Fletcher and colleagues wrote recently,[18] and the rewilding movement's notion favoring the "restoration of wildness" where "humans withdraw from landscapes."[19] And yet, I suspect that they'd share whole-hearted agreement with

Fletcher's sentiment that "We need new ways of engaging with the world around us if we're to live sustainably on this planet."

15.2   GMO to GDO

It is tempting to recoil at overt efforts to infiltrate natural systems with the tools of genetic engineering. And yet, human actions have already disrupted the DNA fabric of countless species without any genetic engineering at all. We have caused evolution through gene frequency change as a result of altered environments that shift the pressure of natural selection, of reduced habitats that depress populations and so increase the influence of genetic drift, of wholesale loss of genomic diversity in the form of species extinctions. And that's not even counting the plant, animal, and microbial species that we've molded through intentional artificial selection in the domestication process. Humans have been engineering the genetics of species for millennia, whether we realized it or not.

It is hard for any one of us to stand on righteous ground to balk at genetic welding of a few strings of DNA in the face of such colossal, if surreptitious, slicing and dicing of the genetic content of ecosystems—that we're already doing right under our own noses. Can we come to terms with that idea? "If we want to sustain and enhance a biodiverse natural world," goes the logic of one answer to this question, as I've quoted previously, then "we might have to be forward looking and embrace the notion of bio-novelty."[20] A newly evolved and extraordinary species that follows the recipe for "how to evolve a dragon," in legal terms, would generally be considered a GMO. So, let's dip our toes into views about GMOs as a point of reference.[21]

GMOs are subject to visceral opposition in some quarters that stem from several distinct rationales. One strain of aversion falls into the quasi-religious cognitive error of the naturalistic fallacy. It falls prey to the fear of the unknown, that "unnatural" or "unknown" things are inherently bad. A second strain of animus derives from anticapitalist sentiments and distrust of for-profit corporations. This sentiment gains inspiration from the hollow public relations efforts of petroleum and tobacco industries, and more directly, from patented single-use seed stocks that feed a conveyor belt of profits to multinational agrochemical corporations on the backs of impoverished farmers and nations. We'll return to these concerns over naturalism and bio-capitalism in a moment.

The third kind of antipathy toward GMOs comes from fears over biological unintended consequences: risks to health or to the environment. For example, people may worry about the possibility of adverse effects of potential mis-expression of a gene near to where a transgene gets inserted into a genome of a crop plant or livestock animal. A 2016 report on genetically engineered (GE) crops by the National Academies of Sciences, however, found "no substantiated evidence that foods from GE crops were less safe than foods from non-GE crops."[22]

For developing GMO crops, part of the rationale is to actually improve health or diminish environmental stresses. For example, improved vitamin A availability brought to nutritionally starved regions from Golden Rice is provided by the engineered metabolic capability to produce beta-carotene,[23] or by resistance of crop plants to insect pests to then reduce excessive application of chemical pesticides. Nonetheless, residual fears often remain simply because of the idea of transgenic "contamination" that triggers a psychologically potent disgust response.[24] Other residual fears reflect the case-dependence of GMO applications. Some worry that "gene escape" through hybridization would yield bad environmental or economic outcomes, for example, if non-GMO crops become "contaminated" or if weeds evolve herbicide resistance due to particular GMO strategies that depend on spraying herbicide-resistant crops with chemicals like glyphosate.

When it comes to the foodstuffs themselves, I think that Dominique Bergmann of the Howard Hughes Medical Institute helps to put in perspective what it means to eat something that has genes that you've never eaten before:[25] "When I talk to people . . . If they're concerned about safety or putting foreign genes in food, I ask them when they first tried quinoa or acai berries. Usually, they say not until adulthood. I then ask why they were willing to take the incredible risk of consuming something with upwards of 20,000 genes they had never before been exposed to."

Bergmann's point highlights a profound disconnect in how we often think about GMOs. On the one hand, strict regulation with deeply felt concern acts to restrict transport of a single genetic element. On the other hand, society fully accepts the transport and trade of entire genomes of non-GMOs to novel environments—tens of thousands of "foreign" genes packaged within the nuclei of organisms. The expression of *Bt* to confer plant resistance to insect herbivory[26] hardly seems in the same league as the ferrying of innumerable novel genes in the genomes of the community of seafaring hitchhikers adhered to the hulls of 300-m-long (1000-foot-long) cargo container ships.

No matter your view, the United Nations has established the Cartagena Protocol on Biosafety as an agreed-upon document for responsible use of GE organisms. It defines an obligation "to subject all living modified organisms to risk assessment prior to the making of decisions."[27] It helps to set out the rules for who gets to decide, and who is responsible, for introducing to the uncontrolled outside world any organisms made in whole or part through genetic engineering. Notably, the United States, Canada, Australia, and Russia have not yet signed on. The Cartagena Protocol is an example application of the precautionary principle, which approaches the laudable goal to "do no harm" by placing the burden of proof on proponents.[28] As we will see, however, this inevitably creates the potential for a status quo bias that one must recognize in any decision-making process.[29]

Gene drives, however, were not on the menu when the Cartagena Protocol was initiated. They are distinct enough that the organisms that contain such GE elements get their own acronym: GDOs, gene drive organisms. Their potential to spread beyond national borders—verboten, if not agreed in advance, according to the Cartagena document—is difficult to ensure, an issue most relevant for imminent plans to introduce suppression gene drives into mosquito populations to control malaria.[30] Given the relative ease that one may perform genetic manipulation with CRISPR-Cas9 techniques, it's also difficult to ensure that biohackers wouldn't take matters into their own hands, irrespective of the restrictions placed on corporations or on public initiatives funded by governments.

With these broad points about gene drives in mind, there are several key distinctions among different types of potential GDOs. The first distinction is between modification drives that can push the evolution of a characteristic through a population so that it becomes an integral new feature of a species—the kind involved in genetic welding—versus suppression drives, the kind designed to depress a species' abundance or even to drive it extinct. As a consequence, genetic welding should pose lower intrinsic risk compared to GDOs intended to drive population suppression. A second distinction differentiates gene drives designed to be globally spreading versus locally restricted in space. Gene drives engineered with a geographically restricted scope pose lower inherent risk if we are worried about global species-wide transmission. Related to these dynamics in space are dynamics in time, with a third distinction among a trio of temporal dynamics: self-propagating gene drives that spread from a low initial abundance versus self-propagating from a high initial abundance (sometimes termed "majority wins") versus self-limiting over time such that they are designed to eventually get purged from the population.[31] Self-limiting and majority-win gene drive designs will pose lower risk than self-propagating drives. These risks may also be mitigated by countermeasures like culling or pesticides, as well as GE countermeasures like secondary gene drives that act to reverse or remove the original gene drive function. Each of these factors, in addition to the biological details encoded by the gene drive cargo, must go into any risk assessment about potential ecological or public health impacts of releasing GDOs into the uncontrolled outdoors.

★

If you are busy soaking up the Sun's rays on a beach in Bali, then you can't simultaneously be cheering for your team from the bleachers in Fenway Park or in the stadium formerly known as the SkyDome. Doing one thing creates an opportunity cost, meaning that you must pay the cost of missing out on doing something else. Resources spent on one thing cannot be expended toward another. This logic of opportunity costs provides one consideration in any decision, including decisions that govern the fate of natural systems.

Critics of rewilding and de-extinction often levy this concern over opportunity costs. The argument holds that money and manpower would be disproportionately siphoned away from other conservation projects that offer more bang for the buck. Devoting maximal effort to retain what exists of the world's biodiversity is the most effective approach, the argument goes, not sucking resources toward the tech sector. The same charge could be laid against an agenda to create entirely new species with the aid of genetic welding. Opportunity cost concerns like this, however, presume that all available funds are already allocated. They assume that there is a zero-sum game inside a fully accounted pocketbook of cash for environmental causes.

Maybe, however, that's a false assumption. Grand visions that capture people's imaginations can spur new enthusiasm and previously inaccessible dollars to then expand the pocketbook, or so goes the counterargument. The bold claims of the Revive & Restore Foundation contend that outside-the-box environmental prospects like big-scale rewilding and de-extinction would create synergistic support for conservation objectives, attracting new proponents and new revenue. "In what universe, I have to wonder," opined Stewart Brand in his critique of de-extinction critics,[32] "would people newly excited about mammoths become suddenly indifferent to imperiled elephants? . . . More likely is that de-extinction will attract significant new sources of funding and interest for conservation."

Deciding on the right compromise, or whether to compromise at all, similarly presents a challenge to decision-making. Perhaps an all-or-none attitude is the prevailing sentiment—which, with respect to nature, translates to "all natural" or "extinct." How many degrees of taintedness does *seminatural* have? Certainly, new species that we evolve with the aid of genetic welding fit snugly at one end of the spectrum. In practical terms, accepting only "all natural" is false faith. All the ecosystems of the world have been touched by humanity.

Is all that is natural all good? Sometimes we traffic in that irrational cognitive leap. That is, we presume an equivalence of "natural" and "good." This so-called naturalistic fallacy is "a kind of covert smuggling operation in which cultural values are transferred to nature and nature's authority is then called upon to buttress those very same values," as Lorraine Daston has written, a psychological predilection that "appears to be ubiquitous and irresistible."[33] It is sometimes thought of as a specific case of eighteenth-century philosopher David Hume's famous "is-ought problem": simply because something *is* (a fact) does not mean that it *ought* to be a certain way (a value). Value judgments require moral justification separate from nonmoral observations of fact. Nature is replete with examples that one might falsely employ to justify all manner of horrors: murder (chimpanzees), infanticide (lions), cannibalism (bullfrogs), rape (dolphins), incest (mongoose), plunder (army ants), poison (mushrooms), and deception (Batesian butterfly mimics). Perhaps it is fortunate that moral justification does not arise spontaneously from the details of nature. It's simply insincere to cherry-pick only those natural examples that support a particular value.

Many people find moral bearings in the scriptures of their religion. So, you might ask, is it playing god, is it an affront to god, to evolve a new species—or to de-extinct a species—to harness genetic welding to manipulate the heritable molecules of life? So, too, we can ask: Is it playing god to drive a species extinct? Is it playing god to shape the molecules of life through selective breeding, as humanity has done since the days predating monotheism, in creating peculiar breeds of domestic animals? Is it playing god to carve deep into the environment to support those peculiar breeds—razed prairies for crops to feed cattle, clear-cuts of rainforest to ranch cattle, processed cattle offal as pet food—shaping the molecules of life for all those wild species that previously inhabited those lands or that come to interact with the domesticated animals and plants?

Either we answer "yes" to every one of those questions, or we answer "no" to every one of them—or we accept the cognitive dissonance of hypocrisy.

Whatever the answer, the question that would confront you next asks whether we hold a moral duty to provide reparations to nature for all the extinction and disruption to ecosystems. The currency of reparations, of course, does not recognize dollar bills, but the richness of species and the richness of the biological processes that drive evolution in ecological systems. As conservation interventionist Stewart Brand famously wrote in 1968's *Whole Earth Catalog*, "We are as gods and might as well get good at it." At least some philosophers weigh in to agree, speaking in favor of de-extinction and evolving new species by pointing out that there is no "asymmetry between 'protecting' biodiversity by conserving species alive today, and 'creating' biodiversity."[34] Even if there is no asymmetry, it still remains true that the efficacy of preventing extinctions is vastly greater than reinventing the wheel with species creation.

★

Another difficulty in linking nature with morality is the fact that the meaning of what is natural mutates over time, and "natural" means different things in different cultures. We've already talked about this idea in terms of Eurocentric views in contrast with some Indigenous perspectives. We've also deliberated over Shifting Baseline Syndrome, how we accept the norm of nature as what our youthful experiences set as our baseline perspective. We see differences from or changes to that baseline expectation as discomfiting perturbations to the status quo. This sense of the status quo presents us with an unconscious bias, an irrational predilection to keep things as-is. This tendency is, in psychological lingo, the status quo bias.

Sometimes status quo bias arises out of the psychological phenomenon of "loss aversion," the tendency to judge the same alternatives differently depending on whether they are framed as losses versus gains. Underhanded pollsters can skew survey responses with this wording trick. Similarly, the so-called endowment effect can produce a status quo bias, leading us to value more those things that we already have in hand that are equivalent to things that we just as likely could have had instead. As ethicists have highlighted, "the popular intuition about the preferability of 'the natural' might in part derive from a status quo bias."[35]

What to do? The first step is acknowledgment, as with any kind of implicit bias. As those same ethicists point out: "Recognizing and removing a powerful bias will sometimes do more to improve our judgments than accumulating or analyzing a large body of particular facts." A way to check ourselves is to perform a thought experiment known as the Reversal Test. For example, if we were successful in evolving a new species, and it subsequently became threatened with extinction, would we fight to save it? If so, then you could argue that, in fact, we have a moral duty to try to usher it into the world in the first place.

No matter how you slice it, the individuals that make up any new animal species *are still animals*. Despite genetic engineering and selective breeding having shaped their genomes, their bodies and their behaviors would have all the fundamental attributes that we normally associate with the respectful recognition of living creatures. They wouldn't know any other way to be or to live.

Society could, of course, decide that the foreign policy principle of noninterference, the political notion that "sovereign states shall not intervene in each other's internal affairs," would be appropriate to apply to the internal affairs of nature. The sci-fi Prime

Directive. But we know only too well how governments abide by that principle in practice. Its violation has spawned the entire genre of spy novels and films. We also know that nature does not negotiate and, also, that nature—shark attacks, wildfires, hurricanes, earthquakes, asteroid impacts, disease epidemics—will not reciprocate in kind.

★

A moment ago, I mentioned how one strand of anti-GMO sentiment arises from anticapitalist perspectives. So, what's wrong with bio-capitalism? It is, after all, perfectly legal for a person, institution, or corporation to patent biological material. Perhaps more suspect, single-celled living organisms in their entirety can be deemed patentable in some countries, for example, when "bugs as drugs" treat gastrointestinal conditions through microbiome transplant techniques.[36] In the United States, as of 2013, however, patents of DNA *must* be genetically engineered in some way.

Patents are intended to reward inventiveness by conferring on the awardee sole discretion over dissemination, and a monopoly on profits, for a specified period of time, say, 20 years. This protection motivates the business models of research and development operations within corporations. They work to develop as much patentable, and potentially profitable, intellectual property as possible. There truly have been wonderful creations through this process, including the recent and rapid deployment of mRNA-based vaccines for COVID-19.

The conflation of financial incentives with research innovation, however, makes many folks bristle. One kind of reaction against such bio-capitalism is a perception of vulgarity in presuming that profit motives are required to decipher the natural world, compared to supposed noble pursuits of knowledge for its own sake. A second point of opposition arises from the history of excessive and harassing enforcement of patents by some large corporations. Thirdly, the self-interest of corporate entities often shows a track-record at loggerheads with public interests, of disingenuous public relations surrounding such conflicts, and a tendency to shirk responsibility when things go wrong. If money is on the line, then marketing hype may overestimate the probability and timeline for successful outcomes and give inadequate service to the likelihood and consequences of failures. Corporations cower behind the shield of shareholder interests at the expense of the public good. The public is left on the hook when corporations go belly-up, as witnessed by cases of chemical company contamination of groundwater.[37] Suspicions of this kind about for-profit de-extinction companies, for example, may not be unwarranted.

★

Hunting presents a particularly thorny case of bio-capitalism as it intersects with ecological conservation and ideas of wilderness. On the one hand, trophy hunting of charismatic big mammals—sometimes endangered and hunted with illegal or norm-violating tactics—creates much outrage at what is viewed by nonhunters as an obscene and immoral activity.[38] This belief is fundamental to the outlook of members of Eurocentric animal welfare organizations. The fact that trophy hunting is an inequitable privilege of the rich, with some trophy hunt licenses costing over a quarter-million US dollars, provides only further fuel for infuriation. Poor enforcement of game quotas, or scientifically unsound quota assessments, could lead to unsustainable depletion of some of the world's most-prized organisms. While trophy hunting generates over US$200 million each year in sub-Saharan countries of Africa, only a small portion of these funds get allocated directly to conservation purposes to support the very foundations of the industry: the existence of amazing creatures.[39]

There is, however, a compelling "on-the-other-hand" perspective to trophy hunting, especially in cash-strapped developing countries. Trophy hunting is an unusually lucrative form of eco-tourism. It creates incentives, curiously enough, for both local communities and countries to value and promote ecosystem conservation. The benefits to conservation goals come both from direct injection of a portion of hunting revenue and from systemic supports that act to counter the opportunity costs of converting wildlife habitat to other less sustainable uses if hunting were barred. Indigenous groups and local communities may view hunting and conservation from very different perspectives relative to Eurocentric environmental outlooks.[40] In Namibia, for example, only 11% of rural inhabitants would support wildlife if a ban on trophy hunting were enacted.[41]

The support of local communities is essential to the success of conservation goals to avoid illegal wildlife trade and overly exploitative hunting. Legal trophy hunting operators may encourage the preservation of larger and less easily accessible areas—a contrast to easy viewing of artificially dense wildlife by traditional ecotourists. It also may encourage reintroductions of previously extirpated species and discourage illegal poaching. The total area of sub-Saharan Africa utilized for trophy hunting actually exceeds the combined area of the region's national parks, making non-park land invaluable habitat.[42] The carbon footprint, in terms of travel and local infrastructure demands, is also lower for the small number of trophy hunters relative to large numbers of traditional ecotourists needed to generate the same income. The potential for trophy hunting to support the retention of biodiversity—if we accept the bloody means to that end—requires, in practice, that it be subject to sound and controlled management.

What link to trophy hunting might inadvertently be created by introducing into the world de-extinct or newly evolved hyperexotic species? Such human intervention in the formation of species, for one thing, does not give license to construct a zoo. "[T]he creation of lab specimens and curiosities," Philip Seddon and Mike King say, "cannot be a defensible end goal for de-extinction."[43] Given the popularity of a plethora of genetically modified pets, like GloFish,[44] I'm not sure that the consensus of society would agree with them on this point. In the broader view of fostering biodiversity, it has been argued, the goal of conservation is not "to save wildlife only so we as humans may continue to use and enjoy them as we see fit."[45] These views may not enjoy perfect consensus but do reinforce the notion that any value beyond public health to applying genetic engineering for wild systems would be to fulfill a role that enhances ecosystem function.

15.3   How to think

What are the rights of animals, as individuals and as collective members of a species? If you recoil at the notion of propping up trophy hunting on African safaris, do you also recoil at institutionalized culling of invasive pythons in the Florida Everglades? Pythons, as invasive predators, take a harsh toll on local wildlife, like wading birds and marsh rabbits.[46] Pythons often garner less sympathy, however, than Namibian leopards and elephants. Rather than paying for the privilege of killing wild game, the Python Elimination Program sponsored by the South Florida Water Management District Governing Board will pay *you*. You'd get $125 for your typical 2.2-m-long (7-foot-long) python.[47] How do these things square with animal welfare? One answer is that we owe it to organisms to make available the conditions that enable them to be self-sustaining. Up to a point, anyway. Feral domesticated species and invasive species, society often deems, lose that privilege, at least in some places.

Another answer to the question about animal welfare is that we owe it to organisms to allow them the opportunity to make their own decisions. This idea is, in essence, the logic that rewilding takes and then amplifies it to the scale of ecosystems. We can also examine the individual scale of animals. The translation of sensory inputs into actions is a form of decision-making and problem-solving, whether physiological responses or behavioral actions. In other words, the ability to convert the realm of the senses into subsequent biological activity is, in a sense, a measure of intelligence.

★

Some organism's decision-making is based on simple logic gates. If A then B, with some propensity for noise in producing response B after detecting A. When you tap the nose of a nematode roundworm that is crawling in a Petri dish, almost invariably, it reverses its locomotion in a stereotypical behavioral response. They have only 302 neurons, after all. In these *C. elegans* roundworms, if you had a lab strain with a dysfunctional copy of a gene known as *osm-9*, however, then you'd find that the creatures no longer respond to a poke to the nose in that stereotypical way.[48] The protein encoded by *osm-9* normally gets expressed in neurons in the worm's nose, contributing to sensation of mechanical and odor stimuli. A defective or absent OSM-9 protein leaves the worm no choice but to ignore otherwise obvious environmental cues.

These behaviors are sometimes called instinctual or reflexes or hard-wired. These kinds of hard-wired responses are genetically deterministic and without much scope for rational thought to alter the biological response. Hard-wired responses work well for common experiences, and we humans also make use of them: pulling our hand away from a hot pot handle, involuntary acceleration of heart rate from a brush with danger to ready us to flee or fight.

Hard-wired responses also can be hacked. By editing the genomes of those *C. elegans* roundworms, genetic engineers at the University of Toronto made the worms' muscle cells responsive to light. Specifically, their muscle cells will contract when exposed to a 473-nm wavelength of blue light. By further overriding the roundworm's own neurons with a hyperpolarizing chemical, the genetic engineers produced biological entities whose locomotion they could control by flashing a programmable sequence of laser beams onto selectively targeted groups of muscle cells inside their transparent bodies. In this way, a given worm was compelled by a computer program to navigate a maze, despite its neurons being paralyzed, crawling by laser-light directed activation of its muscles. They describe any one of these entities as a living robot—a biological marionette—a so-called RoboWorm being an "untethered, highly controllable living soft microbot."[49]

For up to 15 minutes at a time, the ghost gets taken out of the organic machine and replaced with a mechanical machine's ghost.

We can also hack our own responses, intentionally holding a too-hot pot handle, for example, to keep from spilling its bacon grease onto the floor. This more sophisticated cognition imparts the ability to recognize and rationally balance alternative possible outcomes, and the behavioral responses best equipped to achieve those outcomes. A sense of agency, of control; to learn and remember and acclimate; to deal with the unanticipated and solve life's myriad puzzles; to modify one's actions to manipulate the world in predictable ways. To paraphrase Shane Legg and Marcus Hutter, the better an individual is at achieving their goals under a wide range of circumstances, the greater is their intelligence.[50] Of course, setting goals in the first place demands a rather sophisticated mind.

The ability to achieve goals—intelligence—can evolve, just like any other characteristic of an organism. Even within our own species, genetic variants contribute to heritable differences in various aspects of problem-solving abilities.[51] Adaptive evolution favored increased intelligence since our common ancestry with other primates, leading to the unusually large brains of *Homo sapiens*. In humans these days, however, it's not clear that those heritable differences of intelligence metrics relate in any consistent way to differences in reproductive output. As a consequence, and despite strong associations with educational and long-term health outcomes, it may be that the modern world does not actually confer an *evolutionary* advantage to further increases in human intellectual capabilities.

When we talk about intelligent animals, there is a broad range of cognitive abilities beyond simple instinctual responses. Most basic is the capacity for associative and reinforcement learning, the ability to mentally link a positive or negative outcome with a given cue or stimulus that correlates with that outcome. A simple example of associative learning is when an animal comes to expect the presence of food with a specific odor. More sophisticated is observational learning, taking cues from the behavior of others to figure out how to do something oneself.

Humpback whales have devised a clever underwater weapon to assist them in capturing prey: the bubble net. This weapon is one that gets taught from whale to whale. Operating solo or in a cooperative group, the whales will swim in a circle around a large school of fish or krill, shooting columns of bubbles up through their blowholes to create a bubbly curtain around their prey that produces, from the surface, a circle or spiral shape.[52] This corrals and concentrates their prey, trapping them for the whales to efficiently gulp as they lunge up through the center of the bubble net, mouths agape (Figure 15.2). While circling and creating their net of up to 30-m (100- foot) diameter, the whales will trumpet so loud that you can hear it from a ship. That trumpeting is thought to interact with the acoustic properties of the bubbles of the curtain to help create an impenetrable wall of sound around a near-silent interior that enhances the net's efficacy in trapping its seafood inside.[53]

Some animals, even, are capable of creating spatial memories or maps to remember how best to get from place to place, as seen among the impressive feats of octopuses, the smartest of mollusks. Another advanced cognitive task is knowing how to recognize

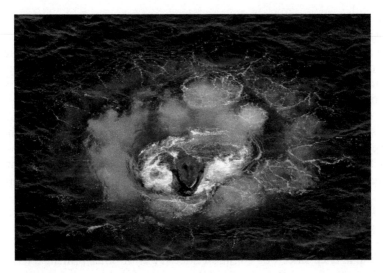

Figure 15.2 *Animal intelligence is the ability to achieve goals under a range of conditions, with the invention and communication by humpback whales of the bubble net hunting technique giving a profound example of intelligent learning. Animals capable of intelligent learning, including humans, show genetic variability for the capacity to perform many intelligence-related tasks.*
Image credit: Christin Khan, NOAA/NEFSC reproduced from public domain.

particular individuals of one's own species or of other species that an individual may encounter repeatedly—as dogs and elephants are famously good at—or to recognize oneself in a mirror, a rare feat that *Pica pica* magpies and some great apes and dolphins can do. This social intelligence sometimes extends to cooperative hunting, as conducted by wolves and some clever cichlid fish, and the ability to put yourself in another's shoes, to mind-read. Then there are the brainiac crows and the famous *Psittacus erithacus* African grey parrot known as Alex, who can use tools in complex ways or can have some sense of numbers and abstract concepts related to shape and color.[54]

It's no surprise that our closest primate relatives are highly intelligent animals: chimpanzees, bonobos, gorillas, and orangutans. And dog owners are well-versed in the smarts of their canines, a trait they share with wolves and pigs.[55] The physically enormous brains of elephants and toothed whales, including bright-minded dolphins, however, does not manage to make them as sharp as some other, smaller, fellow vertebrates. I am speaking, of course, of the crows.[56]

Crow smarts are not just limited to Aesop's fables. For example, not only are New Caledonian crows *Corvus moneduloides* adept at making and using tools; they appear able to use analogical reasoning to use one tool to retrieve another tool to solve problems. In one experiment, each of seven hungry crows was placed in an arena with a tasty treat that they could see and smell but that sat out of reach of their beaks inside a deep, narrow plexiglass box. They had ready access to a short stick, too short to reach the treat. There also was a longer stick, the perfect length for a bright bird to wield in its beak to get the food, but this longer tool lay inside a caged toolbox that was out of reach of their beaks.

The three cleverest crows—named Luigi, Icarus, and Gypsy—figured out on the first try that they could use the short tool to extract the long tool to then obtain the snack. They all figured out the logic in short order, independently arriving at the multitool solution. These bird-brained animals, as it has been said, have "cognitive abilities that rival those of our primate relatives."[57]

But are they sentient? Do they have a sufficiently strong sense of self, ability to reason, capacity to feel emotion that we would recognize them as having minds with consciousness? The authors of the 2012 Cambridge Declaration on Consciousness say, unequivocally, "yes."[58] The sensory systems of other creatures are so distinct from ours, however, that it is difficult to know, exactly, whether or how our human sensibilities could translate to thought and feeling in the minds of other creatures. We differ, even, in our perception of time.[59] How does a crow think and feel? To paraphrase the philosopher's conundrum in knowing what it is like to be a bat—a profound conundrum, given the wide gulf in sensory abilities between us and bats—surely the answer lies in knowing what it is like to be a bird.[60]

★

So far, we've dwelled on different ways that unintended consequences might cause harm. We've thought through a number of possible ethical worries and tangible calamities that genetic welding and directed evolution of new or exotic species might have on ecosystems. Disastrous unintended consequences, however, despite the potential risk, may not actually manifest.

But what about the possibility of *intended* disasters—of wielding genetic welding to wage war, of bioterrorism? Like many technologies, gene drives suggest possibilities for so-called dual use: for good and for ill.[61] A familiar example from physics includes applying nuclear technology for medical procedures and power production versus A-bombs and H-bombs. In genetic engineering, there are, as it turns out, G-bombs.

We've mostly talked about how gene drives could be applied to enhance biological systems in the creation of hyperexotic organisms, to restore ecosystem functions, to eliminate invasive species, or to suppress diseases like malaria. One could, however, imagine bad actors introducing gene drives for disruptive purposes, what we might call a G-bomb. For example, an adversary might engineer with genetic welding a species of mosquito that is endemic to the geographic region of their enemy so that it can deliver a bacterial-derived toxin when female mosquitoes bite people. Or genetic welding could serve to alter the environmental tolerance of the mosquito to allow it to expand its range deeper into enemy territory so as to drive transmission of diseases vectored by the mosquito population. A suppression gene drive might target a species that serves as a national symbol, driving extinct a creature that gives people a point of pride: a genetic debut into psychological warfare.

Food systems or natural resources that an enemy state depends on also could be targets, provided that the animal or plant species reproduce sexually. For example, a rogue state might introduce a suppression gene drive to drive extinct a key fisheries species or set loose a modification gene drive into an insect or fungus to enhance its pestilence—or to confer pesticide resistance—when it attacks a key crop or forestry product. A bioterrorist could even create an innocuous gene drive that nonetheless promulgates a transgenic element through crops or livestock. Such foodstuffs "tainted" with the stain of being GMOs might wreak havoc to rules and regulations in regions like the European Union where farmers and grocers are subject to strict regulation in the possession and use of organisms with genomes subjected to genetic engineering.

These susceptibilities to nefarious applications of genetic welding, and gene drives generally, are indeed alarming. Application of them surely would violate the United Nations Biological Weapons Convention as well as the less-formal Code of Ethics of gene drive researchers.[62] Fortunately, some types of organisms make for inherently hard targets to evade malicious G-bombs. Species that propagate by self-fertilization or asexually, like crops generated through grafting, are not susceptible to genetic invasion because a gene drive depends on the union of gametes from different individuals in order to spread over time. Species with long generation times, like trees grown for timber, similarly would be slow to realize any adverse effects of gene drive genetic invasion. They'd be slow-motion G-bombs, taking centuries to detonate and providing centuries to defuse once detected. Livestock that farmers reproduce in controlled breeding arrangements also would have limited risk, and such species would be amenable to genetic testing to exclude any individuals with the nefarious genetic engineering from contributing to the next generation. Should a weaponized gene drive be detected, it may be possible to engineer so-called reversal gene drives that thwart the spread of the G-bomb. Reversal drives—a last resort that employs a use-a-gene-drive to fight-a-gene-drive approach— have been proposed as an integral part of mitigation strategies in any plan to study even innocuously or positively intentioned gene drives.[63]

★

These are some of the ethical quandaries, the figurative dragons, to consider in developing a biological Artemis mission on Earth. Should we do it because we can? Is it shameful hubris and the height of hedonism to create something grand from the sparks of imagination? What utility does it serve and is "utility" an appropriate benchmark? Our lives are rich even without real, live dragons evolved through human intervention and genetic welding. There are natural wonders right now, many at risk of extinction. Perhaps it would be wasteful to devote substantial resources to creating novel species by genetic welding. Perhaps it is time to embrace humility, with our ambition tempered to simply foster passive rewilding. But grand vision is also a staple of society in building a shared sense of purpose and community. In recent times, science has fostered pride in those shared aspirations, like how we chase space exploration to land humans on the Moon and robots on Mars.

What do I think about all of this? In one word: *conflicted*.

★

The evolution of new hyperexotic species and resurrecting extinct species with the tools of genetic engineering, and genetic welding in particular, raises weighty concerns about ethics and the health of humans and ecosystems. Gene drives hold potential to be weaponized as G-bombs. Some issues of evolving new species overlap with the use of

GMOs, including concerns over bio-capitalism and unintended consequences to ecosystem properties. Opportunity costs in choosing one path to the exclusion of others may be real, or illusory in the presence of synergistic effects. There can be costs to inaction that must be evaluated in risk assessment. In assessing a given issue, it is valuable to disentangle separate concerns and to identify potential unconscious biases, such as the naturalistic fallacy and status quo bias. There are no easy answers to navigating the ethical bureaucracy of interventionist rewilding and genetic welding.

Notes

1. p. 11266 in Davis, M., et al. 2018. Mammal diversity will take millions of years to recover from the current biodiversity crisis. *Proceedings of the National Academy of Sciences USA*. 115: 11262–11267. https://doi.org/10.1073/pnas.1804906115.

2. Aerts, R., et al. 2018. Biodiversity and human health: mechanisms and evidence of the positive health effects of diversity in nature and green spaces. *British Medical Bulletin*. 127: 5–22. https://doi.org/10.1093/bmb/ldy021.

3. Baumgärtner, S. 2007. The insurance value of biodiversity in the provision of ecosystem services. *Natural Resource Modeling*. 20: 87–127. https://doi.org/10.1111/j.1939-7445.2007.tb00202.x; Loreau, M., et al. 2021. Biodiversity as insurance: from concept to measurement and application. *Biological Review*. 96: 2333–2354. https://doi.org/10.1111/brv.12756.

4. http://e360.yale.edu/feature/the_case_for_de-extinction_why_we_should_bring_back_the_woolly_mammoth/2721. Accessed May 19, 2021.

5. Plumptre, A.J., et al. 2021. Where might we find ecologically intact communities? *Frontiers in Forests and Global Change*. 4: 26. https://doi.org/10.3389/ffgc.2021.626635.

6. Cleves, P.A., et al. 2020. Reduced thermal tolerance in a coral carrying CRISPR-induced mutations in the gene for a heat-shock transcription factor. *Proceedings of the National Academy of Sciences USA*. 117: 28899–28905. https://dx.doi.org/10.1073/pnas.1920779117; Howells, E.J., et al. 2021. Enhancing the heat tolerance of reef-building corals to future warming. *Science Advances*. 7: eabg6070. https://dx.doi.org/10.1126/sciadv.abg6070.

7. Quigley, K.M., et al. 2019. The active spread of adaptive variation for reef resilience. *Ecology and Evolution*. 9: 11122–11135. https://doi.org/10.1002/ece3.5616; Caruso, C., et al. 2021. Selecting heat-tolerant corals for proactive reef restoration. *Frontiers in Marine Science*. 8: 632027. https://www.frontiersin.org/article/10.3389/fmars.2021.632027; Anthony, K., et al. 2017. New interventions are needed to save coral reefs. *Nature Ecology and Evolution*. 1: 1420–1422. https://doi.org/10.1038/s41559-017-0313-5; Novak, B.J., et al. 2018. Advancing a new toolkit for conservation: from science to policy. *CRISPR Journal*. 1: 11–15. https://doi.org/10.1089/crispr.2017.0019.

8. Rode, N.O., et al. 2019. Population management using gene drive: molecular design, models of spread dynamics and assessment of ecological risks. *Conservation Genetics*. 20: 671–690. https://doi.org/10.1007/s10592-019-01165-5; Kosch, T.A., et al. 2022. Genetic approaches

for increasing fitness in endangered species. *Trends in Ecology & Evolution*. 37: 332–345. https://doi.org/10.1016/j.tree.2021.12.003.

9. Esvelt, K.M., and N.J. Gemmell. 2017. Conservation demands safe gene drive. *PLoS Biology*. 15: e2003850. https://doi.org/10.1371/journal.pbio.2003850; Moro, D., et al. 2018. Identifying knowledge gaps for gene drive research to control invasive animal species: the next CRISPR step. *Global Ecology and Conservation*. 13: e00363. https://doi.org/10.1016/j.gecco. 2017.e00363; Thresher, R.E., et al. 2019. Stakeholder attitudes towards the use of recombinant technology to manage the impact of an invasive species: sea lamprey in the North American Great Lakes. *Biological Invasions*. 21: 575–586. https://doi.org/10.1007/s10530-018-1848-3; https://www.geneticbiocontrol.org. Accessed September 27, 2021.

10. Katz, E. 2012. Further adventures in the case against restoration. *Environmental Ethics*. 34: 67–97. https://doi.org/10.5840/enviroethics20123416.

11. David, P., et al. 2017. Impacts of invasive species on food webs: a review of empirical data. *Advances in Ecological Research*. 56: 1–60. https://doi.org/10.1016/bs.aecr.2016.10.001.

12. Neve, P. 2018. Gene drive systems: do they have a place in agricultural weed management?: Gene drive and weed management. *Pest Management Science*. 74: 2671–2679. https://doi.org/10.1002/ps.5137.

13. https://www.haudenosauneeconfederacy.com/values; https://www.vox.com/future-perfect/22552963/how-to-be-a-good-ancestor-longtermism-climate-change. Accessed July 5, 2021.

14. Bliege Bird, R., and D. Nimmo. 2018. Restore the lost ecological functions of people. *Nature Ecology and Evolution*. 2: 1050–1052. https://doi.org/10.1038/s41559-018-0576-5.

15. Ducarme, F., et al. 2021. How the diversity of human concepts of nature affects conservation of biodiversity. *Conservation Biology*. 35: 1019–1028. https://doi.org/10.1111/cobi.13639.

16. Ellis, E.C., et al. 2021. People have shaped most of terrestrial nature for at least 12,000 years. *Proceedings of the National Academy of Sciences USA*. 118: e2023483118. https://dx.doi.org/10.1073/pnas.2023483118.

17. As philosopher Richard Watson pointed out, "If we are to treat man as a part of nature on egalitarian terms with other species, then man's behavior must be treated as morally neutral, too." That's a tough pill to swallow. Watson, R.A. 1983. A critique of anti-anthropocentric biocentrism. *Environmental Ethics*. 5: 245–256. https://doi.org/10.5840/enviroethics19835325; quotation in the text from Fletcher, M.-S., et al. 2021. Indigenous knowledge and the shackles of wilderness. *Proceedings of the National Academy of Sciences USA*. 118: e2022218118. https://dx.doi.org/10.1073/pnas.2022218118.

18. https://cosmosmagazine.com/people/culture/indigenous-knowledge-and-the-persistence-of-the-wilderness-myth. Accessed October 19, 2021.

19. Perhaps common ground may be found in appeals to the value of wilderness in having intrinsic beauty, like works of art, as environmental philosopher Janna Thompson pointed out: "the free creativity of nature is the reason for finding wilderness beautiful . . . human beings too are part of nature." Thompson, J. 1995. Aesthetics and the value of nature. *Environmental Ethics*. 17: 291–305. https://doi.org/10.5840/enviroethics199517319; quotation in the text

from Perino, A., et al. 2019. Rewilding complex ecosystems. *Science*. 364: eaav5570. https://dx.doi.org/10.1126/science.aav5570.

20. Quotation from Seddon, P.J., and M. King. 2019. Creating proxies of extinct species: the bioethics of de-extinction. *Emerging Topics in Life Sciences*. 3: 731–735. https://doi.org/10.1042/ETLS20190109.

21. https://www.scientificamerican.com/article/why-people-oppose-gmos-even-though-science-says-they-are-safe. Accessed September 28, 2021.

22. GE and GMO are synonyms in this context.https://www.nap.edu/read/23395. Accessed September 22, 2021.

23. More recently, a tomato that produces vitamin D was genetically engineered with CRISPR-Cas9 to modify a pre-existing duplicate metabolic pathway in the plants. Li, J., et al. 2022. Biofortified tomatoes provide a new route to vitamin D sufficiency. *Nature Plants*. https://doi.org/10.1038/s41477-022-01154-6.

24. Perhaps ironically, "reversal" gene drives could be engineered to remove transgenic material from GMOs that have "leaked" into wild populations.

25. https://www.hhmi.org/bulletin/winter-2015/how-can-scientists-help-ease-societys-fear-gmo. Accessed September 22, 2021.

26. *Bt* crops, most commonly maize and cotton, are genetically engineered to express an insecticidal toxin that is ordinarily produced by the *Bacillus thuringiensis* bacterium, making the crops resistant to insect pests to increase yield and reduce pesticide spraying.

27. https://bch.cbd.int/protocol/text and https://bch.cbd.int/protocol/parties. Accessed September 22, 2021.

28. The precautionary principle, in effect, prioritizes the bioethical tenet of nonmaleficence above the other core tenets of beneficence, respect for autonomy, and equity of justice.

29. Cynics, like Cass Sunstein, point out that "undoubtedly the [precautionary] principle is invoked strategically by self-interested political actors" in order to stymie action. Sunstein, C.R. 2003. Beyond the precautionary principle. *University of Pennsylvania Law Review*. 151: 1003–1058. http://www.jstor.com/stable/3312884.

30. Marshall, J. 2010. The Cartagena Protocol and genetically modified mosquitoes. *Nature Biotechnology*. 28: 896–897. https://doi.org/10.1038/nbt0910-896; Hammond, A., et al. 2021. Gene-drive suppression of mosquito populations in large cages as a bridge between lab and field. *Nature Communications*. 12: 4589. https://doi.org/10.1038/s41467-021-24790-6; Long, K.C., et al. 2020. Core commitments for field trials of gene drive organisms. *Science*. 370: 1417–1419. https://doi.org/10.1126/science.abd1908.

31. Long, K. C., et al. 2020. Core commitments for field trials of gene drive organisms. *Science*. 370: 1417–1419. https://doi.org/10.1126/science.abd1908.

32. http://e360.yale.edu/feature/the_case_for_de-extinction_why_we_should_bring_back_the_woolly_mammoth/2721. Accessed May 19, 2021.

33. The naturalistic fallacy makes "an appeal to some feature of the natural world in order to represent moral, aesthetic, or social norms," but "there is no legitimate inference that can be

drawn from how things happen to be (equated with natural regularities) to how things should be (equated with human norms)." Daston, L. 2014. The naturalistic fallacy is modern. *Isis.* 105: 579–587. https://dx.doi.org/10.1086/678173.

34. Gyngell, C., and J. Savulescu. 2017. Promoting biodiversity. *Philosophy & Technology.* 30: 413–426. https://doi.org/10.1007/s13347-016-0234-2.

35. Bostrom, N., and T. Ord. 2006. The reversal test: eliminating status quo bias in applied ethics. *Ethics.* 116: 656–679. https://dx.doi.org/10.1086/505233.

36. FitzGerald, M.J., and E.J. Spek. 2020. Microbiome therapeutics and patent protection. *Nature Biotechnology.* 38: 806–810. https://doi.org/10.1038/s41587-020-0579-z.

37. https://publicintegrity.org/environment/bankrupt-companies-avoid-more-than-700-million-in-cleanup-costs. Accessed October 20, 2021.

38. Batavia, C., et al. 2019. The elephant (head) in the room: a critical look at trophy hunting. *Conservation Letters.* 12: e12565. https://doi.org/10.1111/conl.12565.

39. Di Minin, E., et al. 2016. Banning trophy hunting will exacerbate biodiversity loss. *Trends in Ecology & Evolution.* 31: 99–102. https://doi.org/10.1016/j.tree.2015.12.006.

40. https://www.theguardian.com/inequality/2017/nov/01/animal-rights-activists-inuit-clash-canada-indigenous-food-traditions. Accessed October 20, 2021.

41. Angula, H.N., et al. 2018. Local perceptions of trophy hunting on communal lands in Namibia. *Biological Conservation.* 218: 26–31. https://doi.org/10.1016/j.biocon.2017.11.033.

42. Lindsey, P.A., et al. 2007. Economic and conservation significance of the trophy hunting industry in sub-Saharan Africa. *Biological Conservation.* 134: 455–469. https://doi.org/10.1016/j.biocon.2006.09.005.

43. Seddon, P.J., and M. King. 2019. Creating proxies of extinct species: the bioethics of de-extinction. *Emerging Topics in Life Sciences.* 3: 731–735.

44. GloFish is a brand of pet fish, genetically engineered to express transgenic fluorescent proteins.

45. Batavia, C., et al. 2019. The elephant (head) in the room: a critical look at trophy hunting. *Conservation Letters.* 12: e12565. https://doi.org/10.1111/conl.12565.

46. McCleery, R.A., et al. 2015. Marsh rabbit mortalities tie pythons to the precipitous decline of mammals in the Everglades. *Proceedings of the Royal Society B.* 282: 20150120. https://doi.org/10.1098/rspb.2015.0120.

47. At the posted rate of $50 plus $25 for each foot above 4 feet long, someone apparently earned $400 for hunting an 18 footer. https://www.sfwmd.gov/our-work/python-program. Accessed September 23, 2021.

48. Colbert, H.A., et al. 1997. OSM-9, a novel protein with structural similarity to channels, is required for olfaction, mechanosensation, and olfactory adaptation in *Caenorhabditis elegans. Journal of Neuroscience.* 17: 8259–8269. https://dx.doi.org/10.1523/JNEUROSCI.17-21-08259.1997.

49. Dong, X., et al. 2021. Toward a living soft microrobot through optogenetic locomotion control of *Caenorhabditis elegans. Science Robotics.* 6: eabe3950. https://dx.doi.org/10.1126/scirobotics.abe3950.

50. To quote, using their lingo, "Intelligence measures an agent's ability to achieve goals in a wide range of environments." Legg, S., and M. Hutter. 2007. Universal intelligence: a definition of machine intelligence. *Minds & Machines.* 17: 391–444. https://doi.org/10.1007/s11023-007-9079-x.

51. Plomin, R., and S. von Stumm. 2018. The new genetics of intelligence. *Nature Reviews Genetics.* 19, 148–159. https://doi.org/10.1038/nrg.2017.104; Montgomery, S.H., et al. 2010. Reconstructing the ups and downs of primate brain evolution: implications for adaptive hypotheses and *Homo floresiensis. BMC Biology.* 8: 9. https://doi.org/10.1186/1741-7007-8-9.

52. Hain, J.H.W., et al. 1982. Feeding behavior of the humpback whale, *Megaptera novaeangliae,* in the western North Atlantic. *Fishery Bulletin.* 80: 259–268. https://www.researchgate.net/publication/255590511.

53. Bryngelson, S.H., and T. Colonius. 2020. Simulation of humpback whale bubble-net feeding models. *Journal of the Acoustical Society of America.* 147: 1126–1135. https://doi.org/10.1121/10.0000746; Leighton, T.G., et al. 2004. Trapped within a 'wall of sound': a possible mechanism for the bubble nets of the humpback whales. *Acoustics Bulletin.* 29: 24–29. https://eprints.soton.ac.uk/10410.

54. Roth, G. 2015. Convergent evolution of complex brains and high intelligence. *Philosophical Transactions of the Royal Society B.* 370: 20150049. https://doi.org/10.1098/rstb.2015.0049.

55. Lampe, M., et al. 2017. The effects of domestication and ontogeny on cognition in dogs and wolves. *Scientific Reports.* 7: 11690. https://doi.org/10.1038/s41598-017-12055-6; Mendl, M., et al. 2010. Pig cognition. *Current Biology.* 20: R796–R798. https://doi.org/10.1016/j.cub.2010.07.018.

56. Roth, G. 2015. Convergent evolution of complex brains and high intelligence. *Philosophical Transactions of the Royal Society B.* 370: 20150049. https://doi.org/10.1098/rstb.2015.0049.

57. Taylor, A.H., et al. 2007. Spontaneous metatool use by New Caledonian crows. *Current Biology.* 17: 1504–1507. https://doi.org/10.1016/j.cub.2007.07.057; Emery, N.J., and N.S. Clayton. 2004. The mentality of crows: convergent evolution of intelligence in corvids and apes. *Science.* 306: 1903–1907. https://dx.doi.org/10.1126/science.1098410.

58. http://fcmconference.org/img/CambridgeDeclarationOnConsciousness.pdf. Accessed July 14, 2021.

59. Healy, K., et al. 2013. Metabolic rate and body size are linked with perception of temporal information. *Animal Behaviour.* 86: 685–696. https://doi.org/10.1016/j.anbehav.2013.06.018.

60. As Thomas Nagel wrote of the so-called mind-body problem, "We must consider whether any method will permit us to extrapolate to the inner life of the bat from our own case. . . . In so far as I can imagine . . . it tells me only what it would be like for me to behave as a bat behaves. But that is not the question. I want to know what it is like for a bat to be a bat. . . . but such

an understanding may be permanently denied to us by the limits of our nature." Forty years prior, Jakob von Uexküll might have asked what it is like for a tick to be a tick, his exemplar animal of choice, with the sensory systems that produce its so-called umwelt being so different from ours. Nagel, T. 1974. What is it like to be a bat? *Philosophical Review.* 83: 435–450. https://dx.doi.org/10.2307/2183914; von Uexküll, J. 1934. A stroll through the worlds of animals and men: a picture book of invisible worlds. In: C.H. Schiller (ed.), *Instinctive Behavior: The Development of a Modern Concept.* New York: International Universities Press. pp. 5–80.

61. Getz, L.J., and G. Dellaire. 2018. Angels and devils: dilemmas in dual-use biotechnology. *Trends in Biotechnology.* 36: 1202–1205. https://doi.org/10.1016/j.tibtech.2018.07.016; https://doi.org/10.1126/science.345.6200.1010-b; https://www.independent.co.uk/voices/editorials/power-good-or-evil-technology-gene-drives-holds-immense-potential-improving-our-life-planet-and-also-great-dangers-10434022.html. Accessed September 23, 2021.

62. Annas, G.J., et al. 2021. A code of ethics for gene drive research. *CRISPR Journal.* 4: 19–24. https://doi.org/10.1089/crispr.2020.0096; https://www.un.org/disarmament/biological-weapons. Accessed September 23, 2021.

63. Esvelt, K.M., et al. 2014. Emerging technology: concerning RNA-guided gene drives for the alteration of wild populations. *eLife.* 3: e03401. https://dx.doi.org/10.7554/eLife.03401; Vella, M.R., et al. 2017. Evaluating strategies for reversing CRISPR-Cas9 gene drives. *Scientific Reports.* 7: 11038. https://doi.org/10.1038/s41598-017-10633-2.

16

When we become dragons

As capable as humans are at engineering ecosystems, we truly excel at engineering our own daily lives. We inject ourselves with live viruses and bacteria as vaccines to stimulate our immune systems and so prevent debilitating infections. We consume drugs extracted from microbes or plants, or that we synthesize from scratch, to combat infections that have taken hold. Some of us will accept transplanted organs from cadavers or from the living bodies of other people, even from other species.[1] We might even willfully remove our own organs as a precautionary measure—such as mammary tissue after a BRCA1/2 genetic diagnosis.[2] We'll intentionally dose ourselves with ionizing radiation or poisonous chemicals, whether to combat cancerous lesions of our own body's making, to get a better look at our bones and teeth, or to just get high. We make these choices in attempts, real or perceived, at improving ourselves. Some of those benefits are life-or-death in gravity, some not so much.

Even when death is not at stake, we engineer ourselves to suit our preferences. Females may dose themselves daily with hormone supplements to prevent pregnancy or choose to terminate a pregnancy. Soon, dimethandrolone undecanoate, among other possibilities undergoing testing, may provide "the pill" to males as a chemical means to contraception. Lacking a baculum, we can take sildenafil.[3] Some of us will cut out or staple together parts of our gastrointestinal tract to become slimmer. Elective cosmetic surgeries can reshape breasts, buttocks, bellies, noses, eyelids, labia, calves, and most other body parts. It's common practice to inject colorful dyes beneath our skin, to glue metal braces to teeth, to slice off a baby's foreskin, or to create permanent punctures in our epidermis and subcutaneous cartilage from which to dangle jewelry, entirely for aesthetic or cultural preference.

We'll readily enhance our natural bodily capabilities. We'll do so to boost performance absolutely or to compensate for disabilities relative to other people, those deficits sometimes originating from birth or from aging or from injury. There are surgeries to the cornea of our eyes to improve vision, microchips to augment our hearing, gizmos implanted next to our hearts to improve cardiac pacing, attachment of mechanical or motorized contraptions to improve our motility or manual dexterity or lifting capacity. Couples may choose in vitro fertilization or solicit donations of eggs or sperm in procedures to circumvent infertility. We may undergo gender affirmation surgeries to

Evolving Tomorrow. Asher D. Cutter, Oxford University Press. © Asher D. Cutter (2023). DOI: 10.1093/oso/9780198874522.003.0016

align our physical bodies with how we think and feel they should be. We take drugs to improve our physical abilities, whether to grow muscles (creatine) or to mitigate pain (ibuprofen). We take drugs to improve our mental acuity,[4] whether intellectual (caffeine, methylphenidate a.k.a. Ritalin, or anti-amyloids against Alzheimer disease) or emotional (lithium or selective serotonin reuptake inhibitors like Prozac and Zoloft).

You can justify some uses for their virtue in elevating one's quality of life. Other performance-enhancing uses, however, may be disparaged as frivolous, selfish, or conferring unfair advantage. The people who earn back their self-esteem or earn their paycheck from partaking of such enhancements, nonetheless, may clap back that they do, in fact, materially benefit from an improved quality of life.

None of these modifications to our bodies are risk-free, and many of them are permanent. As we talk about modifying ourselves, we'll also talk about how to draw the line for what's acceptable and what's fair across the entirety of the global population of humans.

16.1 Heritable human genome editing

With so very many intentional body modifications that we accept—for health, for quality of life, for personal preference—why not modify our genomes? After all, roughly 5% of newborns, one in twenty, experience a genetic disorder of some kind.[5] Genome editing in people is not just a barstool flight of fancy; in fact, it is a proven real-world capability.

In 2018, it was revealed that a woman in China birthed twin girls—pseudonyms Lulu and Nana—that resulted from CRISPR-Cas9 genome editing.[6] The motivating rationale, aside from hopes of profit and glory, was to use such so-called gene surgery as a "wholesome development for medicine." More specifically, He Jiankui, the entrepreneurial scientist responsible, intended to use CRISPR-Cas9 to induce genetic changes in the CCR5 gene that would confer resistance to HIV, a virus that afflicted the father. Despite successfully achieving genome edits to the CCR5 gene, not all were the expected DNA changes; and there were unexpected alterations of DNA to other parts of the genome, as well. He Jiankui ended up fined, fired, jailed, and banned from future work on reproductive technologies in the fallout of what amounted to a litany of unethical breaches of protocol. Nonetheless, CRISPR-Cas9 genome editing in humans remains a demonstrated fact.

★

Beyond the breach of medical conduct by He and his team, the controversy highlighted by Lulu and Nana is that of *heritable* human genome editing. Some applications of CRISPR-Cas9 genome editing are *nonheritable*, acting entirely on nonreproductive cells. For example, genome edits that alter cells in the retina of the eye or T-cells in bone marrow can combat inherited blindness or even lung cancers.[7] CRISPR-Cas9 genome editing, in these cases, simply provides a new way to conduct so-called gene therapy.

These nonheritable kinds of treatments, like standard medical interventions, only act on the body of a single individual and will not be passed on to that person's children.

Nonetheless, one criticism invokes the possibility of unequal access: that they'll only be accessible to the rich. This kind of inequity, unfortunately, plagues numerous other medical interventions already. As problematic as that might be, it does not offer an intrinsic reason for nonheritable genome editing to not be pursued for medical purposes. In my view, nonheritable forms of genome editing aren't especially controversial from an ethical standpoint,[8] provided that their development and provision adheres to existing guidelines for clinical trials and distribution that are required for any new medical procedure prior to widespread adoption.

Heritable human genome editing (HHGE), however, is controversial. It is prohibited by law in Canada and the United States, among 73 other countries. No country in the world explicitly allows HHGE, though five leave the door ajar on their prohibitions (Belgium, Colombia, Italy, Panama, and United Arab Emirates).[9] At the most basic level, there are concerns about inducing off-target mutations or that the embryo might become a "mosaic" that contains both edited and unedited cells. The current legal prohibitions to clinical application of HHGE, therefore, are sound policy until the technical hurdles get surmounted, at the very least. Technical advances will eventually minimize or eliminate such unintended side effects, though, and the health and quality of life justifications to altering a genetic lesion to be less debilitating may begin to outweigh these potential medical risks.

Even beyond the technical considerations, ethical concerns crop up. People may worry about appropriately informed consent by the embryonic subjects of HHGE, which, as a blob of cells, are not in a position to sign on the dotted line. Embryonic consent, however, writes ethicist Françoise Baylis, "is a red herring; embryos do not consent to anything that affects their inherited traits."[10] Parents, after all, are already empowered and responsible for making health and other life-altering decisions for their children and certainly so for embryos, which are not people. Some people hold religious objections to the broader enterprise of using human embryos in research. Religious beliefs vary widely, however, with little justification for one tenet to supersede another to dictate policy in a pluralistic society. Consequently, each of these points of objection to heritable human genome editing—technical hurdles of off-target effects, embryo consent, religion—have potent counterpoints to HHGE being intrinsically verboten. One other area of unease with the prospect of heritably modifying human genomes, however, is thornier.

★

The ethical debates about HHGE are similar to those surrounding preimplantation genetic diagnosis (PGD), the testing that can come hand-in-hand with in vitro fertilization procedures (IVF). With IVF, the medical retrieval of about ten eggs and harvesting of millions of sperm are followed by fertilization in the lab to produce a collection of embryos in a Petri dish. Usually, however, only one or two embryos get transferred to the uterus. How to choose among all of the "maybe babies"? Beyond using criteria about the overall outward appearance of cellular progression of the embryos to the blastocyst stage, you could just flip a coin.

Or you can sacrifice a couple of cells with no harm to the young embryo in order to analyze its DNA. Standard services for PGD currently will only test for one or two out of a selection of several hundred simple genetic disorders, such as karyotype anomalies, Huntington disease, sickle cell disease, and phenylketonuria.[11] There also are prenatal genetic tests of a young fetus following amniocentesis, which takes place during pregnancy after an embryo implants in the uterus and develops an amniotic sac. Such genetic tests are commonplace irrespective of IVF, the results of which may lead a pregnant person to elect to legally abort the fetus.[12]

Modern DNA sequencing technology, however, is capable of screening the entire genome from that tiny sample of embryonic cells to assess "pathogenic variation, tandem repeat, copy number and structural variations."[13] This technique has the ability to confidently ascertain "predicted disease status" based on genetic variants passed down from each parent, as well as new genetic lesions introduced by mutation in the production of gametes that are unique to the embryo. Rather than a simple diagnosis inspired by prior suspicions of genetic risk, this procedure would be a full-scale preimplantation genomic screen (PGS). After requisite genetic counseling, the intended parents could then choose to transfer whichever embryos they want.[14]

This is, of course, a small-scale form of artificial selection. Hopeful parents might choose an embryo that is free of a genetic variant that would lead to a severe disease, such as Huntington disease or cystic fibrosis. Or they might make their selection based on genetic matches that would make the potential child a suitable organ donor—a so-called savior sibling—for a sick older sibling. They might even select an embryo for the *presence* of a disability, such as heritable deafness or dwarfism, a choice that parents-to-be make in 3% of cases.[15]

In principle, however, parents-to-be could select embryos on any trait predisposition that has been linked to genetic factors. You are probably familiar with personal genomics kits that provide you with a report containing genealogical information, health predispositions, and genetic predilections for an array of physical and physiological features. One popular company offers details on over 30 heritable traits, including hair characteristics (texture, thickness, pigmentation, sun bleaching, dandruff, baldness, and widow's peak), facial features (dimples, cleft chin, detached earlobes, unibrow, freckles, and eye color), bunions, asparagus pee detection, mosquito-bite predilection, fear of heights, and hatred of the sound of chewing (misophonia).[16] Based on associations between traits like these and their genetic basis, potential parents could select among their fertilized embryos using a genomic selection index that integrates and weights their preferences across all features of the embryos.

Philosopher Julian Savulescu argues controversially that, not only *could* intended parents choose in this way, but that they *should*.[17] We're familiar with the dictum that pregnant people should avoid excess cocaine use and alcohol consumption during pregnancy and that parents should provide their child with nutritious food and the best education. In a similar way, the principle of "procreative beneficence" says that parents-to-be should select embryos that are expected to give the person who develops the best chance in life, in the parents' estimation, given available information. This provocative idea, however, has a mixed reception.[18]

16.2   The echoes of eugenics

Why does the idea of procreative beneficence strike a nerve despite the seemingly admirable ambition to give a child the best life? In short, it sounds like an echo of eugenics.[19] Sometimes the notion of embryo selection gets labeled with the alarming phrase *designer babies* or the euphemistic phrases *genetic enhancement* and *reproductive freedom*. Preimplantation genomic scans that then get applied to embryo selection by individuals would, indeed, represent a form of what is termed by bioethics scholars as *liberal eugenics*.

I approach this loaded topic of eugenics nervously. It is important to make clear that I, and I hope everyone, denounce in the strongest terms all coercive eugenic views, especially those rooted in racist attitudes. Because race and ethnicity are defined by social criteria not genomic criteria,[20] I reject the false genetic determinism that misattributes their basis to a simplistic caricature of genetic differences. As an evolutionary geneticist, I'm ashamed of how genetic and evolutionary principles have been co-opted in misguided ways as pretext for imposing atrocities on people. I reject, as we all must, any system that imposes genetic decision-making on others.

With this backdrop, I think it is important that we have open discussion about the ethics and the science of noncoercive individual selection of genetic properties with respect to the birth of humans. It is hard to talk about, and it is hard to disentangle from the abhorrent social backdrop of racism and bigotry throughout our history of imposed genetic sanctions. That history reveals our deep shame in having allowed forced sterilization, genocide, and antimiscegenation to have taken place as government policy, what is termed *authoritarian eugenics*. It is hard to talk about these issues because of how easy it is to overinflate the role of genetic differences in aspects of ourselves, to put undue emphasis on genetic determinism of traits that arise, in large part, because of environmental and social factors.

Now, what is meant by "eugenics,"[21] exactly? In the broadest of terms, "eugenics is the attempt to improve the human gene pool."[22] Integral to this definition is intention on the part of people. You'll note that there is a lot riding on the word *attempt*, and, as we shall see, there is more ethical debate than you might have guessed over the dividing line between which kinds of practices are morally acceptable and which are unacceptable. Philosophers sometimes distinguish between "positive eugenics" (efforts to increase preferred heritable attributes) and "negative eugenics" (efforts to decrease unpreferred heritable attributes). More important for our discussion is distinguishing between voluntary (freedom of parental choice) and involuntary (coerced, forced, or imposed) genetic decision-making. This distinction is what distinguishes "liberal eugenics" from "authoritarian eugenics."

As disquieting as the word *eugenics* is today, "liberal" or otherwise, I want to talk through how individual choice in decisions about one's offspring is qualitatively distinct from government-imposed authoritarian eugenics. Let's explore the individualist angle with some further nuance.

Like it or not, each of our own implicit biases already create a soft form of liberal eugenics as we go about choosing partners to procreate with. Those features we find attractive that have a genetic basis will bias the likelihood that our offspring will share those features. We engage in assortative mating—discrimination in the form of like choosing like—for heritable characteristics ranging from height to weight to personality, and most strongly so for intelligence.[23] Variability in all of these features is heritable to different degrees. As much as we abhor racism, even racists are permitted by society to discriminate among prospective mates based on people's outward and genetically heritable features. As Walter Veit and colleagues put it, "all human societies engage in a variety of practices that are both widely accepted and plainly eugenic."[24]

Our implicit biases come closer to the surface for people who make use of a sperm bank or egg donor bank as part of IVF. You choose which features are important to you in contributing DNA to your offspring, whether by convincing a relative to donate their sperm or in selecting a donor based on the comprehensive profile in sperm bank databases. Both options imply a preference of some heritable features over others. If you were to peruse such a prospective sperm donor's portfolio, you would learn about his genetic disease predispositions, as well as physical (e.g., eye color, complexion, handedness, teeth condition, and vision) and social characteristics (e.g., ethnic ancestry, educational attainment, preferred academic subjects, personality traits, and life goals). So strong is our impetus to select based on particular traits that sperm banks may even be sued for failing to disclose donor information.[25]

Discrimination, unconscious or overt, in the production of our offspring by preimplantation genomic screening is therefore not a difference in kind from what we already do. It is, however, a way of discerning that gets advised more directly by genetic information. Our selection of partners by eye provides only an inkling of their invisible genetic status.

DNA sequence features, however, are not the infallible blueprints of determinism that you might think. First, genetic predispositions are analogous to a weather forecast: a 30% chance of rain means that there's still a good chance it will end up being sunny. The environment of our childhood and adult lifestyle shapes our genetic predispositions to an astonishing degree. This is why your doctor will first recommend adjusting your lifestyle toward exercise and nutritious meals when you present them with an elevated risk of heart disease or diabetes. It's also why we provide public education to everyone to enhance their mental acuity. The social, economic, and geographic circumstances of families are extremely heritable: the wealth of parents predicts the wealth of children, the geographic location of parents predicts the locations of children. This inheritance, however, arises nongenetically, in a way separate from genetic inheritance of some kind of "gene" that makes you rich or a "gene" for having a particular hometown. Educational attainment is another example trait. You can directly see the profound effects of home environment on educational attainment by comparing adopted and non-adopted children. It turns out that home environment explains half of the predictability in education between parents and children that gets found in otherwise-genetic studies.[26] One's environmental circumstances matter an awful lot for what features we develop.

There is a second reason that genetic determinism isn't the end-all and be-all to define what traits we express. It's a bit technical but boils down to the problem of "garbage in, garbage out." It's also an important caveat to all the news that talks of genetic associations for diseases, personality traits, intelligence, and other measurable aspects of our bodies.

Researchers face a hard problem these days in human population genetic studies. They aim to decode the differences in traits that people have from the complex genetic contributions of variants at thousands of genes and their interactions with environmental factors and historical contingencies. Big datasets with millions of people and millions of genetic differences are a real boon. But they also are susceptible to the problem of biases in the composition of people and genotypes and geographic regions that got measured in the first place. Researchers take great pains in using sophisticated statistical methods to correlate variability in traits, like height or risk of schizophrenia, with variability in genomes. One way to summarize all the information in an integrative way is with a so-called "polygenic score." A polygenic score is a metric that uses the abundances of millions of different genetic variants and how well they associate with the incidence of a given feature, even if no gene individually is predictive in a significant way. For instance, the sum total of genetic variants for someone in a particular population might be able to estimate a 10% elevated risk of heart disease, even though any given gene alone might only contribute a negligible 0.000001% elevated risk. A polygenic score, at its essence, is equivalent to a genomic selection index, except not created with the goal of artificial selection in mind.

The reliability of a polygenic score in predicting the feature that will develop in an individual, however, is only as good as the data and analysis that went into it. These days, there remains much uncertainty in polygenic scores that summarize trait predispositions. In particular, the panels of individuals used in today's largest studies correspond to a non-representative sampling of humans on Earth. As a result, the DNA markers that purport to tell you about your genetic predispositions, for many traits, will be poor or even spurious predictors for many people.[27] Even if you wanted to use a pre-implantation genomic screen to select among IVF embryos, that "best available information" in the form of a polygenic score, for many traits, will not be very good or even downright wrong.

16.3   How much meddling is too much meddling?

As interventionist as preimplantation genomic screening might be, it simply permits the sifting among genotypes that chance events in meiosis and fertilization provide. It does not proactively alter DNA. Heritable human genetic editing *does* proactively alter DNA. The prospect of actually altering DNA in embryos through HHGE represents a major escalation in how much we already meddle with the genetic potential of our offspring.

The forces of evolution have produced "vast natural inequality" among humans as a byproduct of genetic disorders and chronic diseases that emerge with a heritable component of risk.[28] Altering an embryo's DNA to reduce or eliminate the likelihood of a heritable disease makes for the most obviously laudable goal of HHGE. It can reduce suffering and improve the welfare of recipients. Genetic diseases that manifest irrespective of lifestyle, after all, get thrust upon the bearer through no fault of their own.

He Jiankui, the infamous source of Lulu and Nana, declared in an emotional appeal in his defense of performing human genome editing that "everyone deserves freedom from genetic disease."[29]

The editing-away of inherited diseases is easiest to think about for those disorders caused by a single genetic difference with a large effect, termed *Mendelian diseases.* There are more than 4100 genes conferring Mendelian diseases. Examples of such Mendelian diseases include cystic fibrosis, Duchenne muscular dystrophy, Rett syndrome, and dystrophic epidermolysis bullosa.[30] In these cases, there is a nearly 100% association between the risk of disease and just one or two genetic factors. Children born of HHGE to remedy the genetic lesion, in such cases, would undoubtedly experience a higher quality of life than if they had not had their DNA edited as embryos. "Letting nature decide, when we can intervene," so argue Julian Savulescu and Guy Kahane,[31] "is foolish."

Interestingly, HHGE in these cases also would confer benefits on others. Given the cost to families and health care systems of the long-term burdens of experiencing and treating those conditions—not to mention the psychological and emotional tolls—removing those opportunity costs could contribute to improved equity of health resources.[32] "In a world of limited resources, taking a more expensive therapy has the opportunity cost of preventing the treatment of someone else's disease," writes one team of bioethicists, "Justice requires we choose the most cost-effective option, other things being equal. If we do not invest in the most cost-effective option, we harm others who could use these resources."[33] Such moral obligations in favor of HHGE, moral philosopher Russell Powell further argued, also are a matter of intergenerational justice to reduce suffering of future generations.[34]

What of future generations? "[H]umans uniquely modify the environment in ways that minimize the consequences of acquired genetic afflictions," writes Michael Lynch, which comes with "an expected genetic deterioration in the baseline human condition."[35] He goes on to say that "the price will have to be covered by further investment in various forms of medical intervention."

Perhaps counterintuitively, HHGE could help to demedicalize human lives. It could make us less dependent on the chronic use of pharmaceutical drugs, corrective surgeries, and physical aids like spectacles, hearing aids, and pacemakers. These medical dependencies are susceptible to disruption by natural disasters, wars, corporate bankruptcies, travel, power outages, and other supply chain interruptions. HHGE represents a one-time cost, albeit a potentially big one-time cost. Such a single preventative "genetic treatment" contrasts with the ongoing lifetime of costs associated with many diseases.

★

Provided that CRISPR-Cas9 genome editing for HHGE is technically safe, reliable, and relatively inexpensive to make its access an equitable choice, potential parents choosing HHGE for their offspring would fulfill the four pillars of bioethical values. They would fulfill the tenets of nonmaleficence to avoid harm, beneficence to reduce suffering, respect for autonomy in decision-making, and equity of justice across society and later generations. In principle, ethical arguments in support of HHGE that we've covered do not discriminate based on the likelihood that a genetic disease predisposition will manifest, whether a baby will be born with 100% certainty of displaying the disorder or acquiring the disease, versus a 50% chance or 10% or 1%. These ethical arguments also do not depend on whether the baby is born with the condition, or if the disease will only arise later in life. Genetic ethicist Henry Greely states, when posing the question of whether HHGE is inherently bad, "No. At least, I don't think so."[36] I'm inclined to agree, though I find it very hard to draw a line between the kinds of genome edits that might be permissible and those that aren't.

We may feel that one could stand on moral high ground for using CRISPR-Cas9 HHGE to alter a genetic variant of, say, the HEXA (beta-hexosaminidase A) gene that confers Tay Sachs disease on infants into a variant that allows the body to break down fats properly and so prevent near certain death by the age of five years. After all, this severe disease has no known treatment. Perhaps we'd even be comfortable with editing embryos that contain broken versions of the BRCA1 gene that give a person a 72% chance of developing breast cancer later in life.[37]

But what about a more subtle gene variant associated with cardiac arrhythmia and, say, a 10% higher incidence of heart attack? Or lactose intolerance, responsible for the cramps, flatulence, and diarrhea that results from the dialing down of expression of the LCT lactase enzyme after early childhood? Or those variants of the CCR5 gene that make a person susceptible to HIV infection should they ever encounter the HIV virus?[38] Or inherited color blindness? Whatever threshold we draw into the gray zone, it will be difficult to defend the boundary of what might seem like a laudable divide between those genetic edits that are morally justifiable and those that are not. How would you draw the line between treatment and enhancement, between medical necessity and personal preference?

★

It is hard to impugn the idea of reducing the possibility that someone would suffer from a debilitating heritable disease. Disease matters in that it influences our wellbeing.[39] We can frame this same idea, however, in a way that might make us more squeamish. Genome edits to prevent disease act to increase the capacity of an individual. How much increased capacity—performance enhancement—are we comfortable with?

Some inherited diseases impinge on the cognitive capacity of the affected person. Consequently, HHGE to counteract the disorder would increase the intelligence of the person who develops from such an embryo relative to what they would have had otherwise. Does this then permit HHGE to increase *any* embryo's predisposition to higher intelligence? After all, as the Nuffield Council on Bioethics reported in 2018, "If it is accepted that some capacity-increasing interventions are morally permissible, the difficulty then arises of distinguishing those from others that are impermissible."[40]

One criterion to use for deciding whether a particular kind of genome editing is permissible is the "normal range human enhancement" criterion. This is the idea that HHGE could be used only to alter a capability or predisposition—whether to disease or any other feature—to a level that lies within the bounds imposed by what already exists as genetic variability within humans. A practical variation on this theme is to use only the "most prevalent" genetic variants as replacement templates in HHGE, which ethicist John Evans concluded is one of the only "structurally sound" ethical barriers to a slippery slope of HHGE.[41]

Implicit in these "normal range" perspectives, however, is that they pronounce that what is normal defines what is acceptable. Disability rights activists point out its intrinsic discrimination and inequity. For example, some people who experience heritable deafness consider themselves not to comprise a group with a shared disability, but to represent a "Deaf-World" ethnic group, with a shared culture and language. One argument holds that attempts to reduce the number of deaf births, as could occur by replacing via HHGE any gene variants that confer hereditary deafness with alternate versions that permit hearing, would place "the Deaf-World in jeopardy of ethnocide and even genocide."[42] Regardless of your opinion of this particular case, it is nonetheless true that medical applications of HHGE more generally would typically act to reduce genetic variability and to reduce human diversity. Such so-called normativity is, perhaps, a recipe for intolerance and monoculture, a recipe for the dystopian world of Kurt Vonnegut's *Harrison Bergeron*.

From an evolutionary perspective, we know the value of retaining genetic variation to provide the raw material for adaptive evolution. Genetic variability is the source material to counteract pathogens as, for example, the so-called Δ32 genetic variant of the CC chemokine receptor type 5 gene (CCR5) is able to confer resistance to infection by HIV, the virus that causes AIDS.[43] How many gene variants that one might rationally choose to eliminate by editing them into a standard variant might provide resistance to newly emerging infectious diseases?

★

From our earlier discussion of the speed bumps to adaptive evolution, we also know about pleiotropic effects of genetic changes: a single mutation may alter multiple characteristics, and only some aspects of those alterations may be beneficial or desirable. A danger to genome editing of genetic variants with benefits in one way is that they may confer detriments, known or unknown, in other ways. For example, increased height confers economic and social advantages to tall individuals, but the tall also suffer greater risk of ailments like cancer and heart disease.[44] This hazard of unintended consequences in genetically correlated features may prove especially challenging to manage in cases of highly polygenic traits. Genetic predispositions to high intelligence, for example, appears to involve thousands of gene variants that overlap with predispositions to severe autism.[45] Embryonic selection or genome editing that increases the incidence of such variants would be expected to increase the likelihood that both high intelligence and severe autism manifest in the person who develops from the resulting embryo.

Part of the reason that it is so hard to draw a clean line in the sand is that the distinction is blurry between genome edits that would remove a welfare-*diminishing* feature and those that would add a welfare-*improving* feature. They are two sides to the same coin. They both enhance the potential performance of the individual that develops from an HHGE embryo compared to how that same embryo would develop without having undergone HHGE. One way to think about this conundrum is with the thought experiment of a Double Reversal Test.

Imagine a case in which a woman was exposed to a chemical that would depress the cognitive ability of her offspring. With no known treatment other than HHGE, doctors perform HHGE to alter the DNA of her embryos to confer on her children normal levels of intelligence to then live out lives of their own making. Later, it's discovered that the woman's grandchildren are endowed with cognitive enhancement by virtue of the combination of both an inherited edited genome and lack of parental chemical exposure. Should such individuals be given drugs or have their genomes edited to *reduce* intelligence back to the normal range? "Surely, it would be absurd to do so," argue ethicists Nick Bostrom and Toby Ord in accepting the consequences of this Double Reversal Test.[46]

If we were to have, instead, rejected the consequences of the Double Reversal Test in this thought experiment, it would imply that somehow the average level of intelligence across humanity is the optimal level: that both lower and higher cognitive abilities are suboptimal. When we have a gut reaction of reluctance to editing genomes for the purposes of performance enhancement, it is our worry about unfairness talking. We are attuned to think about differences in relative and, therefore, competitive terms. We spurn the possibility that others might gain a disproportionate advantage. We tend to think about it solely in terms of a zero-sum game of winners and losers.

Performance enhancements, however, hold the potential to produce absolute benefits to society writ large. After all, we employ public sanitation to reduce disease burdens and public education to raise the problem-solving ability and intellectual capacity of everyone. In addition to the plausible societal benefit to having citizens with a predisposition to high intelligence, there are plausible societal benefits to other features that one might target for genome editing. For example, both individual and societal benefits would accrue as a result of decreasing the incidence of inherited diseases, increasing the incidence of resistance to transmissible diseases, or acquiring the ability to biosynthesize vitamins and so prevent nutritional deprivation that currently requires vitamin consumption.

If that all sounds too good to be true, there may indeed be a serious downside. The alignment of interests by individuals and society might predispose states to mandating their uptake—a decidedly undesirable transition from the liberal eugenics of individual choice to the detestable circumstances of coerced or imposed authoritarian eugenics. There is another way that such authoritarian eugenics might slide onto the scene, under the guise of equity promotion. To prevent inequitable access to HHGE, public health care policies might be obliged to cover the procedure for everyone. That is, government agencies would be tasked with determining which HHGE procedures ought to be promoted for reasons of equity of access as well as public health or social welfare. There is genuine debate about which traits, exactly, promote human flourishing. It's then a short leap to coercive or imposed authoritarian eugenic applications.

Private health care systems also might not be immune to coercive eugenic applications of HHGE. In particular, the marketplace of medical procedures is keenly sensitive to capitalist incentives and the ad-agency marketing that goes along with it. Effective advertising and marketing tap into our psychological and emotional vulnerabilities as a socially acceptable kind of coercion. Promotional advertising of HHGE, especially when coupled with deference to medical authority figures, could easily generate social coercion that subverts rational individual choice. Different prospective parents may be permitted to make different decisions about what kinds of genetic predispositions they care about, though some folks still hold the legitimate worry that social norms and expectations may represent an unofficial form of coercion that leaves individual choice as a superficial veneer.

Heritable germline editing raises the specter of exacerbating existing inequity. If accessible to only a subset of people, then it creates yet another social injustice, and potentially economic and racial injustices. Heritable germline editing also raises the specter of societies actualizing our worst eugenic fears. More specifically, we all repudiate authoritarian eugenics as embodied in the atrocities committed by the Nazi regime of the twentieth century, as well as the authoritarian eugenic laws like those that led to forced sterilization of marginalized people in Canada, the United States, and many other countries.[47] We fear that misguided genetic notions that undergird racist discrimination might lead to coerced genetic choices that governments impose or mandate on us, its citizens. Imposed genetic controls usurp the autonomy of individual reproductive choice. This is the voice of the legitimate fear of "slippage down to the dystopian bottom of pervasive eugenic enhancement" as depicted in books and films like *Brave New World* and *Gattaca*.[48]

*

In their call for a moratorium on HHGE, Eric Lander, Françoise Baylis, and their co-signatories pointed to a series of societal, ethical, and moral considerations.[49] They raised the concern of increased discrimination experienced by people with genetic differences or disabilities, and worried that children that result from HHGE might experience mental illness as a consequence. They pointed out that peer pressure or marketing could coerce people into using HHGE. Françoise Baylis further has highlighted that power differentials and any direct harms of HHGE procedures themselves, which depend on IVF-related treatments, would be felt disproportionately by women.[50] The proponents of a moratorium suggested that inequitable access to HHGE could exacerbate existing social inequities. They even proposed that the changes to human genomes could exert permanent and potentially harmful effects on the human gene pool or, in the long term, even lead to distinct subspecies of humans.

One important consideration in weighing legitimate concerns like these is how much they depend on an essentialist sense of genetic determinism.[51] Unfortunately, it is easy to unintentionally hold an overinflated notion that organisms are fixed, with their genetic make-up firmly predetermining their fate, and to falsely conclude that one's genetic make-up is sufficient to define social groups or moral distinctions. The state of one's genome, however, is but one piece among the confluence of factors that determine who a person is and how society perceives them. Environmental circumstances and social influences interact with genetics in complicated ways that often mitigate against genetic predispositions.

A second consideration is the uniqueness of these ethical concerns about HHGE relative to other practices, such as adoption, egg donation, and PGD. Are the concerns truly intrinsic to something distinctive about HHGE, or are they broader concerns about medical and social practices in general? Unless they are distinctive to HHGE, it's hard to invoke them to bar only HHGE.

A third consideration is one of timelines and incidence rates. Remember that a new mutation, as might arise from off-target effects of HHGE, has a probability of fixation that is proportional to the reciprocal of the population size. For humans, this probability implies a new mutation will have a chance of surviving stochastic elimination from the human gene pool on the order of 0.0001% or less, with that likelihood of stochastic loss being about 37% in just the first generation following HHGE.[52] True, the gene pool might be different, but only barely.

These ethical concerns and considerations, among others, are important to weigh thoughtfully as all of humanity now contemplates whether or not to pursue HHGE and, if so, under what conditions. In deliberating this topic, and how much to invest into its potential, it's important to keep in mind that "[p]eople's genetic essentialist biases make it easy for them to assume that genes are the ultimate solution to social problems,"[53] even though most of society's problems, medical and otherwise, may have nongenetic remedies that would prove more timely, effective, and cost-efficient. But do the sum total of these kinds of concerns add up to something that is sufficient for a blanket ban on heritable genetic editing? Many bioethicists actually say "no." Some ethicists instead argue that some applications of human germline genome editing "are not merely morally permissible but are moral imperatives."[54] Clearly, ethical unanimity isn't easy to come by, and may take a long time for us to reach societal consensus.

★

There is, also, a more extreme view toward performance enhancement through CRISPR-Cas9 HHGE. It's the view of so-called transhumanists. People who adhere to the transhumanist perspective envision people editing their genomes to gain altogether new capabilities. The DNA might be transgenes, obtained from sequences of other species, or even altogether novel genetic elements that we synthesize ourselves. Just as we imagined modifying the genome of a *Gallus* founding stock to contain more opsin genes to confer greater visual acuity across the light spectrum or more chemoreceptor genes to confer improved sense of smell, the same technique could be applied to *Homo sapiens*. These and other kinds of genomic manipulations are seen by transhumanists as helping our species to achieve its full potential through evolutionary change.

Where does genetic welding fit in? Genetic welding involves that self-propagating incarnation of genome editing in the form of a gene drive. You could imagine designing a gene drive to proactively edit out those genetic lesions that would otherwise confer an inherited disease.[55] For example, it might overwrite the broken version of BRCA1 to encode a version of the BRCA1 protein that can properly repair damaged DNA and so prevent the onset of breast cancer. Such a gene drive would spread throughout the population to preemptively alter human genomes so that we'd never need to see BRCA1-induced cancer again. Such genetic welding, in the long term,[56] would act to prevent the societal burden of treating the possibility of this horrible disease, precluding preventative mastectomies, avoiding anxiety over cancer testing, and redistributing limited healthcare resources toward other ailments.

To apply genetic welding in this way would serve the public good. But it also would undermine respect for autonomy as the gene drive spreads a genetic feature through the genomes of all humanity. Gene drives in humans would perform genome edits in every generation, irrespective of the volition of the parent who is the genetic carrier, or of their partner. It would perform genome edits in every reproductive episode that a carrier undertakes, not just with the episode of consent that could be given with PGD or nondrive HHGE. The genome editing would occur irrespective of the wishes or laws of governments, as well, and human procreation is not constrained by political boundaries. BRCA1 gene drives might have our sympathy, but what about gray-zone alterations? Once started, to stop gene drive spread in humans would require either release of another gene drive, a reversal gene drive,[57] or sweeping government-mediated genomic surveillance that would determine whether individuals were permitted to reproduce: authoritarian eugenic practice.

Consequently, genetic welding in humans with gene drives, even in the seemingly virtuous way of preemptively "treating" genetic disorders, is fundamentally subversive. Gene drives in humans would produce a new form of eugenics, surreptitiously promoting certain genotypes across the species, generation after generation. HHGE with gene drives would be distinct from both liberal and authoritarian strains of eugenic selection, but close in spirit to genetic coercion. Genetic welding in humans would be a sneaky, hidden—if long-term—way of shaping the genetic make-up of our entire species. Genetic welding in humans would represent an insidious new strand of eugenics that we might term "guerrilla eugenics."[58]

From "gene surgery" prevention of disease to quality-of-life genetic enrichment to transhuman performance enhancement to guerrilla eugenics, one can see why John Evans talks about HHGE as a slippery slope. This is why we need education to improve the literacy of our populace in both scientific topics and in ethics. This is why we need open discourse about both science and ethics. In this case, it is essential to improve literacy about CRISPR-Cas9 genome editing, as well as literacy surrounding the core bioethical tenets of beneficence, nonmaleficence, respect for autonomy, and equity of justice.[59] We already have wide gulfs in public opinion about HHGE that depend, in part, on gulfs in familiarity with scientific principles.[60] The informed views of the general public are crucial to ensuring that consensus decisions about genetic sovereignty are made in an ethically deliberate, scientifically informed, and socially acceptable way.

HHGE with CRISPR-Cas9, after all, isn't a blue-sky notion. There are humans living in the world today that are the result of this technique, in the physical form of Lulu and Nana. The technique itself is neither good nor bad, and the facts of nature, as we have learned, are no guide to deciding what is good. The decisions of if, when, and how we ourselves become dragons is up to us.

*

Humans are adept at engineering our own daily lives out of medical necessity and out of personal or cultural preference. The line is ethically indistinct between disability compensation and functional enhancement. Humanity now holds the demonstrated power to perform heritable human genome editing. The most serious ethical concerns intrinsic to HHGE center on the risk of negative pleiotropic effects of genetic changes and on the ease of misappropriation for coercive eugenic applications. Application of genetic welding to humans through HHGE with gene drives, even with noble intent, would usher in an era of guerrilla eugenics.

Notes

1. Most recently, using the heart from a genetically-modified pig. https://www.cnn.com/2022/01/10/health/genetically-modified-pig-heart-transplant. Accessed January 11, 2022.

2. Some gene variants of the BRCA1 and BRCA2 genes lead to very high risk of developing cancers of the breasts and ovaries at some point in life.

3. Recall from Chapter 11 our discussion of baculum bones in the penises of many species of mammal. You might know sildenafil by its first trade name of Viagra as a drug that promotes male erection. Recent clinical trials with a different drug, dimethandrolone undecanoate, encourage further development of it as a male contraceptive. Thirumalai, A., et al. 2019. Effects of 28 days of oral dimethandrolone undecanoate in healthy men: a prototype male pill. *Journal of Clinical Endocrinology & Metabolism*. 104: 423–432. https://doi.org/10.1210/jc.2018-01452; Thirumalai, A., and J.K. Amory. 2021. Emerging approaches to male contraception. *Fertility and Sterility*. 115: 1369–1376. https://doi.org/10.1016/j.fertnstert.2021.03.047.

4. Mehlman, M.J. 2004. Cognition-enhancing drugs. *Milbank Quarterly*. 82: 483–506. https://doi.org/10.1111/j.0887-378X.2004.00319.x.

5. Verma, I.C., and R.D. Puri. 2015. Global burden of genetic disease and the role of genetic screening. *Seminars in Fetal and Neonatal Medicine*. 20: 354–363. https://doi.org/10.1016/j.siny.2015.07.002.

6. https://www.science.org/content/article/untold-story-circle-trust-behind-world-s-first-gene-edited-babies and https://www.nature.com/articles/d41586-020-00001-y and https://www.youtube.com/channel/UCn_Elifynj3LrubPKHXecwQ. Accessed October 28, 2021.

7. https://www.labiotech.eu/best-biotech/crispr-technology-cure-disease. Accessed October 28, 2021.

8. I acknowledge that this opinion is not shared universally.

9. Canada's 2004 Assisted Human Reproduction Act states that "No person shall knowingly ... alter the genome of a cell of a human being or in vitro embryo such that the alteration is capable of being transmitted to descendants," with violations punishable by 10 years in prison and a $500,000 fine. https://laws-lois.justice.gc.ca/eng/acts/a-13.4. Accessed October 28, 2021; Baylis, F., et al. 2020. Human germline and heritable genome editing: the global policy landscape. *CRISPR Journal*.3: 365–377. https://doi.org/10.1089/crispr.2020.0082.

10. Baylis, F. 2019. *Altered Inheritance*. Cambridge, MA: Harvard University Press. p. 108.

11. Preimplantation genetic diagnosis today also enables parents-to-be to select embryos on the basis of sex. Some sex selection takes place for health reasons, as for sex-linked disorders, whereas others may aim to equalize the sex composition of the children in the family, and cultural norms and pressures may lead still others to select disproportionately one sex or the other. Sex selection of embryos, however, is not unique to preimplantation genetic tests. Sex-selective abortion following fetal testing provides the path most responsible for the sex bias at birth in a number of countries around the world, albeit with those countries' sex biases generally expected to decline over the coming decades. Sex selection in people is illegal for nonmedical use in some countries, like Canada and China, but is legal in most others, like the United States. In livestock, sex selection can prevent the energenic, economic, and animal welfare drawbacks to postnatal culling in species like chickens and cows, where only the female offspring may be desired for egg laying and milk production. There's now a CRISPR-Cas9 technique to produce litters of just one sex or the other, at least in mice. Douglas, C., et al. 2021. CRISPR-Cas9 effectors facilitate generation of single-sex litters and sex-specific phenotypes. *Nature Communications*. 12: 6926. https://doi.org/10.1038/s41467-021-27227-2; Chao, F., et al. 2019. Systematic assessment of the sex ratio at birth for all countries and estimation of national imbalances and regional reference levels. *Proceedings of the National Academy of Sciences USA*. 116: 9303–9311. https://dx.doi.org/10.1073/pnas.1812593116.

12. Grossman, T.B., and S.T. Chasen. 2020. Abortion for fetal genetic abnormalities: type of abnormality and gestational age at diagnosis. *AJP Reports*. 10: e87–e92. https://doi.org/10.1055/s-0040-1705173.

13. Murphy, N.M., et al. 2020. Genome sequencing of human in vitro fertilisation embryos for pathogenic variation screening. *Scientific Reports*. 10: 3795. https://doi.org/10.1038/s41598-020-60704-0.

14. Extra embryos can get cryopreserved for subsequent cycles of IVF, donated to other intending parents, donated for research purposes, or destroyed.

15. Baruch, S., et al. 2008. Genetic testing of embryos: practices and perspectives of US in vitro fertilization clinics. *Fertility and Sterility*. 89: 1053–1058. https://doi.org/10.1016/j.fertnstert.2007.05.048; https://www.nytimes.com/2006/12/05/health/05essa.html. Accessed October 28, 2021.

16. https://www.23andme.com/en-ca/dna-reports-list. Accessed October 30, 2021.

17. Savulescu, J., and G. Kahane. 2017. Understanding procreative beneficence. In L. Francis (ed.), *The Oxford Handbook of Reproductive Ethics*. Oxford: Oxford University Press, pp. 592–622. https://dx.doi.org/10.1093/oxfordhb/9780199981878.013.26.

18. De Melo-Martín, I. 2004. On our obligation to select the best children: a reply to Savulescu. *Bioethics*. 18: 72–83. https://doi.org/10.1111/j.1467-8519.2004.00379.x.

19. In lay usage, eugenics sometimes is loaded with a moral judgment of condemnation because of its association with some of humanity's most horrific abuses, including by the Nazi regime. As we'll consider further shortly, I will use "eugenic" in a morally neutral way so that we may discuss the ethics of particular practices that relate to human genetic change. Despite the high emotion and politicization that surrounds the term *eugenics*, I feel that it is a more

honest approach to use the term and talk about its nuance than to sweep the term under the carpet.

20. Yudell, M., et al. 2016. Taking race out of human genetics. *Science*. 351: 564–565. https://doi.org/10.1126/science.aac4951.

21. Historically, it was Francis Galton, the cousin of Charles Darwin, who coined the term *eugenics* in 1883 and first promoted the idea that society should actively apply the logic of artificial selection to humans ourselves with the goal of improving humanity.

22. Wilkinson, S., and E. Garrard. 2013. *Eugenics and the Ethics of Selective Reproduction*. Keele, UK: Keele University.

23. Assortative mating in humans, with up to 11-fold increased odds of partners sharing a trait relative to random pairings, also occurs with respect to psychiatric conditions like attention-deficit/hyperactivity disorder, autism, schizophrenia, obsessive-compulsive disorder, and substance abuse. Plomin, R., and I. Deary. 2015. Genetics and intelligence differences: five special findings. *Molecular Psychiatry*. 20: 98–108. https://doi.org/10.1038/mp.2014.105; Nordsletten, A.E., et al. 2016. Patterns of nonrandom mating within and across 11 major psychiatric disorders. *JAMA Psychiatry*. 73: 354–361. https://dx.doi.org/10.1001/jamapsychiatry.2015.3192.

24. Some of the more familiar practices include genetic counseling of carriers of heritable disorders, elective abortion following the results of a fetal genetic test, consideration of heritable traits of sperm or egg donors, and incest taboos. Veit, W., et al. 2021. Can "eugenics" be defended? *Monash Bioethics Review*. 39: 60–67. https://doi.org/10.1007/s40592-021-00129-1.

25. For just one sperm bank's example: https://www.repromed.ca/sperm_donor_catalogue; https://www.ctvnews.ca/canada/families-sue-ontario-sperm-bank-allege-they-were-misled-on-donor-background-1.5123116. Accessed October 29, 2021.

26. Cheesman, R., et al. 2020. Comparison of adopted and nonadopted individuals reveals gene-environment interplay for education in the UK Biobank. *Psychological Science*. 31: 582–591. https://doi.org/10.1177/0956797620904450.

27. Duncan, L., et al. 2019. Analysis of polygenic risk score usage and performance in diverse human populations. *Nature Communications*. 10: 3328. https://doi.org/10.1038/s41467-019-11112-0.

28. Gyngell, C., et al. 2019. Moral reasons to edit the human genome: picking up from the Nuffield report. *Journal of Medical Ethics*. 45: 514–523. http://dx.doi.org/10.1136/medethics-2018-105084.

29. https://www.youtube.com/channel/UCn_Elifynj3LrubPKHXecwQ. Accessed October 28, 2021.

30. https://www.omim.org/statistics/geneMap. Accessed November 1, 2021.

31. Savulescu, J., and G. Kahane. 2017. Understanding procreative beneficence. In L. Francis (ed.), *The Oxford Handbook of Reproductive Ethics*. Oxford: Oxford University Press, pp. 592–622. https://dx.doi.org/10.1093/oxfordhb/9780199981878.013.26.

32. As has been pointed out for the case of PGD and PGS to use in embryo selection in IVF, "The ongoing emotional and psychological burden born by the parents and the monetary cost of support from a healthcare system for caring for an affected individual is vastly greater than the cost of a genome sequencing test." For many inherited diseases, the same logic would apply to HHGE. Murphy, N.M., et al. 2020. Genome sequencing of human in vitro fertilisation embryos for pathogenic variation screening. *Scientific Reports*. 10: 3795. https://doi.org/10.1038/s41598-020-60704-0.

33. Gyngell, C., et al. 2019. Moral reasons to edit the human genome: picking up from the Nuffield report. *Journal of Medical Ethics*. 45: 514–523. http://dx.doi.org/10.1136/medethics-2018-105084.

34. Powell, R. 2015. In genes we trust: germline engineering, eugenics, and the future of the human genome. *Journal of Medicine and Philosophy*. 40: 669–695. https://doi.org/10.1093/jmp/jhv025.

35. Lynch further points out that "the one truly exceptional human attribute, brain function, may be particularly responsive to mutation accumulation . . . . A fitness decline of a few percent on the timescale of a century is on the order of the rate of global warming, and that is part of the problem." Lynch, M. Mutation and human exceptionalism: our future genetic load. *Genetics*. 202: 869–875. https://doi.org/10.1534/genetics.115.180471.

36. Greely, H. 2021. *CRISPR People*. Cambridge, MA: MIT Press. p. 203.

37. Kuchenbaecker, K.B., et al. 2017. Risks of breast, ovarian, and contralateral breast cancer for BRCA1 and BRCA2 mutation carriers. *JAMA*. 317:2402–2416. https://dx.doi.org/10.1001/jama.2017.7112.

38. Interestingly, the Δ32 variant of the protein CCR5 actually impairs a piece of the immune system and appears to more than triple the likelihood of a fatal outcome of influenza but does so in such a way as to evade the typical route of infection by HIV that targets CCR5. Falcon, A., et al. 2015. CCR5 deficiency predisposes to fatal outcome in influenza virus infection. *Journal of General Virology*. 96: 2074–2078. https://doi.org/10.1099/vir.0.000165.

39. The self-reported quality of life by people who experience a condition can differ markedly, however, from what outsiders assume. Barata, A., et al. 2017. Do patients and physicians agree when they assess quality of life? *Biology of Blood and Marrow Transplantation*. 23, 1005–1010. https://doi.org/10.1016/j.bbmt.2017.03.015.

40. Quotation from p. 71 in Nuffield Council on Bioethics. 2018. *Genome Editing and Human Reproduction: Social and Ethical Issues*. London: Nuffield Council on Bioethics. https://www.nuffieldbioethics.org/publications/genome-editing-and-human-reproduction.

41. Evans, J.H. 2021. Setting ethical limits on human gene editing after the fall of the somatic/germline barrier. *Proceedings of the National Academy of Sciences USA*. 118: e2004837117. https://dx.doi.org/10.1073/pnas.2004837117.

42. Lane, H. 2005. Ethnicity, ethics, and the Deaf-World. *Journal of Deaf Studies and Deaf Education*. 10: 291–310. https://doi.org/10.1093/deafed/eni030.

43. The Δ32 variant, however, disables the CCR5 protein. Given the protective role of a functional form of CCR5 in other aspects of immune function, taking a long-term view, it would

be unwise for everyone's genome to encode the Δ32 variant. Lederman, M.M., et al. 2006. Biology of CCR5 and its role in HIV infection and treatment. *JAMA*. 296:815–826. https://dx.doi.org/10.1001/jama.296.7.815.

44. Nunney, L. 2018. Size matters: height, cell number and a person's risk of cancer. *Proceedings of the Royal Society B*. 285: 20181743. https://doi.org/10.1098/rspb.2018.1743; Raghavan, S., et al. 2022. A multi-population phenome-wide association study of genetically-predicted height in the Million Veteran Program. *PLoS Genetics*. 18: e1010193. https://doi.org/10.1371/journal.pgen.1010193.

45. Crespi, B.J. 2016. Autism as a disorder of high intelligence. *Frontiers in Neuroscience*. 10: 300. https://doi.org/10.3389/fnins.2016.00300.

46. This is another way of testing the status quo bias. Bostrom, N., and T. Ord. 2006. The reversal test: eliminating status quo bias in applied ethics. *Ethics*. 116: 656–679. https://dx.doi.org/10.1086/505233.

47. Patel, P. 2017. Forced sterilization of women as discrimination. *Public Health Reviews*. 38: 15. https://doi.org/10.1186/s40985-017-0060-9.

48. Evans, J.H. 2021. Setting ethical limits on human gene editing after the fall of the somatic/germline barrier. *Proceedings of the National Academy of Sciences USA*. 118: e2004837117. https://dx.doi.org/10.1073/pnas.2004837117.

49. Lander, E., et al. 2019. Adopt a moratorium on heritable genome editing. *Nature*. 567: 165–168. https://www.nature.com/articles/d41586-019-00726-5.

50. For an extended consideration of diverse ethical issues related to heritable human genome editing, I point you to the 2019 book *Altered Inheritance* by esteemed bioethicist Françoise Baylis.

51. Dar-Nimrod, I., and S.J. Heine. 2011. Genetic essentialism: on the deceptive determinism of DNA. *Psychological Bulletin*. 137: 800–818. https://doi.org/10.1037/a0021860.

52. Recall our discussion of the molecular clock in Chapter 3. This calculation of 0.0001% presumes a new genetic variant that is selectively neutral, given our population size of roughly 8 billion and a present-day genetically effective population size of well over 1 million. We would expect natural selection to purge harmful mutations within tens of generations, though, clearly, they would represent a burden in the interim. Keinan, A., and A.G. Clark. 2012. Recent explosive human population growth has resulted in an excess of rare genetic variants. *Science*. 336: 740–743. https://doi.org/10.1126/science.1217283.

53. Dar-Nimrod, I., and S.J. Heine. 2011. Genetic essentialism: on the deceptive determinism of DNA. *Psychological Bulletin*. 137: 800–818. https://doi.org/10.1037/a0021860.

54. Gyngell, C., et al. 2019. Moral reasons to edit the human genome: picking up from the Nuffield report. *Journal of Medical Ethics*. 45: 514–523. http://dx.doi.org/10.1136/medethics-2018-105084.

55. Nestor, M.W., and R.L. Wilson. 2020. Beyond Mendelian genetics: anticipatory biomedical ethics and policy implications for the use of CRISPR together with gene drive in humans. *Bioethical Inquiry*. 17: 133–144. https://doi.org/10.1007/s11673-019-09957-7.

56. Remember from Chapter 7 that gene drive genetic variants increase in abundance on a timescale measured in generations, which for humans is now approximately 30 years. For a gene drive to break past 50% of people, from having been inserted into 10 people initially, would take 900 years or so. Even if the gene drive were inserted into 1000 people right now, it would still take nearly 700 years to hit 50% incidence in the global population.

57. Esvelt, K.M., et al. 2014. Emerging technology: concerning RNA-guided gene drives for the alteration of wild populations. *eLife*. 3: e03401. https://dx.doi.org/10.7554/eLife.03401; Heffel, M.G., and G.C. Finnigan. 2019. Mathematical modeling of self-contained CRISPR gene drive reversal systems. *Scientific Reports*. 9: 20050. https://doi.org/10.1038/s41598-019-54805-8.

58. Cutter, A.D. 2022. Guerrilla eugenics: gene drives in heritable human genome editing. *SocArXiv*. https://doi.org/10.31235/osf.io/wqjsd.

59. Veatch, R.M. 2020. Reconciling lists of principles in bioethics. *The Journal of Medicine and Philosophy*. 45: 540–559. https://doi.org/10.1093/jmp/jhaa017.

60. https://www.pewresearch.org/science/2018/07/26/public-views-of-gene-editing-for-babies-depend-on-how-it-would-be-used. Accessed October 29, 2021.

17

A midnight coterie of transcendent fauna

Genetic welding and the forces of evolution are at our fingertips to direct evolution and to witness heritable change in the living world. We've done just that throughout the Anthropocene, albeit without the technological innovation afforded by genetic engineering. Evolutionary change is not just a natural force that operates in the hinterlands of the wild. For millennia, we humans have shaped the evolution of animals and plants and microbes and viruses according to our own whimsy. That whimsy has given us fancy pigeons for amusement and fancy dogs for companionship, fancy grasses for cob corn and fancy fungal yeasts for wine, fancy microbes for sanitation and fancy viruses for vaccines.

That whimsy—launched with the aid of genetic welding—could give us new ecosystem engineers, species capable of reconstituting some of the gaps that resulted from our ancestors having caused past extinctions. Should society endorse the rationale, human whimsy could create entire communities, even, of new and hyperexotic species, to interact in a landscape of entirely novel ecosystems, each species with an origin that emanated from evolutionary engineering. "[O]nce the taboo of releasing genetically engineered species into the wild is thoroughly broken, albeit for the best of conservation reasons," writes Leslie Paul Thiele in describing such a world as Nature 4.0, "there may be no stopping, or even slowing down, the undiscriminating proliferation of synthetic life-forms."[1] We could walk through that new incarnation of nature's landscape at noon or at midnight to witness those tangible facts of evolution in the flesh.

★

"Earth is now nowhere pristine," it has been said, on good authority.[2] Analyses of climate change point to the world already being past the point of no return. The possibility of reined-in greenhouse gas emissions no longer provides a sufficient solution to catastrophic climatic disruption in the coming decades.[3] The impact of 2 °C global warming on ecosystems and biodiversity is expected to be abrupt and severe. As a consequence, serious research is exploring extreme remedies, like so-called solar geoengineering. Solar geoengineering is as eyebrow-raising as it sounds: the deliberate injection of aerosolized

Evolving Tomorrow. Asher D. Cutter, Oxford University Press. © Asher D. Cutter (2023). DOI: 10.1093/oso/9780198874522.003.0017

sulfur compounds throughout the atmosphere around the globe, belched into the sky from sortie after sortie of high-altitude rockets, jets, dirigibles, and balloons.[4] These stratospheric aerosols would reflect the Sun's light back into space to offset, temporarily, the warming influence of greenhouse gasses until large-scale emissions cuts and carbon sequestration could be implemented (Figure 17.1). In saltier words, it would, at this late date, buy us time to get our shit together. At a direct cost of US$10 billion per year, it is considered cheap, given the scope of the problem.[5] If anyone still thought that proclaiming this age as the Anthropocene were premature, then solar geoengineering should prove a convincing demarcation point while watching the especially gorgeous sulfur-infused sunsets.[6]

Solar geoengineering is as desperate a climate countermeasure—and as intense of a human intervention into nature—as one might imagine.[7] If not accompanied by strong reductions in carbon dioxide emissions, as well as active extraction of atmospheric CO_2, however, then life as we know it will simply become addicted to solar geoengineering. In yet another of the Anthropocene's great ironies, given the climate problem created by dumping chemicals into the sky, we'd come to depend on the continual infusion of chemicals into the sky. Premature termination of those infusions would lead to "climate

Figure 17.1 *Elevated concentrations of sulfur-based molecules in Earth's atmosphere act to reflect infrared heat energy from the Sun, leading to climate cooling. Volcanoes are a natural source of atmospheric sulfur, as from the Pavlof Volcano in Alaska that was visible from space in 2013 (the gray streak arcing through the clouds in this photo). Solar geoengineering is a human-mediated version of sulfur-injection into the atmosphere, and it has been proposed as a feasible, if ironic, means to temporarily offset the effects of climate change attributed to human-mediated carbon-injection into the atmosphere.*

velocities more than double in speed" and unconscionable threats to ecosystems and biodiversity, and to us humans who depend on them.[8] The increasing plausibility of worst-case scenarios has prompted calls from the pages of the *Proceedings of the National Academy of Sciences* for directed scientific study of global catastrophe and human extinction in light of unlikely but possible climate change outcomes.[9] When set alongside such seemingly outrageous interventions and risks to global ecosystems as these, genetic welding of new and hyperexotic species almost seems quaint.

17.1   Lost world

This morning, my news feed told me that the US federal government had announced plans to formally recognize 23 additional species as extinct, a demotion from their more hopeful prior status of critically endangered.[10] This admission single-handedly tripled the number of species designated extinct since the 1973 adoption of the Endangered Species Act. With the sincere risk of truly calamitous biodiversity loss in my lifetime and in my daughters' lifetimes, I wonder if the sentiments of society may shift closer toward the view that humanity should proactively generate biodiversity in addition to staving off further extinction. An attitude that prints a welcome sign on a placard, *Enter the Novafauna*, to usher into the world a diversity of new species evolved through the contributions of genetic welding to perform a suite of ecosystem functions. To echo Seddon and King's words yet again,[11] "If we want to sustain and enhance a biodiverse natural world we might have to be forward looking and embrace the notion of bio-novelty," they tell us somberly, "rather than backward looking and seeking to try and recreate lost worlds." The world of today, it seems, already is a lost world.

A telling sign of this being a lost world: dinosaurs today swarm the Earth, all 50 billion of them, outnumbering us sixfold. They just might not be the dinosaurs of your imagination: birds.[12] In outlining a recipe for how to "evolve a dragon," I traced the merits of chickens as a founding stock to provide the basis for selective breeding and genetic welding. I emphasized practical considerations and relatively uncomplicated visible characteristics. In generating a "novafauna," one might cast a wider net for those initial founding stocks. One might start with representatives across the tree of life, from mammals and lizards, insects and crustaceans, plants and fungi and bacteria.

To pass muster from legal, political, and ethical standpoints for release into uncontrolled environments, however, there would need to be clear justification for the role that the evolutionarily engineered organisms would play in an ecosystem. The most popular support might get behind recently extinct species that proffer crucial ecosystem engineer activity or that express capabilities suitable to counteracting adverse effects of human actions. Easier still, rather than wholesale evolutionary engineering of new species, modification gene drives might genetically weld beneficial features into existing species: Disease resistance; temperature tolerance; countermeasures to inbreeding depression; regional color morphs. This would, in effect, be the genetic domestication of wild animals and plants. They might not be cuddly, but they'd have been genetically tamed.

Fulfilling roles occupied by species that went extinct longer ago might prove a harder sell. This is something that de-extinction advocates have faced, in no small part owing to our collective syndrome of shifted baselines and bias toward the status quo. Right now, the most effective way to bolster the Earth's species richness is to stave off extinction, that much is true. If the world's extinctions mount high enough, however, then perhaps biodiversity creation will itself provide ample justification for evolutionary engineering of biological novelty. Perhaps job postings of that bleak future would advertise for "wildlife evolutionary engineers" as a warranted and sought-after occupation.

★

The creation of "curiosities" may be touted as an insufficient reason to evolve a new species, but I admit that I am indeed curious. The notion of hyperexotic creatures in the flesh is captivating, tantalizing, awesome. We all want more awesome in our lives.[13] For me, dragons got their start in the acres of woods back behind my childhood home. We lived at the end of a dirt road in a rustic town outside of Boston. There was a horse farm at the top of the road and a walking path at our end that led over to some cranberry bogs, but most of the area was covered in New England forest. The mix of hardwoods and conifers had sprung up over the Earth through the decades and centuries, a landscape dimpled and rolling from past glacial action or perhaps long-forgotten sand-mining dugouts. Some of the depressions into the forest floor were huge, rounded bowls, bigger across than a house, layered in inches of rusty red pine needles and moss. My eight-year-old self envisioned them having been formed from the weight of a dragon as it curled down to make its nest, each time circling its tail for a few revolutions the way a dog prepares its cushion for a nap. I imagined the mammoth beast having flown back weary from the east, tired from a full day of soaring over the Atlantic to fish for whales to eat. The fancies of a little boy don't disappear completely, even decades on.

To bring life to the idea of genetically engineering a new species, however, is also to bring about loss: a loss of a kind of biological innocence, a loss of that self-respect that you earn through self-restraint. It's a particular kind of melancholy. Like many people, I yearn for experiences with some conception of authentic nature. I love to camp in the backcountry with just my hiking boots and the contents of a backpack; I love traipsing through off-the-path rainforests in Australia and South America and, one day I hope, southern Asia and Africa. Surely, it is wrapped up in nostalgia from my youth, a nostalgia that set my own baseline for what is wild.

Also in my youth, the man who turned out to be something of a physical doppelganger of my older self, Patrick Stewart as Captain Jean-Luc Picard, helped me cement and internalize the principle of noninterference toward nature, what Trekkies know as the Prime Directive.[14] As an eager high school student, I feverishly gathered biological evidence of vernal pool habitat in the woods of Massachusetts to file those ephemeral springtime ponds for certification and protection under the Wetlands Protection Act.[15]

That's classic habitat preservation at the grass roots, promoting noninterference by humans in the innerworkings of ecosystems.

And yet, it is hard to square noninterference alone as enough to contend with the profound impacts that humanity has already made and will continue to make. The alternatives of proactive restoration ecology and rewilding, consequently, come across as especially compelling. I dipped my toe into small-scale restorative conservation in a college project on saltmarsh grasses along the banks of an estuary near a former construction site and, later, by investigating insect biodiversity in former cattle paddock that had been revegetated with rainforest plants native to the Atherton Tablelands of Australia.[16] As satisfying as those days were amid swamp and jungle, the world is aching for much bigger-scale measures.

De-extinction holds potential as a big-scale means to a rewilding end. For me, it is also seductive in its appeal to my techno-curiosity. I'm no stranger to genetic machinations. In my hands, though, they are contained and highly controlled, dissociated from the wilds of the outdoors and the siren song of the Anthropocene. In the cinder block confines of my university laboratory, my research team creates marvels of genetic engineering and carries out experimental evolution to watch genetic change accrue across generations. I am comfortable with directing evolution in the *laboratory*. And yet, I can't help but tremble at the prospect of infusing *nature* with a newly evolved species, genetically engineered with hyperexotic features soldered into its DNA with genetic welding. The idea of evolutionary engineering is both invigorating—the tantalizing prospect of demonstrating in the flesh one's commanding knowledge about the forces of evolution—and deflating, an anticlimax to be dumped with a middle-manager's responsibility, a slog of a job to manage nature rather than letting nature go, trusting it to run smoothly on life's intrinsic properties.

17.2   Defenses of fauna

We can genetically engineer novel or distinctive features, using modification gene drives to perform genetic welding, to induce evolutionary change in a species in the wild. We can use those same tactics to evolve new and extraordinary species. But should we? Isn't the wonder in nature—that we can seek out and soak in right now—enough? Many species, after all, have evolved impressive defenses to take care of themselves.

Cnidarians—jellyfish and their kin—are one of the beneficiaries of the Anthropocene, seeming to have expanded their numbers. They are also potion masters. Among cnidarians, the medusas, or jellyfish, enchant us the most: the seemingly innocuous drape of tendrils from a helpless gelatinous orb, or if you cock your head the other way, the dangling lethal string-curtains of a malevolent undersea hippie's blobby doorway. Each beaded tentacle strand gets coated in toxic stinging cells called cnidocytes, "the most sophisticated cellular gadgetry and weaponry found in the animal kingdom."[17] Any sense of malevolence, though, is in our own minds. After all, we bestowed it with the name of the Gorgon monster, Medusa. The jellyfish, one must admit, is just fishing for its supper in the open-water market of the ocean. The cnidocyte is its poison-tipped harpoon with a hair-trigger. Each stinging cell contains a spring-loaded tubule that pierces into a target's skin through a hole punched by mechanical force with fierce and sharp stylets, injecting a complex suite of bioactive toxins along the way.

One such cellular harpoon wouldn't be so bad; they are single-use devices. But a single jellyfish may fire off millions at once, like the Australian sea wasp *Chironex fleckeri*, a box jellyfish with 3-m-long (10-foot-long) tentacles. Depending on the species, the poisons can exert neurotoxic effects on ion channels or rupture cells, and the venomous cocktail can differ among cnidocyte cells located in different parts of the jellyfish body for optimal functionality in defense, prey capture, and digestion. The lion's mane jelly, *Cyanea capillata*, at over 30 m (100 feet) long, with 1000 tentacles, has a lot of cnidocytes to manage. Even small jellyfish can pack a punch. Stings from the four tentacles that trail beneath the 1-cm (1/2-inch) transparent bell of Irukandji jellyfish like *Carukia barnesi* cause "excruciating pain, sweating, nausea and vomiting, hypertension and a feeling of impending doom ... involuntary grunting, shivering and teeth chattering, a creepy skin feeling, and, in some cases, priapism (prolonged erection)."[18] Also, sometimes, death.

★

Not all species are so invasive in administering their complex chemistry. Unlike the active, wounding delivery of toxin by venomous creatures, the majestically colored *Dendrobates* dyeing dart frogs are simply poisonous (Figure 17.2). They secrete pumiliotoxins from their skin as a defensive weapon, advertising their toxicity with brilliant hues, a carefree dare and dire warning. I was unprepared for their nonchalance when I shared their jungle one November. Each beautifully deadly streak of blue-and-yellow was a miniature surprise as the *Dendrobates tinctorius* hopped through their happy-go-lucky lives by the dozen across the rainforest floor of inner French Guiana, the little piece of France in South America that rests just above the equator.

On the other side of the world, in the hills of Papua New Guinea, there is a black-and-orange songbird. It's an old-world oriole that looks like a new world Baltimore oriole that tried to turn itself into a raven. Its name is the hooded pitohui, and it is poisonous. The locals knew to avoid it, that it was no good for eating. But field biologist John Dumbacher found out the hard way when catching birds in the forest.[19] He got itchy eyes and, worse, a tingly and burning sensation in his mouth for hours after having licked the scratches on his hands incurred from handling the birds. From his years of experience, he knew that this kind of reaction was not normal. He knew that something was very peculiar about that pitohui. It turns out that the bird's skin and feathers are covered in batrachotoxins, the same family of incredibly potent alkaloid neurotoxins that poison dart frogs secrete from their skin.

Batrachotoxins target protein channels in the membranes of animal cells, irreversibly opening the channels and causing sodium ions to flood the cells, leading to muscle paralysis and blocked transmission of nerve signals. The bird cells evolved the ability to ignore these devastating effects. The molecular mechanism for the bird's tolerance is not known for certain, but it may come from so-called toxin sponge proteins. The idea is that such

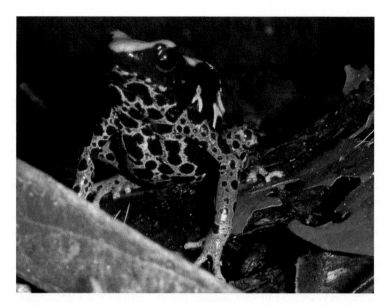

Figure 17.2 *Animals, plants, and microbes have evolved diverse mechanisms for defending themselves against antagonists. This dying poison dart frog,* Dendrobates tinctorius, *from the Nouragues Nature Reserve of French Guiana, secretes extremely hazardous pumiliotoxins from its brightly colored skin that provides warning coloration.*
Photo by the author.

proteins can sop up the poison to prevent the bird's ion channels from succumbing to the adverse influence of batrachotoxin.[20] In fact, the pitohui doesn't make the batrachotoxin at all, but acquires it from its diet, most likely from eating melyrid beetles that contain the poison; that also is how poison dart frogs get their batrachotoxin.[21] A benefit to the birds of this topical poison is in killing off parasites: the pitohui feathers mortally repel lice.

The African crested rat uses a different technique to make itself poisonous. Looking like a rodent version of a skunk-porcupine chimera, these animals gnaw the roots and bark of the poison bush *Acokanthera schimperi*, a plant that produces cardiac glycosides akin to the famed traditional arrow poison, ouabain.[22] The *Lophiomys imhausi* crested rats then slather this masticated spittle onto the specialized long crest of hairs that give the rats their name. These special hairs have a high wicking ability to soak up the poison. When assaulted, the crested rat flares its fur, and, if bitten, the mouthful of poison makes the attacker suffer uncoordinated movement, frothing at the mouth, and, potentially, death by heart failure. This behavior of painting oneself with poison shows the many paths to better living through chemistry.

★

Among creatures that make their own poisons, the complex chemical mixtures of the bioactive cocktails from venom glands can do more than kill a foe. They can mediate communications among friends or antagonism between competitors, subdue prey, offer protection in the care of offspring, and provide topical antiseptic treatment against microbes or parasites, seduction in mating, or even terraforming the landscape to create suitable habitat.[23] Sometimes they can even be sexy. Male scorpions of the species *Euscorpius alpha* of Switzerland don't just sting to subdue prey or in defense. They sting their mates as part of courtship behavior.[24]

Even quiet platypus get in on the venomous action in the name of sex. During the breeding season, the literally horny males assert dominance by elbowing one another with hollow spurs made of keratin on their hind legs. These spurs, like ominously arcing fangs made out of fingernails that poke out behind their ankles, connect to a gland containing a complex soup of venomous proteins.[25] These proteins include some constituents that are similar to, but independently evolved from, the toxic peptides made by some reptiles.[26]

Spiders and many wasps deploy a save-it-for-later feeding strategy with their venom. Most macabre, the venom from an emerald jewel wasp sting acts to manipulate the brain of its targets, inducing peculiar behaviors and eventually making the victim compliant, allowing the wasp to direct it to an underground lair.[27] To achieve this end, the wasp actually employs three distinct envenomations to sting carefully defined parts of the victim's central nervous system that induce distinct behavioral effects. First, it attacks with a sting that induces a brief paralysis so that the wasp can skillfully line up a pair of precise jabs into the brain that cause the prey to compulsively groom itself while the wasp prepares its burrow, after which it steers the subdued creature inside. The story doesn't end there. The beautifully iridescent green wasp then lays a single majestic egg, which hatches and eats the envenomated and entombed creature alive from the inside out. Take comfort that the zombies that get consumed come from only a single species: American cockroaches.

To avoid a wasp's sting when stealing their nest, I'd advise snatching while they sleep. That was the trick I learned as a graduate student from my friend Barrett Klein while assisting him in collecting *Polistes flavus* paper wasps from the underside of a bridge in the dark of the Arizona desert. Perched on a step ladder at night with a nest at eye level, the wasps of the colony clung quietly from their gray sponge-looking saucer of open cells, their spittle-paper nests facing down and dangling from a thin petiole adhered to the concrete underside of the overpass. Gently and decisively, we'd encircle each nest with a bag and pinch off the petiole, the nest's stalk of an anchor, entrapping the colony in the bag to bring back to the lab unstung. Once cared for in the laboratory, Barrett quantified circadian rhythms with the aid of infrared cameras to visualize muscle heating and cooling, seeking in these insects the telltale behavioral and physiological signatures of sleep. As an artist-scientist, Barrett couldn't help himself but infuse some beauty into his experiments: some colonies were treated to colored construction paper as their source of nest-making fiber. What resulted was a wild and elegant rainbow, accreting slowly each day as the wasps chewed up and spit out the brilliant colors of the construction paper, a comb of cells for their brood fashioned into a *papier-mâché arc-en-ciel.*[28]

★

Sometimes the best defense is a good offense. You know about skunks and spitting cobras. Camels spit, too. When agitated, *Camelus bactrianus* may spew a stream of spit and foul-odored fluid from the rumen of their gut at a target up to 3 m (10 feet) away.[29] Thankfully, I wasn't on the receiving end while riding a cousin of the two-humped camel, the one-humped dromedary *Camelus dromedarius*, through the dunes of the Sahara in eastern Morocco one October. Nonetheless, I couldn't help but note the alarming profusion of noisy gas emitted from both ends of the animal as we sauntered over the morning sand. Camels, it turns out, aren't the only creatures to employ vomit as a defensive weapon. Fulmars, a kind of petrel, sea birds that superficially look like gulls and nest on rocky cliffs, will spit defensively starting as nestlings. Their name derives from the Old Norse for "foul gull," and it is apt. The nasty-smelling yellow-orange spit is actually made of stomach oils, normally used as an energy reserve, which the fulmars can projectile vomit accurately to 2 m (6 feet) in multiple volleys.[30] Would-be predatory eagles and other birds can get coated, their feathers soaked in a heavy mat so devastating that they can no longer fly properly or repel water. This becomes a death sentence for predators that dive into the sea to fish in these subarctic and subantarctic latitudes. Death by spewed puke is surely one of the more inglorious deaths.

★

Other times a good defense is the best defense. The evolution by grasshopper mice to be impervious to the pain of a scorpion sting has nothing to do with external armor and everything to do with ion channel proteins in the cell membranes of their neurons.[31] *Onychomys* grasshopper mice in American deserts are little carnivores with a predilection for scorpion flesh. This presents a problem, given the exceedingly painful and potentially fatal sting by some of their prey like bark scorpions. Bark scorpion venom is neurotoxic to mammals, binding to sodium ion channel proteins in neural and muscle cells to activate pain pathways and disrupt lung function. The protein sequence of these ion channels in *Onychomys*, however, have evolved changes causing the toxin to bind to the protein in a way that, instead, blocks pain signals rather than transmitting them. These mice barely flinch when stung.

Similar to grasshopper mice, *Thamnophis* garter snakes have evolved resistance to the normally incapacitating effects of tetrodotoxin made by their newt prey.[32] This physiological response of the snake to the newt's toxic defense has a long evolutionary history. More specifically, it reflects a co-evolutionary arms race: newt makes toxin as defense against predator; predator evolves toxin tolerance; newt evolves higher concentration of toxin to overpower predator tolerance; predator evolves still stronger resistance. This evolutionary back-and-forth is responsible for producing a titer of tetrodotoxin in a single little *Taricha granulosa* rough-skinned newt in Oregon sufficient to kill 20 humans. It's amazing what wonders evolution puts into the world.

17.3   The change to come

As potent as an organism's weaponry may be, poisons are not the kind of defense a species needs to stave off the effects of climate change and habitat disruption. Maybe the ecosystems of the world won't be disrupted so radically as anticipated by climate change exceeding the 2 °C warming threshold, or by solar geoengineering efforts to contain it. Maybe the Earth's natural wonders will continue to be accessible to us. The "30 × 30" movement to set aside, by the year 2030, 30% of the world's surface for conservation certainly seems a noble effort to increase the odds.[33] The motivations behind rewilding activities, whether for passive rewilding or interventionist trophic rewilding, seek a public good. They aim to retain self-sustaining ecological and evolutionary processes so that we may continue to be humbled by nature. Surely, that would be something to brag about. The denizens of social networks seem to have mastered the art of humblebrag. In the real world, though, I just don't know what the right balance is between being humble and promoting grand ambition for biological systems that the Anthropocene so desperately needs.

I've explained to you in this long argument my own internal conflict, and those of others, about how humanity will go about evolving tomorrow. With certainty, we will influence the evolution of countless species on Earth by accelerating extinction, perturbing selection pressures and contributing to genetic drift as a byproduct of our ecosystem engineering, and by continuing our multimillennial tradition of artificial selection in shaping domesticated animals and plants. I'm disappointed by the role of *Homo sapiens* in the extinction of other species and admit that we've given the world an extinction debt that still remains to get paid. I am comfortable with selective breeding and genome editing of crops and livestock and pets. I can get behind evolving new species with new combinations of features, at least under controlled ecological circumstances. In spite of my initial reflex to recoil, I've surprised myself to discover that I am even willing to accept some forms of genome editing of our own species' genome.

As for wielding genetic welding or suppression gene drives to domesticate any wild organism of our choosing? I'm not so sure that I'm on board with that. Despite all my enthusiasm for proof-in-the-pudding tests of evolutionary forces—as well as my fascination with biotechnological techniques, my gratitude for the lives and companionship of domesticated creatures, my compassion for the world's health and environmental inequities, my recognition of the astonishingly great magnitude of global problems— I still have trouble consenting to humans playing kingmaker over the entirety of the living world.[34] Some folks, however, are less circumspect than I am about applying genetic engineering as an intervention into wild organisms. As ancient DNA expert Beth Shapiro wrote recently,[35] "We need to direct evolution so that species adapt more quickly to today's world." I appreciate the sentiment, but to me, "*need*" is a strong word, too strong.

We have sequestered a subset of organisms to shape as tools, using evolutionary forces to enhance them for our own needs as food and resources. I accept and support this carving out of a portion of the living world to modify in support of our own survival and proliferation and education. As living things, after all, we are participants with nature. I object to the idea, however, of treating the entirety of nature with a utilitarian outlook such that it needs to behave in a certain way, with particular sets of species needing to be present or absent or with properties that serve only to benefit us.

Despite my reticence, I understand how people may have a clear conscience about evolutionary engineering outside of the confines of the laboratory. They may even have a clear conscience when the eradication of an entire species is at stake. It is the outlook that human welfare is priority number one. Millions of people die prematurely each year through no fault of their own, after all, simply as a consequence of where in the world they happen to have been born. In the unrecorded Q&A session of a recent workshop on gene drives that I attended,[36] one panelist indicated with a melancholy look in his eye that intentional extinction of a handful of mosquito species with suppression gene drives was an ethical price they were willing to pay in order to reduce the inequity and health burden of human malaria in the world. Another panelist was entirely willing to develop and deploy modification or suppression gene drives to control tick-borne disease transmission or to eradicate invasive species, so long as the outcome of open discussions with local community members also supported that path (they didn't). Open dialog with society and communities about new technological approaches, of course, is essential.

With all the implicit biases and cognitive dissonance and short-term thinking that we humans are prone to, is support through community consensus enough? The acceptable answer for the purposes of due diligence is "yes." Perhaps a cynical answer, though, is that it doesn't matter: sometime, somewhere, community support will give a green light to evolutionary engineering in the wild, and the baseline for social acceptance elsewhere will shift. Shifting baseline syndrome, genetic welding edition. Genome editing in its myriad incarnations with CRISPR-Cas9 techniques is just so easy and has so many individually defensible rationales. With exuberant resignation, I see it as inevitable that humanity will intervene with genome editing through evolutionary engineering to manipulate organisms living in wild ecosystems. Genetic welding is the new evolutionary force that can domesticate all of nature. Only the asexuals are off-limits.

★

Some people will tell you how life on Earth started four billion years ago. I am here to tell you that the world today began this morning. It all changed overnight with every little death and every little birth to produce the genetic state in which each species finds itself, each population of individuals spread across their particular geography. I ask of you to bask in the evolution all around you, in the works in progress that we see in the form of each characteristic of each species as an ongoing evolutionary outcome. Enjoy it today, enjoy it while you can, think about what you want to enjoy tomorrow, and what the next generation will enjoy. Life today, like every day, began fresh and full of the promise that there is more change to come.

Notes

1. This chapter is titled in homage to the 2013 *Saturday Night Live* sketch "The Midnight Coterie of Sinister Intruders," a movie trailer spoof that envisions a horror flick directed by Wes Anderson. Quotation from Thiele, L.P. 2020. Nature 4.0: assisted evolution, de-extinction, and ecological restoration technologies. *Global Environmental Politics*. 20: 9–27. https://doi.org/10.1162/glep_a_00559.

2. Donlan, C.J., et al. 2006. Pleistocene rewilding: An optimistic agenda for twenty-first century conservation. *American Naturalist*. 168: 660–681. https://doi.org/10.1086/508027.

3. Not only would total and immediate elimination of fossil fuel emissions be required, but Clark and colleagues say that "meeting the 1.5° and 2°C targets will likely require extensive and unprecedented changes to the global food system." Starker still, Randers and Goluke conclude, "the world is already past a point-of-no-return for global warming." Clark, M.A., et al. 2020. Global food system emissions could preclude achieving the 1.5° and 2°C climate change targets. *Science*. 370: 705–708. https://doi.org/10.1126/science.aba7357; Randers, J., and U. Goluke. 2020. An earth system model shows self-sustained thawing of permafrost even if all man-made GHG emissions stop in 2020. *Scientific Reports*. 10: 18456. https://doi.org/10.1038/s41598-020-75481-z; Trisos, C.H., et al. 2020. The projected timing of abrupt ecological disruption from climate change. *Nature*. 580: 496–501. https://doi.org/10.1038/s41586-020-2189-9; Duffy, K.A., et al. 2021. How close are we to the temperature tipping point of the terrestrial biosphere? *Science Advances*. 7: eaay1052. https://doi.org/10.1126/sciadv.aay1052.

4. Sulfur dioxide (SO_2), which you know from the smell of a lit match, and volatilized sulfuric acid (H_2SO_4) are the leading contenders for solar geoengineering, though, at higher cost, diamond dust also would do the trick. Solar geoengineering, however, would not counteract ocean acidification. Irvine, P.J., et al. 2016. An overview of the Earth system science of solar geoengineering. *WIREs Climate Change*. 7: 815–833. https://doi.org/10.1002/wcc.423; Grasso, M. 2019. Sulfur in the sky with diamonds: an inquiry into the feasibility of solar geoengineering. *Global Policy*. 10: 217–226. https://doi.org/10.1111/1758-5899.12646; Dai, Z., et al. 2018. Tailoring meridional and seasonal radiative forcing by sulfate aerosol solar geoengineering. *Geophysical Research Letters*. 45, 1030–1039. https://doi.org/10.1002/2017GL076472.

5. Harding, A., and J.B. Moreno-Cruz. 2016. Solar geoengineering economics: from incredible to inevitable and half-way back. *Earth's Future*. 4: 569–577. https://doi.org/10.1002/2016EF000462.

6. Kravitz, B., et al. 2012. Geoengineering: whiter skies? *Geophysical Research Letters*. 39: L11801. https://dx.doi.org/10.1029/2012GL051652.

7. An alternative mind-blowing strategy involves orchestrating a shield of bubbles at a Lagrange point in outer space between the Earth and the Sun to deflect the Sun's rays, https://senseable.mit.edu/space-bubbles. Accessed July 13, 2022.

8. Trisos, C.H., et al. 2018. Potentially dangerous consequences for biodiversity of solar geoengineering implementation and termination. *Nature Ecology and Evolution*. 2: 475–482. https://doi.org/10.1038/s41559-017-0431-0.

9. Kemp, L., et al. 2022. Climate endgame: exploring catastrophic climate change scenarios. *Proceedings of the National Academy of Sciences*. 119: e2108146119. https://doi.org/10.1073/pnas.2108146119.

10. https://www.reuters.com/world/us/us-declare-23-species-including-ivory-billed-woodpecker-extinct-ap-2021-09-29. Accessed Septem-ber 29, 2021.

11. Seddon, P.J., and M. King. 2019. Creating proxies of extinct species: the bioethics of de-extinction. *Emerging Topics in Life Sciences*. 3: 731–735. https://doi.org/10.1042/ETLS20190109.

12. And this number doesn't even factor in the 68 billion chickens grown each year. Callaghan, C.T., et al. 2021. Global abundance estimates for 9,700 bird species. *Proceedings of the National Academy of Sciences USA*. 118: e2023170118. https://doi.org/10.1073/pnas.2023170118; https://ourworldindata.org/meat-production#number-of-animals-slaughtered. Accessed June 3, 2021.

13. If any indication comes from aquarium fish with transgenic modifications to make them fluoresce, the more than $1.5 million in sales of GloFish pets in 2017 show that wanting more awesome in one's life is a common desire. Another company is slated to offer bioluminescent plants that produce their own light by co-opting an autoluminescence pathway of four genes introduced transgenically from fungi. Mitiouchkina, T., et al. 2020. Plants with genetically encoded autoluminescence. *Nature Biotechnology*. 38: 944–946. https://doi.org/10.1038/s41587-020-0500-9.

14. For a thoughtful and humorous discourse on the links between biology and Star Trek, I recommend Mohamed Noor's 2018 book *Live Long and Evolve*.

15. The group launched by inspiring high school biology teacher Leo Kenney is still running, three decades later: https://www.vernalpool.org. Accessed September 29, 2021.

16. King, J.R., et al. 1998. Response of rainforest ant communities in Australia's humid tropics to disturbance: validation of the functional group model. *Biodiversity and Conservation*. 7: 1627–1638. https://doi.org/10.1023/A:1008857214743.

17. Quotation from p. 351 in Tardent, P. 1995. The cnidarian cnidocyte, a high-tech cellular weaponry. *Bioessays*. 17: 351–362. https://doi.org/10.1002/bies.950170411.

18. Quotations from pp.1, 4 in Gershwin, L., et al. 2013. Biology and ecology of Irukandji jellyfish (*Cnidaria: Cubozoa*). *Advances in Marine Biology*. 66: 1–85. https://doi.org/10.1016/B978-0-12-408096-6.00001-8.

19. Dumbacher, J.P. et al. 2009. Skin as a toxin storage organ in the endemic New Guinean genus *Pitohui*. *Auk*. 126: 520–530. https://doi.org/10.1525/auk.2009.08230.

20. Abderemane-Ali, F., et al. 2021. Evidence that toxin resistance in poison birds and frogs is not rooted in sodium channel mutations and may rely on "toxin sponge" proteins. *Journal of General Physiology*. 153: e202112872. https://doi.org/10.1085/jgp.202112872.

21. Dumbacher, J.P., et al. 2004. Melyrid beetles (*Choresine*): a putative source for the batrachotoxin alkaloids found in poison-dart frogs and toxic passerine birds. *Proceedings of the National Academy of Sciences USA*. 101: 15857–15860. https://dx.doi.org/10.1073/pnas.0407197101.

22. Kingdon, J., et al. 2011. A poisonous surprise under the coat of the African crested rat. *Proceedings of the Royal Society B: Biological Sciences*. 279: 675–680. https://doi.org/10.1098/rspb.2011.1169.

23. Ants and jellyfish are natural masters of venom biochemistry, applying their self-made potions in an especially diverse number of capacities. Schendel, V., et al. 2019. The diversity of

venom: the importance of behavior and venom system morphology in understanding its ecology and evolution. *Toxins*. 11: 666. https://dx.doi.org/10.3390/toxins11110666.

24. Sentenská, L., et al. 2017. Sexual dimorphism in venom gland morphology in a sexually stinging scorpion. *Biological Journal of the Linnean Society*. 122: 429–443. https://doi.org/10.1093/biolinnean/blx067.

25. I've never been pricked by platypus spurs, but platypus have pricked me with surprise. While tromping through the snow up on Mount Field in Tasmania one Friday in July, I managed to disrupt the quiet solitude of a lone platypus in a narrow stream along the trail. As it emerged from underfoot, I scrambled from my momentary shock into a surreal chase of a platypus across the white of the snow, angling for an up-close photograph. It's a bit blurry and black-and-white, but, frozen in my frame, I snapped proof of that wintertime monotreme.

26. Whittington, C.M., and K. Belov. 2007. Platypus venom: a review. *Australian Mammalogy*. 29: 57–62. https://doi.org/10.1071/AM07006; Whittington, C.M., et al. 2008. Defensins and the convergent evolution of platypus and reptile venom genes. *Genome Research*. 18: 986–994. https://dx.doi.org/10.1101/gr.7149808.

27. Arvidson, R., et al. 2019. Parasitoid jewel wasp mounts multipronged neurochemical attack to hijack a host brain. *Molecular & Cellular Proteomics*. 18: 99–114. https://doi.org/10.1074/mcp.RA118.000908.

28. The spectrum of the rainbow could have been broader than he thought, as it turns out that parts of the nests for some species of *Polistes* wasps also fluoresce green or blue when exposed to UV light. Daney de Marcillac, W., et al. 2021. Bright green fluorescence of Asian paper wasp nests. *Journal of the Royal Society Interface*. 18: 20210418. https://doi.org/10.1098/rsif.2021.0418.

29. Tahzib, F. 1984. Camel injuries. *Tropical Doctor*. 14: 187–188. https://doi.org/10.1177/004947558401400417.

30. Warham, J. 1977. The incidence, functions and ecological significance of petrel stomach oils. *Proceedings (New Zealand Ecological Society)*. 24: 84–93. http://www.jstor.org/stable/24064251.

31. Rowe, A.H., et al. 2013. Voltage-gated sodium channel in grasshopper mice defends against bark scorpion toxin. *Science*. 342: 441–446. https://dx.doi.org/10.1126/science.1236451.

32. Brodie, E.D., III, and E.D., Brodie, Jr. 1999. Predator-prey arms races: asymmetrical selection on predators and prey may be reduced when prey are dangerous. *BioScience*. 49: 557–568. https://doi.org/10.2307/1313476.

33. https://www.hacfornatureandpeople.org/science-and-reports. Accessed June 14, 2022.

34. Admittedly, this perspective might butt heads with the traditional Judeo-Christian view that the nonhuman world is a gift to humanity intended for our benefit and use. The twenty-eighth line of the *Bereishit* and *Book of Genesis*, after all, invites us "to fill the earth and subdue it."

35. Quotation from p. 280 in Beth Shapiro's fascinating 2021 book *Life as We Made It*.

36. Interdisciplinary Workshop on Synthetic Gene Drives, June 21–24, 2021. https://flodebarre.github.io/genedrive2021. Accessed October 27, 2021.

Index

For the benefit of digital users, indexed terms that span two pages (e.g., 52–53) may, on occasion, appear on only one of those pages.

Note: references to figures are indicated by *f* after the page number.